UCL INSTITUTE OF ARCHAEOLOGY PHD SERIES

VOLUME 4

On the Hunt for Medieval Whales

Zooarchaeological, historical and social perspectives on cetacean exploitation in medieval northern and western Europe

YOURI VAN DEN HURK

BAR INTERNATIONAL SERIES 2998 | 2020

BAR PUBLISHING

Published in 2020 by
BAR Publishing, Oxford

BAR International Series 2998

UCL Institute of Archaeology PhD Series, 4
On the Hunt for Medieval Whales

ISBN 978 1 4073 5720 1 paperback
ISBN 978 1 4073 5721 8 e-format

DOI https://doi.org/10.30861/9781407357201

A catalogue record for this book is available from the British Library

COVER IMAGE *Selection of whale, dolphin, and porpoise bones analysed as part of this study. Photos by author*

BAR
PUBLISHING

BAR titles are available from:

BAR Publishing
122 Banbury Rd, Oxford, OX2 7BP, UK
EMAIL info@barpublishing.com
PHONE +44 (0)1865 310431
FAX +44 (0)1865 316916
www.barpublishing.com

UCL INSTITUTE OF ARCHAEOLOGY

UCL INSTITUTE OF ARCHAEOLOGY PHD SERIES

Series Editor: Ruth Whitehouse (Chair of the IoA Publications Committee)

The aim of this series is to offer rapid publication of largely unaltered PhD theses. It covers the full range of subject areas studied at the Institute, including world archaeology, archaeological science, cultural heritage, conservation and museum studies. It makes data, analyses and interpretation available promptly to a wide audience.

The UCL Institute of Archaeology is one of the oldest, largest and most prestigious archaeology research facilities in the world. Its extensive publications programme includes the best theory, research, pedagogy and reference materials in archaeology, cultural heritage and cognate disciplines. Through its publications, which also include a General Series and a Critical Cultural Heritage Series, and a number of associated journals, the Institute brings together key areas of theoretical and substantive knowledge, improves archaeological and heritage practice and brings archaeological findings to the general public, researchers and practitioners.

Volumes in the Series

Pattern and Process in the Material Culture of Anglo-Saxon Non-elite Rural Settlements
Hana Lewis

Oxford, BAR Publishing, 2019 — BAR British Series **649**
UCL Institute of Archaeology PhD Series, 1

Ancient Maya Settlement and the Alacranes Bajo
Landscape and Communities in the Three Rivers Region, Northwestern Belize
Gail A. Hammond

Oxford, BAR Publishing, 2019 — BAR International Series **2956**
UCL Institute of Archaeology PhD Series, 2

A Consideration of Gender Roles and Relations in the Aegean Bronze Age Interpreted from Gestures and Proxemics in Art
Susan E. Poole

Oxford, BAR Publishing, 2020 — BAR International Series **2980**
UCL Institute of Archaeology PhD Series, 3

On the Hunt for Medieval Whales
Zooarchaeological, historical and social perspectives on cetacean exploitation in medieval northern and western Europe
Youri van den Hurk

Oxford, BAR Publishing, 2020 — BAR International Series **2998**
UCL Institute of Archaeology PhD Series, 4

Of Related Interest

Les Ovibovini, Caprini et Ovini (Mammalia, Artiodactyla, Bovidae, Caprinae) du Plio-Pléistocène d'Europe, Volumes I and II
Systématique, évolution et biochronologie
Évelyne Crégut-Bonnoure

Oxford, BAR Publishing, 2020 — BAR International Series **2975**

Animaltown: Beasts in Medieval Urban Space
Edited by Alice M. Choyke and Gerhard Jaritz

Oxford, BAR Publishing, 2017 — BAR International Series **2858**

Archaeozoological Analysis of the Fortified Settlement of Sand (10th Century AD, Lower Austria)
Early Medieval faunal remains from Sand an der Thaya
Konstantina Saliari

Oxford, BAR Publishing, 2018 — BAR International Series **2892**

In the land of the Ichthyophagi
Modelling fish exploitation in the Arabian Gulf and Gulf of Oman from the 5th millennium BC to the Late Islamic period
Mark J. Beech

Oxford, BAR Publishing, 2004 — BAR International Series **1217**

Fishing and Managing the Trent in the Medieval Period (7th–14th Century)
Excavations at Hemington Quarry (1998–2000), Castle Donington, UK
Lynden P. Cooper and Susan Ripper with contributions by Matt Beamish, Jennifer Browning, Nicholas J. Cooper, Robert Howard, Patrick Marsden, Angela Monckton, Anita Radini and Deborah Sawday

Oxford, BAR Publishing, 2017 — BAR British Series **633**

La parure en coquille à Sayula, Occident du Mexique
Approche techno-stylistique et rôle dans la dynamique socioculturelle entre 450 et 1000 apr. J.-C.
Elodie Mas

Oxford, BAR Publishing, 2018 — BAR International Series **2900**

"The more I consider this mighty tail, the more do I deplore my inability to express it. At times there are gestures in it, which, though they would well grace the hand of man, remain wholly inexplicable. In an extensive herd, so remarkable, occasionally, are these mystic gestures, that I have heard hunters who have declared them akin to Free-Mason signs and symbols; that the whale, indeed, by these methods intelligently conversed with the world. Nor are there wanting other motions of the whale in his general body, full of strangeness, and unaccountable to his most experienced assailant. Dissect him how I may, then, I but go skin deep; I know him not, and never will. But if I know not even the tail of this whale, how understand his head? Much more, how comprehend his face, when face he has none? Thou shalt see my back parts, my tail, he seems to say, but my face shall not be seen. But I cannot completely make out his back parts; and hint what he will about his face, I say again he has no face."

Moby-Dick; or, The Whale by Herman Melville (1851, 357)

Contents

List of Figures

List of Tables

Acknowledgements

I would like to thank my parents John and Margreet van den Hurk, and my sister Iris for supporting me in writing this book. Moreover, my parents are thanked for putting on nature documentaries when I was a kid, exposing me to killer whales ripping apart several seals and penguins. Most kids would have been horrified by such a scene, but for me it triggered an extreme interest and fascination for cetaceans, which eventually led to me down to road to pursuing a PhD in the zooarchaeology of cetaceans.

I am also grateful to Erik Stoter who endured my endless talks regarding cetaceans. Eline Tabak is thanked for being my best friend since secondary school and pursuing similar dreams. I am truly pleased that she will be able to start her PhD this year.

Funding for the project, from its inception at University College London (UCL) to its final stages was made available by the Arts and Humanities Research Council – London Arts and Humanities Partnership (AHRC LAHP). This made it possible for me to pursue my dream in performing research on the archaeology of cetaceans. The Ocean's Past Platform (OPP), the Australian Archaeology Association (AAA), the Association for Environmental Archaeology (AEA) and the Scottish Archaeological Research Framework (ScARF) are thanked for allowing me to attend and present my work at their conferences and funding my travel expenses.

I am extremely grateful to Dr Louise Martin and Prof Andrew Reynolds, who have been very supporting supervisors for the past four years and have helped me in any way they could. Furthermore, Dr Canan Çakırlar is also thanked for being a mentor since I got interested in zooarchaeology. Without her support none of this would have been possible and her passion and work ethics are something that greatly inspire me. Dr James Barrett and Dr Vicki Szabo are thanked for serving as examiners and providing valuable comments to improve the study as well as give me suggestions for future directions.

The Institute of Archaeology, University College London is thanked for providing an excellent research environment to perform my research. The people of the Institute are thanked, and special thanks go to Liz Farebrother for welcoming me to London. Goodenough College is also thanked for being a home for me in London, and for introducing me to Michael Borsky, Scott Myers, and Luis Benitez, three of the best friends someone can wish for and being always very supportive. Special thanks to Cristina Brito, Nina Viera, and Anne Karin Hufthammer.

Furthermore, the Groningen zooarchaeology group is thanked for their support, including Esther Scheele, Francesca Slim, Jildou Kooistra, Francis Koolstra, Fleur Dijkstra, Nynke de Boer, Dimitris Filioglou, Safoora Komijani, Yannic Rabou, Rosanne van Bodegom, Rachel Winter, Willemien de Kock, Wietske Prummel, Hans Christian Küchelmann, as well as all the other people of the Groningen Institute of Archaeology, University of Groningen. Laurens Middag, Huub Winters, Anouk de Regt, Stephanie Reitsma, Liesbeth Schut, Bastiaan de Wolde, Yu-tz Tung, Florence Kok, Peter Kane Buchowiecki, Pinju Lee, Viola Wiegand, Kyunghee Kook, Zachary Ke, Jordan Thorpe, Ben Purvis, and Adam Horsley are thanked as well.

I would like to thank Veronica White for allowing me to stay with her for seven months in Washington DC and giving me a home there, and Peter and Gudule Theunissen whom made me feel like I had in family away from family. I truly had a wonderful during my stay in DC.

Furthermore, I am thankful to have been able to perform research at the Smithsonian Institution, through the AHRC International Placement Scheme. I have never seen a more outstanding collection of natural history. Thanks go to Nick Pyenson for being my supervisor at the Smithsonian, as well as to John Ososky and Paul Bohaska for creating a pleasant work environment and making the many hours in spent in the "Whale Warehouse" a joy.

The AHRC is also thanked to provide me with the additional scholarship to visit the Zoological Museum, Statens Naturhistoriske Museum, Copenhagen, Denmark and perform research on their collection. Morten Tange Olsen and Daniel Klingberg Johansson of the Naturhistoriske Museum are thanked for their help.

Henk Aandewiel is thanked for sharing his photos of the grey whale mandible present at the Poldermuseum in Heerhugowaard. Klaas Post, Arthur Oosterbaan, Adrie Vonk, and Ineke Vonk of Ecomare are also thanked for their insights.

Several other museums were visited as part of this study and Christine Lefèvre of the Muséum National D'Histoire Naturelle, Paris, France; Severine Hannam of the Auckland War Memorial Museum, Auckland, New Zealand; Olivier Pauwels of the Royal Belgian Institute of Natural Sciences, Brussels, Belgium; Jerry Herman, Martin Goldberg, Jim Wilson, Zena Timmons, Craig Angus, and Fraser Hunter of the National Museums of Scotland, Edinburgh, Scotland; Klaas Post and Bram Langeveld of the Natuurhistorisch Museum Rotterdam, Rotterdam, The Netherlands; and Pepijn Kamminga of the Naturalis Biodiversity Centre are all thanked.

Furthermore, several people should be thanked for providing specimens that were analysed as part of this study. These people include Frits Laarman of the Cultural Heritage Agency, Amy Kuiper and Jelle Schokker of the Noordelijk Archaeology Depot, Mark Philippeau of the Provincial Archaeology Depot Zuid Holland, Joost van den Berg of the Zeeland Archaeology Depot in Middelburg, Martin Veen, Joanneke van den Engel-Hees, and Kees Zwaan of the Provincial Archaeology Depot North Holland, Ronald Louer and Martin Meffert of the Provincial Archaeology Depot North Brabant, Lisette Kootker of ArcheoPlan Eco, Ronald 't Hart of RAAP Oost, Jeroen Loopik of ADC ArchaeoProjecten, and Jørn Zeiler of ArchaeoBone.

Moreover, all the zooarchaeologists I have bothered over the past year in my quest to locate as many medieval cetacean remains spread over northern and western Europe as possible are thanked. It is truly amazing to be part of the global zooarchaeology community and I hope to be able to spend several more years on reconstructing past whaling practices and understanding our dynamic relationship with these amazing and mysterious animals.

Abstract

Medieval cetacean exploitation has been connected to various societies, including the Basques, Norse, Normans, Portuguese, Frisians, and Flemish. The extent to which active whaling was practiced however remains unclear. The possibility has even been raised that many of these cultures relied on the opportunistic exploitation of stranded cetaceans in order to acquire the meat, bone, oil, and other resources that could be stripped from the carcasses. Furthermore, primarily for the ninth to the twelfth centuries AD, it has been argued that the symbolic significance of cetaceans surpassed their utilitarian value and that their consumption was restricted to the social elite. For many European regions, laws were set in place ensuring that any stranded cetacean was the property of the social elite ("wreck of sea right"). Both the nobility and clergy of medieval Europe are frequently associated with the consumption of cetacean meat, presumably the result of Christian dietary restrictions, as well as the development of salting as a method of preservation, making cetacean meat more easily accessible.

Zooarchaeology offers the possibility to reconstruct which species were exploited and whether the social elite did indeed monopolize this. However, little research has been conducted on zooarchaeological cetacean remains. The identification of cetacean fragments to the species level is hard and specimens are frequently merely identified as "whale", "cetacean" or even "marine mammal" resulting in a poor understanding of human-cetacean interaction in the past.

As part of this study, a literature review was undertaken for which medieval sites from Northern and Western Europe where cetacean remains have been found were assessed. A total of 406 medieval sites with cetacean remains have been identified, with a total of 5528 cetacean specimens (NISP), representing at least eighteen different species. The context and the taphonomy of these specimens were assessed and a select number of remains were re-examined and Zooarchaeology by Mass-Spectrometry (ZooMS) was performed in order to identify the remains to species level.

This study indicated that especially the harbour porpoise (*Phocoena phocoena*), North Atlantic right whale (*Eubalaena glacialis*), and the common bottlenose dolphin (*Tursiops truncatus*) are well represented in the medieval archaeological record, indeed suggesting that many cultures performed active hunting on cetaceans. This was especially the case for the period of AD 1000-1350 and for this period cetacean remains are frequently recovered from high-status and ecclesiastical sites, suggesting that the exploitation and consumption of cetaceans was restricted to the social elite. However, cetacean remains have also been recovered from "rural" site types, signifying peasant efforts to undermine elite control of stranded cetaceans.

As a result of the collapse of the North Atlantic right whale population in the fourteenth century, many cultures ceased their whaling activities. Several specimens of the grey whale (*Eschrichtius robustus*) were also identified, suggesting that active whaling might also have resulted in the extirpation of this species from the North Atlantic. The Medieval Period can therefore be seen as the dawn of a period of centuries of relentless and decimating whaling activities.

Samenvatting

De exploitatie van walvisachtigen (walvissen, dolfijnen, en bruinvissen) is in connectie gebracht met verschillende middeleeuwse culturen zoals de Basken, Noormannen, Normandiërs, Portugezen, Friezen, en Vlamingen. In hoeverre actieve walvisvaart heeft plaatsgevonden blijft onbekend. Het is gesuggereerd dat veel culturen afhankelijk waren van opportunistische exploitatie van gestrande walvisachtigen om toegang te krijgen tot het vlees, bot, olie, en andere grondstoffen die verkregen kunnen worden van de karkassen. Tevens, voornamelijk voor de negende tot en met de twaalfde eeuw AD, wordt gedacht dat de symbolische betekenis van walvisachtigen belangrijker was dan utilitaristisch waarde en dat de consumptie alleen toegestaan was voor de sociale elite. Voor grote delen van Europe werden wetten ontworpen die gestrande walvisachtigen een monopolie van de sociale elite maakte. Zowel de adel als de geestelijkheid van middeleeuws Europa worden frequent geassocieerd met de consumptie van walvisvlees. Dit is waarschijnlijk het gevolg van Christelijke dieet restricties en de ontwikkeling van het zouten van vlees als een preservatiemethode. Hierdoor werd walvisvlees makkelijker te verkrijgen.

Archeozoölogie biedt de mogelijkheid om te achterhalen welke walvisachtigensoorten er werden geëxploiteerd en of de sociale elite inderdaad een monopolie had op de consumptie. Echter is er weinig onderzoek uitgevoerd naar archeozoölogische walvisrestanten. De identificatie van walvisfragmenten tot de soort is moeilijk en leidt er vaak toe dat deze geïdentificeerd worden als "walvis", "walvisachtige", or slechts "zeezoogdier". Dit heeft geleid tot weinig kennis betreffende mens-walvisachtigen interactie in het verleden.

Als onderdeel van dit onderzoek, is een literatuurstudie uitgevoerd naar middeleeuwse archeologische sites in Noord en West Europe waar restanten van walvisachtigen zijn aangetroffen. In totaal zijn er 406 sites middeleeuwse sites geïdentificeerd, waarbij in totaal 5528 walvisachtigenrestanten (NISP) zijn aangetroffen. Deze representeren ten minste achttien verschillende walvisachtigensoorten. De context en de tafonomie van deze vondsten werd geanalyseerd en aantal fragmenten werd opnieuw bekeken en *Zooarchaeology by Mass-Spectrometry (ZooMS)* werd uitgevoerd om de restanten tot diersoort te identificeren.

Dit onderzoek heeft aangetoond dat de bruinvis (*Phocoena phocoena*), noordkaper (*Eubalaena glacialis*), en de tuimelaar (*Tursiops truncatus*) sterk gerepresenteerd zijn in het middeleeuws archeologisch materiaal. Dit suggereert dat veel culturen actieve walvisvaart uitvoerden. Dit was met name het geval voor de periode AD 1000-1350, waarvoor veel vondsten uit aristocratische en kerkelijke contexten komen, waardoor er inderdaad gesuggereerd wordt dat de consumptie van walvisachtigen alleen was toegestaan door de sociale elite. Echter, vondsten van restanten van walvisachtigen uit rurale contexten toont ook aan dat de horigheid regelmatig de regels van sociale elite ondermijnde.

Nadat de noordkaper populatie in Europe in de veertiende eeuw ineenstortte, stopten veel culturen met hun walvisvaartactiviteiten. Enkele vondsten van de grijze walvis (*Eschrichtius robustus*) zijn ook geïdentificeerd. Dit oppert de mogelijkheid dat de middeleeuwse walvisvaart mogelijk ten dele verantwoordelijk was voor de extirpatie van de soort uit de Noord Atlantische Oceaan. De middeleeuwen kunnen daarom gezien worden als het begin van eeuwen aan genadeloze en decimerende walvisvaart activiteiten.

Zusammenfassung

Der Jagd mittelalterlicher Cetacea (Wale, Delfine und Schweinswale) wurde mit verschiedenen Gesellschaften in Verbindung gebracht, darunter Basken, Nordländer, Normannen, Portugiesen, Friesen und Flamen. Inwieweit aktiver Walfang betrieben wurde, bleibt jedoch unklar. Es wurde sogar die Möglichkeit erwägt, dass viele dieser Kulturen auf die gelegentliche Ausbeutung gestrandeter Wale angewiesen waren, um Fleisch, Knochen, Öl und andere Ressourcen zu erwerben, die den Kadavern entnommen wurden. Darüber hinaus wurde vor allem für das 9. bis 12. Jahrhundert n. Chr. argumentiert, dass die symbolische Bedeutung der Wale ihren eigentlichen Nutzwert übertraf und dass ihr Konsum auf die soziale Elite beschränkt war. Für viele europäische Regionen wurden Gesetze erlassen, die gestrandete Wale zum Eigentum der sozialen Elite machten („Wrack des Meeresrechts“). Sowohl der Adel als auch der Klerus des mittelalterlichen Europas werden häufig mit dem Verzehr von Walfleisch in Verbindung gebracht, vermutlich aufgrund christlicher Fastengewohnheiten sowie der Einführung des Salzens als neue Konservierungsmethode, was den Zugang zu Walfleisch erleichterte.

Die Zooarchäologie bietet die Möglichkeit, die Frage zu erörtern, welche Arten ausgebeutet wurden und ob die soziale Elite diese Ausbeutung tatsächlich monopolisierte. Bisher wurden jedoch nur wenige Untersuchungen zu zooarchäologischen Walresten durchgeführt. Die Artenidentifizierung von Walfragmenten ist schwierig, und Überrestproben werden häufig lediglich als „Wal“, „Cetacea“ oder als „Meeressäuger“ identifiziert, was in der Vergangenheit nur ein schlechtes Verständnis der Interaktion zwischen Menschen und Wal erlaubte.

Im Rahmen dieser Studie wurde eine Literaturrecherche durchgeführt, in der die mittelalterlichen Stätten aus Nord- und Westeuropa bewertet wurden, in denen Walreste gefunden wurden. Insgesamt wurden 406 mittelalterliche Stätten mit Walresten identifiziert, mit 5528 Walproben (NISP) über mindestens achtzehn verschiedene Arten. Der Kontext und die Taphonomie dieser Exemplare wurden bewertet, und eine ausgewählte Anzahl von Überresten wurde untersucht, um mit der Durchführung der Zooarchäologie durch Massenspektrometrie (ZooMS) die genaue Art der Überreste zu identifizieren.

Diese Studie zeigt, dass insbesondere der Gewöhnliche Schweinswal (*Phocoena phocoena*), der Atlantische Nordkaper (*Eubalaena glacialis*) und der Große Tümmler (*Tursiops truncatus*) in den mittelalterlichen archäologischen Aufzeichnungen gut vertreten sind. Dies deutet darauf hin, dass viele Kulturen Wale aktiv jagten, insbesondere von 1000 bis 1350 n. Chr. Für diesen Zeitraum wurden Walreste häufig aus kirchlichen und anderen hochrangigen Stätten geborgen, was darauf hindeutet, dass die Nutzung und der Konsum der sozialen Elite vorbehalten war. Walreste wurden jedoch auch in „ländlichen“ Standorten entdeckt, was als eine gewisse bäuerliche Untergrabung dieser Beschränkungen interpretiert werden kann.

Infolge des Zusammenbruchs des Bestands des Atlantischen Nordkapers im 14. Jahrhundert stellten viele Kulturen ihre Walfangaktivitäten ein. Die Identifikation mehrerer Exemplare des Grauwals (Eschrichtius robustus) deutet darauf hin, dass der aktive Walfang auch zur lokalen Ausrottung dieser nordatlantischen Art geführt haben könnte. Das Mittelalter kann daher als der Beginn der artenbedrohenden Walfangaktivitäten gesehen habe, die Jahrhunderte andauert.

Resumen

La explotación medieval de cetáceos (ballenas, delfines y marsopas) ha estado relacionada con varias sociedades, incluyendo los vascos, nórdicos, normandos, portugueses, frisones y flamencos. Sin embargo, la medida en que se practicó la caza activa de ballenas todavía no es del todo clara. Incluso se ha planteado la posibilidad de que muchas de estas culturas dependían de la explotación oportunista de cetáceos varados para adquirir la carne, el hueso, el aceite y otros recursos que podrían ser despojados de los cadáveres. Además, principalmente durante los siglos IX y XII d.C., se ha argumentado que el significado simbólico de los cetáceos superó su valor utilitario y que por consiguiente su consumo se limitó a la élite social. En muchas regiones europeas, se establecieron leyes que garantizaban que cualquier cetáceo varado fuera propiedad de la élite social ("naufragio del derecho marítimo"). Tanto la nobleza como el clero de la Europa medieval se asocian con frecuencia al consumo de carne de cetáceo, presumiblemente el resultado de las restricciones dietéticas cristianas, así como el desarrollo de la salazón como método de conservación, haciendo que la carne de cetáceo fuera más fácilmente accesible.

La zooarqueología ofrece la posibilidad de indicarnos qué especies fueron explotadas y si la élite social monopolizó esta práctica. Sin embargo, se han realizado pocas investigaciones sobre restos de cetáceos zooarquiseológicos. La identificación de fragmentos de cetáceos al nivel de la especie es complicada y los especímenes se identifican con frecuencia simplemente como "ballena", "cetáceo" o incluso "mamífero marino", lo que deriva en una limitada comprensión de la interacción humano-cetácea del pasado.

Como parte de este estudio, se llevó a cabo una revisión de la literatura para evaluar los yacimientos medievales de Europa Septentrional y Occidental donde se han encontrado restos de cetáceos. Se han identificado un total de 406 yacimientos medievales con restos de cetáceos, con un total de 5,528 especímenes de cetáceos (NISP), que representan al menos dieciocho especies diferentes. Se evaluó el contexto y la taponomía de estos especímenes y se reexaminaron un número selecto de restos y se realizó zooarqueología por espectrometría de masas (ZooMS) con el fin de identificar los restos a nivel de especie.

Este estudio indicó que especialmente la marsopa portuaria (*Phocoena phocoena*), la ballena franca del Atlántico Norte (*Eubalaena glacialis*), y el delfín nariz de botella común (*Tursiops truncatus*) están bien representados en el registro arqueológico medieval, lo que sugiere que muchas culturas realizaron caza activa en cetáceos. Este fue especialmente el caso para el período de 1000-1350 d.C durante el cual os restos de cetáceos se recuperaron con frecuencia en sitios eclesiásticos y de alto estatus, lo que sugiere que la explotación y el consumo de cetáceos se limitaba a la élite social. Sin embargo, los restos de cetáceos también han sido recuperado de los tipos de sitios "rurales", lo que evidencia los esfuerzos de los campesinos para socavar el control de los cetáceos varados de la élite.

Como resultado del declive en la población de ballenas francas del Atlántico Norte en el siglo XIV, muchas culturas cesaron sus actividades de casa de cetáceos. Varios especímenes de la ballena gris (*Eschrichtius robustus*) también fueron identificados, lo que sugiere que la caza de ballenas activa también podría haber dado lugar a la extirpación de esta especie en el Atlántico Norte. Por lo tanto, el Período Medieval puede verse como el comienzo de un período de siglos de actividades implacables y diezmadas de caza de ballenas.

Résumé

L'exploitation médiévale des cétacés (baleines, dauphins et marsouins) a été le fait de diverses sociétés, notamment basques, norvégiennes, normandes, portugaises, frisonnes et flamandes. L'étendue exacte de la chasse baleinière reste toutefois incertaine. La possibilité que plusieurs de ces cultures s'appuyaient plutôt sur l'exploitation opportuniste des carcasses de cétacés échoués afin d'en prélever la viande, les os, l'huile et d'autres ressources a même été évoquée. En outre, il a été avancé que, en particulier entre le 9e et le 12e siècles, la valeur symbolique des cétacés dépassait leur valeur d'usage et que leur consommation était restreinte à une élite sociale. Dans plusieurs régions européennes, des lois furent adoptées afin que les carcasses de cétacés échoués soit la propriété de l'élite sociale (le « droit sur les épaves »). La noblesse et le clergé de l'Europe médiévale sont souvent associés à la consommation de viande de cétacé, sans doute en raison de restrictions alimentaires chrétiennes et du développement du salage comme méthode de préservation rendant la viande de cétacé plus accessible.

La zooarchéologie permet d'établir quelles espèces étaient exploitées et si l'élite sociale en exerçait effectivement le monopole. Toutefois, peu de recherches ont été conduites sur les restes zooarchéologiques de cétacés. L'identification des fragments de cétacés à des espèces précises est difficile, et les spécimens sont ainsi souvent identifiés comme appartenant à des « baleines », des « cétacés » ou même des « mammifères marins », débouchant sur une faible compréhension des interactions passées entre êtres humains et cétacés.

Dans le cadre de cette recherche, un état de la question a été mené sur les sites médiévaux d'Europe septentrionale et occidentale où des restes de cétacés ont été trouvés. Un total de 406 sites médiévaux ont été identifiés, contenant 5528 spécimens de cétacés (NISP), représentant au moins 18 différentes espèces. Le contexte et la taphonomie de ces spécimens ont été établis, une sélection d'entre eux ont été réexaminés et une zooarchéologie par spectrométrie de masse (ZooMS) a été conduite pour identifier les restes à leur espèce spécifique.

Cette recherche indique que les marsouins communs (*Phocoena phocoena*), les baleines franches de l'Atlantique Nord (*Eubalaena glacialis*) et les grands dauphins (*Tursiops truncatus*) sont très représentés dans les archives archéologiques médiévales, suggérant ainsi que plusieurs cultures pratiquaient en effet une chasse baleinière active. Ceci était particulièrement le cas entre les années 1000 et 1350, période durant laquelle les restes de cétacés sont fréquemment retrouvés dans des sites associés à l'aristocratie et au clergé, suggérant que l'exploitation et la consommation de cétacés relevaient bien de l'élite sociale. Toutefois, des restes ont aussi été retrouvés dans des sites de type « rural », indiquant des efforts de la paysannerie pour miner le contrôle de l'élite sur les carcasses de cétacés échoués.

Suite à l'effondrement de la population de baleines franches de l'Atlantique Nord au 14e siècle, plusieurs cultures ont cessé leurs activités de chasse baleinière. Plusieurs spécimens de baleine grise (*Eschrichtius robustus*) ont aussi été identifiés, suggérant que la chasse baleinière ait aussi pu entrainer l'extirpation de ces espèces de l'Atlantique Nord. Le Moyen-Âge peut ainsi être perçu comme le crépuscule d'une période multiséculaire de chasse baleinière insatiable et destructrice.

Introduction

While zooarchaeological cetacean material is frequently recovered from medieval sites, the study of it has been lacking in comparison to their terrestrial mammalian counterparts. The field of zooarchaeology has often even ignored cetacean material, resulting in a poor understanding of past cetacean exploitation. The situation is further construed by a lack of expertise and methods dealing with these animals in the field of zooarchaeology. The lack regarding research on cetacean exploitation can presumably be ascribed to the continuous influence of evolutionary thought in archaeology which perceives hunting in farming societies as a remnant of a backward evolutionary stage (Zvelebil, 1992, 8).

Several studies have attempted to tackle the lack of research being performed on cetacean exploitation in pre-modern Europe. The first major study concerned with the archaeology of whales and their relationship with humans was conducted by Clark (1947). Clark focused on Prehistoric Europe as a whole and created a database of archaeological sites where cetacean remains had been discovered. Clark was able to collect data for 79 archaeological sites in Northern and Western Europe. The main purpose of his study was to find out to what extent cetaceans played an economic role in Europe and since most of the sites were found in Scotland, most of the attention went to that region in combination with the rock engravings depicting whaling in northern Scandinavia. This study was undertaken 73 years ago, and a lot of new information has been acquired since. Furthermore, the archaeological discipline has changed as well during these years, not solely focused on the "economic" aspects of zooarchaeology anymore, but also analysing social and cultural aspects of past foodways.

Though Clark's study is still the most extensive one for Europe, it is limited in its scope, generalizes a lot and made uncritical use of sources of evidence (Szabo, 2008, 15). New extensive studies were undertaken almost 40 years later. In 1997 Gardiner published his study "The Exploitation of Sea-Mammals in Medieval England: Bones and their Social Context" which, though not specifically a zooarchaeological study, used historical sources to argue that in the High Medieval Period, cetaceans (especially porpoises) were seen as a luxury food and were claimed by royal figures or religious complexes.

Following this, new research was conducted by Mulville (2002a, 2002b). Her research can be seen as one of the first true modern zooarchaeological studies in North-Western Europe concerned with cetaceans and focused primarily on the Hebrides in Scotland. Furthermore, she has argued that archaeological remains are getting increasingly more attention in the modern whaling debate. Zooarchaeological remains are often used in it to argue that whaling has been a vital part of cultures of specific groups for a long time and therefore these people should be allowed to practice whaling now as well (Mulville, 2002b). In this way, zooarchaeology of cetaceans is connected to the debate in regard to modern whaling practices.

More recent studies focusing on the archaeology of cetaceans, focused especially on species identification, have more regularly used molecular-based analysis. The application of mtDNA analysis on cetaceans has been conducted by Foote *et al.* (2013) and has shown excellent results for cetacean species identification. However, DNA analysis remains an expensive technique and as a result it has only occasionally been undertaken.

This financial issue has limited identification of zooarchaeological cetacean remains up until recently, but the development of a new method tackled this problem. This is the method of Zooarchaeology by Mass Spectrometry (ZooMS) of collagen fingerprinting, which allows an efficient and low-cost possibility of species identification (Buckley *et al.*, 2014). It has been applied on zooarchaeological cetacean remains and has proven to be an excellent identification method. However, it is less precise in comparison to aDNA, not always allowing identification to the species level.

In comparison to zooarchaeological studies, historical studies concerned with medieval whaling practices, have been undertaken more frequently. These studies have however also been limited to several cultural groups. One of these groups are the Normans and historical sources concerned with this group have been extensively studied for decades. Examples are studies conducted by Lestocquoy (1948), Musset (1964), Lebecq (1997), and Guizard (2011, 2018), and have provided a wealth of information regarding the history of cetacean exploitation in Normandy and bordering regions. Even though zooarchaeological remains have been discovered at medieval contexts in France, no extensive study has been undertaken comparing the zooarchaeological record to the historical record.

A similar situation is evident for the Basque Country (both the Spanish and the French parts). Historical sources have been studied by Jenkins (1921), Aguilar (1981; 1986), Goyheneche Farnie (1984), Proulx (1986), Kurlansky (1999), Loewen (2009), and Laist (2017), but zooarchaeological remains appear to be rare and understudied (personal communication Grau-Sologestoa, 2016).

To date, the most extensive historical study concerned with medieval cetacean exploitation is the book "Monstrous Fishes and the Mead-Dark Sea" by Szabo (2008). This comprehensive work focused on the Medieval Period of Northern Europe by considering historical sources and sagas. Limited zooarchaeological remains are considered, and Szabo (2008) highlights that zooarchaeological remains can provide a wealth of new information in regard to the reconstruction of medieval whaling endeavours.

Historical sources appear to suggest a pattern in medieval cetacean exploitation. Gardiner (1997) has proposed that whale populations along the English coast were declining in the late twelfth and early thirteenth century. The records of Battle Abbey suggest that strandings appear to occur less frequently from the mid-twelfth century onwards (Gardiner, 1997). This has been proposed for the English Channel coastline for France as well by Musset (1964). The historical records for that region suggest that whaling was most frequently practiced during the eleventh and first half of the twelfth century but declined soon after. Whaling in the Bay of Biscay peaked during the twelfth and thirteenth centuries, but after that, Basque whalers ventured to other regions in pursue of whales (Fischer, 1881, 24; Kurlansky, 1991).

This suggested pattern based on historical sources, has led to Gardiner (1997) proposing a three-phase system for England. In the first phase, during the Anglo-Saxon period until the eleventh century (AD 410-1066), cetacean exploitation was limited to coastal communities. Active whaling was occasionally undertaken, but opportunistic scavenging of stranded individuals was the most common source of procurement. Cetacean meat did not travel far inland, and it was not restricted to the social elite.

The second phase started in the early eleventh century and lasted until AD 1300, in which the King, nobility and clergy were interested in cetacean consumption and tried to monopolize its consumption. It was during this period that porpoises were occasionally exploited, and whale meat was imported from France to England. Active whaling was undertaken in Normandy during this period and for England, as well as several other European countries, stranded cetaceans were from this period onwards a royal and seigneurial right.

The third phase started at AD 1300 and as mentioned, the whale population appears to have been in decline from this period onwards. This led to less whale meat being available to the social elite and it fell out of favour. Stranded cetaceans remained a royal and seigneurial right but claiming of these stranded cetaceans by the social elite was less frequently undertaken. Porpoise meat however continued to be sold as a high-status food.

This system was set out over twenty years ago, was based on historical sources and limited zooarchaeological data, and was restricted to England. Up until this point no extensive study has attempted to combine zooarchaeological and historical sources in order to reconstruct medieval whaling practices. Many studies have focused on the one discipline and used some arguments or sources from the other discipline, but a truly interdisciplinary study, connecting historical and zooarchaeological sources, has not been conducted and has limited our understanding of past cetacean exploitation. On top of this, even though many medieval historical sources hold valuable data in regard to human-animal interaction, many zooarchaeological reports concerned with historical periods, refer to historical texts only in anecdotal or factual manners instead of interpretative (Ervynck, 2004).

Moreover, though more research has been performed on medieval whaling from a historical perspective than from an archaeological one, historical sources concerned with cetaceans are also still understudied, without a comprehensive overview of European medieval whaling practices being created. Lindquist (1997) noted that studies regarding Norse whaling and cetacean exploitation are limited and not comprehensive. It has even been suggested that "no detailed comprehensive presentation of Scandinavian medieval whaling and whale utilisation" exists (Schnall, 1992). This appears to not only be the case for the Norse, but for many other medieval cultural groups as well.

This study aims to combine the historical sources and zooarchaeological sources in order to reconstruct medieval whaling practices, leading to the main research question: **What are the Social Implications of Cetacean exploitation in Medieval Northern and Western Europe?** This question is primarily based on an assessment of Gardiner's proposed three-phases, though the analysis will encompass a larger region than Gardiner was concerned with, namely northern and western Europe (see figure 1, for all the countries assessed as part of this study, and table 1 for time periods considered).

To answer this question, it will be attempted to reconstruct whether cetacean exploitation was limited to particular social strata. It will be analysed whether the undertaking of whaling itself, the scavenging of stranded cetaceans, the consumption of cetacean meat, and the utilisation of raw resources (e.g. bone, baleen, and teeth) was limited to the social elite. Moreover, the species exploited will be identified through the study of both the historical and

Table 1. Time periods considered in this study. *For Scandinavia this includes the Germanic Iron Age (fifth to eighth centuries AD) and the Viking Age (late eighth – mid eleventh century AD)

Time period	Dates
Early Medieval*	AD 400 - 1066
High Medieval	AD 1066 - 1300
Late Medieval	AD 1300 - 1500
Post Medieval	AD 1500 - 1600

Figure 1. Map of Europe, with countries and regions considered as part of this study in dark grey.

zooarchaeological sources and it will be attempted to find out in which regions and periods active whaling was undertaken, and in which opportunistic scavenging of stranded individuals was the main source of procurement.

In this study a clear distinction between "whaling" and "cetacean exploitation" is made. "Whaling" in this study is considered active hunting of cetaceans in its widest sense (e.g. hunting using spears or harpoons, driving them to shore, poisoning them, trapping them in bays of inlets, etc.). "Cetacean exploitation" on the other hand both encompasses actively caught individuals through "whaling" or the (opportunistic) exploitation of stranded individuals.

Material and Methods

In order to answer the main question "**What are the Social Implications of Cetacean exploitation in Medieval Northern and Western Europe?**", this study will be subdivided into chapters that deal with several sub-questions.

The first chapter is a background chapter, focusing on the biological and ecological aspects of the cetaceans present within the North Atlantic in order to answer the question "Which cetacean species were present in the eastern North Atlantic Ocean during the Medieval Period?". A short overview of cetacean evolution will be provided as well, primarily focusing on the changing osteology of the animals from terrestrial mammals to fully aquatic mammals. Furthermore, in order to understand cetacean exploitation, this chapter will provide a general overview of the 35 cetacean species present within the North Atlantic. Distribution, size, weight, estimate of current world and North Atlantic population, pelagic/coastal lifestyle, pod size, prey, maximum speed, lifespan, as well as whether the animal sinks or floats after it has been killed, are contributors to potential active whaling activities.

Another important aspect to consider regarding the ecology of cetaceans is the tendency to live strand. Though now often caused by anthropogenic factors (e.g. entanglement in nets, ship strikes, or the swallowing of plastic), natural strandings also occur frequently and happened in the past as well. Coastal communities might have exploited the carcasses, eliminating the need to venture out on the sea and pursuing the animals. Furthermore, the products that can be exploited from cetacean carcasses will be discussed in this chapter. A wide variety of products including meat, baleen, bone, and blubber can be extracted, but this varies for the various species.

The second chapter focuses on the medieval historical sources regarding cetaceans and will answer the question: "What do medieval sources mention about human-cetacean interaction and how should these be interpreted?". Several communities, cultures and societies have been associated with cetacean exploitation. Historical sources associated with these cultures will be discussed in order to get an idea of in which geographic regions and temporal settings cetacean exploitation might have been practiced.

Furthermore, medieval cetacean exploitation and consumption has often been associated with social elite. An overview of the medieval diet and the variation between the three orders (peasantry, nobility, and clergy) will be provided. This is of importance in associating zooarchaeological data with historical data. Additionally, an overview of medieval marine resource exploitation will be provided, as cetacean exploitation might be linked to the exploitation of other marine resources.

The third chapter will focus on zooarchaeological analysis concerned with cetaceans and is concerned with "How can zooarchaeological cetacean remains be studied?". First of all, the taphonomic factors that affect the survival of cetacean remains, the identification of the specimens and the publicizing of the data will be discussed. Especially the identification of cetacean remains has been problematic and little research has been undertaken in that field. This can primarily be attested to the poor preservation of the cetacean remains and the relatively few instances their remains have been encountered in medieval European contexts.

Moreover, the techniques used to identify cetaceans remains, including aDNA and ZooMS, as well as identification based on morphology will be discussed. The lack of extensive osteological reference collections as well as the inexistence of a comprehensive osteological manual have hampered the identification of cetacean remains from the archaeological record.

In addition, the distinction between bones deriving from actively caught individuals or opportunistically exploited live stranded individuals is hard if not impossible. Numbers of medieval cetacean remains are generally low, often leading to the interpretation that the identified material merely derived from stranded individuals that were opportunistically exploited. This was for example the case for the site of Barvas, Scotland in which "Occasional whale bones suggest the periodic exploitation of strandings, although in general marine mammals are surprisingly scarce on the site" (Amit, 1996). The same was the case for the site of Green Shiel, Lindisfarne, England for which "The remains of some marine mammals - seals and a whale - which may have been stranded on the nearby shore" were uncovered (O'Sullivan and Young, 1995). While in other instances the zooarchaeological remains of cetaceans have been interpreted to have been simply used as a raw material for the creation of tools and artefacts. This was for example the case for the site of Church Street, Seaford, England, for which "The occurrence of two fragments of whale bone might perhaps be related to its use commercially, and is certainly not evidence of a food joint" (Brothwell, 1979).

These cases clearly indicate the contradicting reasoning that is active in the field of zooarchaeology, and interpretation of zooarchaeological cetacean remains are based on these contradicting arguments. On the one hand cetacean material is interpreted to represent remains of opportunistically exploited stranded cetaceans used for dietary purposes, while on the other hand the material is interpreted to have been merely used for the creation of artefacts or tools. These cases clearly show the confusing reasoning zooarchaeologists are dealing with. Even though zooarchaeological cetacean remains will probably always be hard to interpret, this study aims to give a better understanding of medieval cetacean exploitation and the accompanied social aspects, based on extensive zooarchaeological and historical analyses.

It is however clear that in order to prove the existence of active whaling, besides zooarchaeological remains, secondary sources are required as well. These include historical sources, ethnographic accounts, hunting tools and gear, and archaeological shipwrecks. Lastly, this chapter will discuss the potential zooarchaeology can have on the interpretation of social aspects of medieval cultures, especially the identification of a high-status diet and dietary differences between the three orders.

For chapter four, an extensive zooarchaeological study was conducted in order to answer the question "At which medieval sites were cetacean remains found?". As part of this study zooarchaeological cetacean remains from medieval contexts were assessed. No new material was studied and therefore this is a literature review or Big-Data analysis of previously published reports. This study aims to seek and understand the changing cultural perception of cetaceans in medieval Europe, through the analysis of zooarchaeological remains. This was accomplished by examining the species, skeletal elements, number of identified specimens (NISP), age of the individuals, signs of butchery present on the specimens, and the contexts the remains were found in and compare those to the historical record. This resulted in a better understanding of medieval cetacean exploitation, where active whaling was practiced, who performed it, which species were exploited, and who had access to the consumption of cetacean meat.

As both the temporal as well as the geographic scope of this study is considerable, several in depth case studies were undertaken as well. Two case studies are undertaken based on geography and include a Netherlands and Flanders case study and an England case study. These case studies offer the possibility to identify differences between the geographical region by analysing the zooarchaeological remains and identify those to the species level and connect those to the historical sources concerned with those areas. These case studies, therefore, perfectly make use of all the sources previously discussed and interpret them on a smaller scale.

In addition, another case study is based on the exploitation of a single species. The species considered is the grey whale (*Eschrichtius robustus*). This species is thought to have gone extinct in Europe somewhere during the Medieval Period. It is still not precisely understood what caused the extinction of this species from the North Atlantic. Remains of this species have not frequently been identified in archaeological contexts, but as part of this study four or potentially five grey whale specimens from Roman and medieval contexts in the Netherlands were identified, suggesting that the species might have been hunted as well.

By addressing all the sub-questions, as well as looking at several case studies in more detail, this study will create a comprehensive database based on both zooarchaeological and historical sources, that eventually leads to a better understanding of the social implications of cetaceans in Northern and Western Europe. It allows for the reconstruction of medieval whaling practices and the social emphasis placed on the exploitation on the cetaceans (both opportunistically stranded and actively caught), the consumption of cetacean meat, which species were exploited, and the using of secondary resources extracted from the carcass.

1

Cetaceans and the North Atlantic Ocean

In this background chapter, first, a short discussion regarding the disputed phylogenetic classification of cetaceans will be provided, as well as a short overview of the evolution of cetaceans and the transition from a terrestrial lifestyle to an eventually fully aquatic lifestyle undertaken by their ancestors and how this affected their osteological features.

Furthermore, the biological and ecological aspects of the cetaceans present in the eastern North Atlantic will be discussed. This is however based on modern data. In the past several species might have occupied different regions, but centuries of excessive whaling might have forced them to retreat from particular areas. Moreover, this might have allowed for other species to take over those grounds, in this way decreasing the distribution of the one species and increasing the distribution of the other. As a result, the distribution of several species is likely to have been very different to their distribution in the past.

Additionally, it is important to consider the natural environment, in the case of this study: the eastern North Atlantic. The Atlantic is indented by numerous smaller seas, gulfs and bays which provide a variety of environments for the different cetacean species. Humans have adapted the coastline significantly, but in the Medieval Period, large areas were still unpopulated and unaltered. Information regarding bathometry, gyres, water characteristics and food availability for all the separate regions will be provided as all these aspects have an influence on the cetaceans present within each region.

Numerous cetacean species are present within the North Atlantic and all these cetacean species will be discussed separately and information regarding their ecology (range, size, diet, speed, etc.) will be given. Their osteological features (number of vertebrae, number of phalanges, etc.) will also be provided as this is important when considering the different species within a zooarchaeological context. This data will be provided for the 34 species currently present in the European North Atlantic, as well as one species that disappeared from the region during the Medieval Period: the grey whale (*Eschrichtius robustus*).

Another ecological aspect will be discussed separately: cetacean strandings. These events happen regularly, and it can be assumed that these events happened frequently in the past as well, if not more frequently as the population size of most species would have been higher prior to the industrial whaling era, which could have led to opportunistic scavenging of the carcasses. Modern stranding data will be analysed and will be compared to zooarchaeological data discussed later in this study. However modern strandings are often caused by anthropogenic factors and should therefore be considered carefully, as this was not the case during the Medieval Period and might have resulted in different stranding patterns.

Furthermore, the various exploitable products coming from cetaceans will be discussed as well. These products included meat, baleen, bone, teeth, ivory, oil, spermaceti, ambergris, and blubber. These products can be used for various purposes and are of importance when considering cetacean exploitation, as different species can provide different products.

1.1. Origin and Evolution of Cetaceans

There are six separate extant mammal groups that have adapted to a (semi-)marine lifestyle and are therefore called "marine mammals". Improving foraging efficiency was the main driver for an adaptation to a (semi-)marine lifestyle and has proven to be extremely successful as it has resulted in the evolution of the largest animal to have ever lived on earth: the blue whale (*Balaenoptera musculus*), as well as numerous other species (Perrin, Würsig and Thewissen, 2009). Cetaceans are by far the most successful of the marine mammal group, but they have undergone millions of years of evolution to accomplish this.

1.1.1. Marine Mammals

The six groups of extant marine mammals include the Cetacea infraorder, the Sirenia order (including the dugong and manatees), the Pinnipedia clade (including the true seals, sea lions, fur seals and the walrus), the polar bear (*Ursus martimus*), the sea otter (*Enhydra lutris*) and the marine otter (*Lontra feline*) (Tedford, Barnes and Ray, 1994, 11-32; Jefferson, Webber and Pitman, 2008, 37-44; Canto *et al.*, 2008-09, 918-922).

Fossil remains suggest that in the past there were even more mammal lineages that had a (semi)-marine lifestyle, these include the Desomostylia order (extinct relatives to both the Sirenia order, as well as the Proboscidea order, which includes elephants), the *Thalassocinus* (an aquatic sloth) the *Kolponomos* (a marine bear), and two members of the *Eutriconodonta* order (a group of early mammals that existed during the Jurassic and Cretaceous period): the *Ichthyoconodon* and the *Dyskirtodon* (Perrin *et al.*, 2009).

While, especially the members of the Sirenia order show similarities to the members of the Cetacea order, various large-scale molecular analyses have confirmed that the six current marine mammal lineages all adopted a marine lifestyle parallel to each other, which is a classic example of convergent evolution (Foote *et al.*, 2015, 272). The Cetacea order switched to an aquatic lifestyle around 55 million years ago (mya), while the Sirenia order followed 10 million years later, around 40 mya. The other 4 groups followed later and have developed semi-marine lifestyles. The Pinnipedia clade made this switch around 20-25 mya, followed by the sea otters around 2 mya, while the polar bear did this only around 130,000 to 110,000 years ago.

Of the six lineages, only three can be found in North Atlantic waters. The infraorder Cetacea, includes 88 known extant species of whales, dolphins and porpoises, of which 35 can be found in the North Atlantic area. The infraorder includes the two parvorders Mysticeti (also known as baleen whales) and Odontoceti (also known as toothed whales; (Rice, 2009c, 236). These two parvorders of Mysticeti and Odontoceti can be further subdivided into families, with Mysticeti containing four families and Odontoceti ten families (figure 2). The four modern Mysticeti families are: Balaenopteridae, Balaenidae, Eschrichtiidae and Neobalaenidae and contain at least 15 known species. Of these four families, only the former two can be found in the North Atlantic, while the third was present until the seventeenth- or eighteenth-century AD.

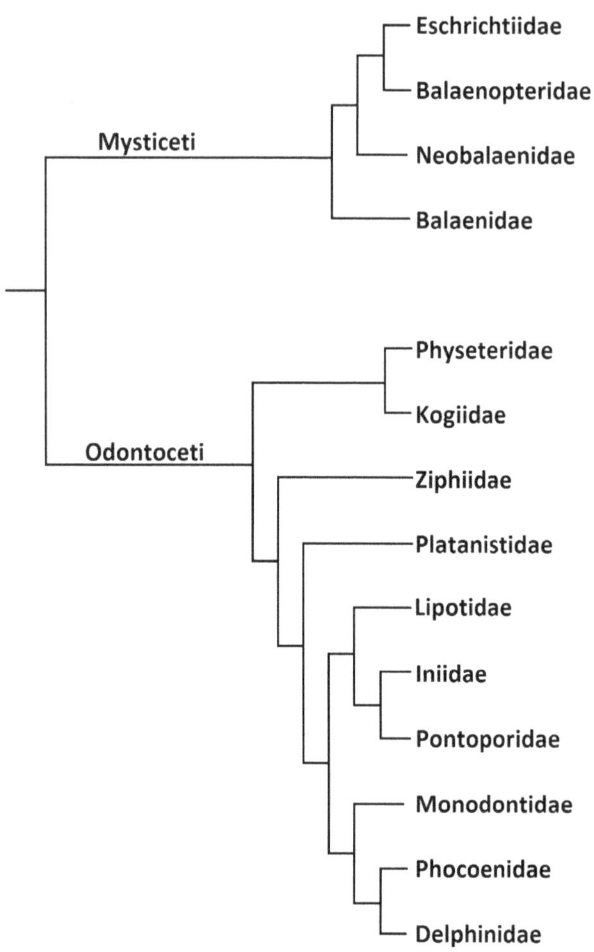

Figure 2. Phylogenetic classification of the infraorder Cetacea and its families, based on Hooker (2009) and Deméré *et al.* (2008), created by author.

The ten Odontoceti families are the Delphinidae, Monodontidae, Phocoenidae, Iniidae, Pontoporiidae, Plataniistidae, Lipotidae, Kogiidae, Physeteridae, and Ziphidae families and contain at least 73 known species (Rice, 2009c, 235). Of these, the Platanistidae, Pontoporiidae, Iniidae, and Lipotidae families are river dolphins. The baiji/Chinese river dolphin (*Lipotes vexillifer*), the only member of the Lipotidae family, has not been seen since 2006 and has been presumed extinct ever since, leaving only nine Odontoceti families still extant. Of these nine families, only member of the Delphinidae, Monodontidae, Phocoenidae, Kogiidae, Physeteridae, and Ziphidae are present in the North Atlantic.

The baiji is the very first cetacean species to have gone extinct by human activity (Turvey *et al.*, 2007, 53-540; Zhou, 2009, 71-76). It however still has a "Critically Endangered" IUCN status. Additionally, the vaquita (*Phocoena sinus*) and the narrow-ridged finless porpoise (*Neophocaena asiaoientalis*) also have a "Critically Endangered" status. Numbers of vaquita fell to merely 30 individuals in November 2016, making this species the most endangered species of marine mammal since the extinction of the baiji (Taylor *et al.*, 2016).

The Pinnipedia clade is also represented in the North Atlantic by the common/harbour seal (*Phoca vitulina*), grey seal (*Halichoerus grypus*), harp seal (*Pagophilus groenlandicus*), hooded seal (*Cystophora cristata*), bearded seal (*Erignathus barbatus*), ringed seal (*Pusa hispida*), Mediterranean monk seal (*Monachus monachus*) and the walrus (*Odobenus rosmarus*). Additionally, the polar bear can be found in the waters of the North Atlantic as well, while the other three lineages all inhabit more southern waters (Shirihai and Jarrett, 2011, 260-359).

1.1.2. Phylogenetics

The phylogenetic classification of cetaceans is still a matter of debate. Cetaceans were originally thought to be closely related to the extinct mesonychids (of the order Mesonychia or Acreodi). The mesonychids were wolf-like creatures with five toes that bore hoof-like claws. Their close relationship was accepted and the two were put together in the order Cete by McKenna and Bell (1997, 631).

However, when molecular analysis was applied to study the classification of cetaceans, another theory arose. The molecular studies suggested that cetaceans are the sister-taxon of the family Hippopotamidae and are part of the Artiodactyla order (even-toed hoofed mammals) that includes the Bovidae, Moschidae, Cervidae, Antilocapridae, Giraffidae, Trangulidae, Suidae, Tayassuidae, and Camelidae families (Gatesy *et al.*, 2013, 487). Since it is thought that these families and cetaceans share a common ancestor, it was suggested to change the name from Artiodactyla in Cetartiodacyla (Cetacea + Artiodactyla), though this name change is also still a matter of debate (Price, Bininda-Edmonds and Gittleman, 2005, 452).

As stated above, of the six lineages of extant marine mammals, the Cetacea infraorder is the group that developed the earliest and its evolution and adaptation to an aquatic lifestyle can be traced back to 55 million years ago. Around this time the very first "pre-whales" developed, the ancestors of the two extant parvorders of Cetacea, the now extinct third parvorder: Archaeoceti (Thewissen, 2009, 46).

1.1.3. Evolution of Modern Cetaceans

The evolution, and adaptation to an aquatic lifestyle, took cetaceans millions of years. This evolution also changed numerous aspects of their body, including, and most relevant to this study, their osteology. Their skulls, teeth, vertebral column, and extremities, all changed, and a short overview will be given in order to understand their osteology.

The Archaeoceti are a group of ancestors to modern whales, that lived in the Eocene, approximately 55 to 34 million years ago (Thewissen, 2009, 46). Five families of Archaeoceti are recognized, each representing an important step in evolution: Pakicetidae, Ambulocetidae, Remingtonocetidae, Protocetidae and Basilosauridae (Williams, 1998, 1-28).

The Pakicetidae, were digitigrade hoofed animals that lived in the Eocene, 50 million years ago and lived in what is now Pakistan (Lazarus, 2006, 10). Though the Pakicetidae were still terrestrial mammals, they possessed the first adaptations to live partially in fresh water (Thewissen, 2009, 46). Their bones were unusually thick (caused by osteosclerosis), which made the animal heavier to counteract the buoyancy of the water. This allowed the animals to wade more effectively through the water (Gray *et al.*, 2006, 638-653). Pakicetids were additionally able to listen underwater, but not to an extent that provides directional hearing. Their fossil remains have only been discovered in shallow freshwater deposits, suggesting that Pakicetidae were not good swimmers and did not venture into marine environments (Thewissen, 2009, 46).

The next stage in the evolution of cetaceans were the Ambulocetidae, which lived approximately 49 million years ago. Ambulocetidae lived in both fresh as well as saltwater, tolerating a wide range of salt concentrations (Thewissen and Fish, 1997, 482-490). Furthermore, Ambulocetidae had an adaptation in the nose enabling them to swallow whilst being underwater and they were additionally able to hear underwater as well. Their teeth are also comparable to the modern toothed whales and they probably displayed a lifestyle comparable to modern crocodiles. They additionally had strong hind legs which they used for pelvic paddling (swimming using the hind limbs for propulsion) in combination with caudal undulation (using the undulations of the vertebral column for propulsion), swimming in a similar way as otters (Thewissen and Fish, 1997, 482-490). These adaptations however took a toll, making the Ambulocetidae slow walkers as well as slow swimmers (Thewissen, 2009, 47).

Following the Ambulocetidae in the line of evolution, are the Remingtonocetidae, that lived approximately 49 to 43 million years ago. They were even more adapted to an aquatic lifestyle than their ancestors which is suggested by the fact that their fossils were retrieved from a wide variety of coastal marine deposits (Thewissen, 2009, 47). Their limbs were reduced in size, though they were still able to support their weight on land, and their semi-circular canals (which are used for balance in terrestrial mammals) were also reduced in size. They used caudal undulation for swimming and were well adapted to it. The inner ears of Remingtonocetidae were larger than those of its ancestors and are set further apart, possibly increasing directional hearing (Thewissen, 2009, 47).

The Protocetidae are the next family in line of cetacean evolution. The Protocetidae family is a heterogeneous and diverse group that lived around 48 to 35 million years ago with some of the members being fully aquatic, while others were probably still able to support their weight on land. This cetacean group was the first to leave the Indian subcontinent and their remains have been uncovered from coastal deposits in Africa, Europe, and North America as well (Thewissen, 2009, 47).

Following the Protocetidae family, the Basilosauridae emerged, which can be subdivided into two subfamilies: Basilosaurinae and Dorudontiae. These two families lived around 41 to 35 million years ago and are the oldest most fully aquatic cetaceans. The two groups are closely related, and both lack the melon organ modern Odontoceti whales use for echolocation. Additionally, the two families both have relatively small brains suggesting both were solitary and did not live in large groups with a complex social structure. The blowhole is higher up the snout, closer to the position of the modern cetaceans. Furthermore, both families still had external hind limbs, though these were very reduced in size and probably were not involved in locomotion (figure 3). Their pelvic bones were no longer attached to the vertebral column (Fordyce and Barnes, 1994; Uhen, 2009a, 91-94).

The Cetacea infraorder eventually split into the Mysticeti and Odontoceti suborders. Molecular data has suggested that this happened around 27 million years ago. However, the oldest fossils date to about 34 million years ago and the actual date of the split is still a matter of debate. At this point, after millions of years of evolution, their hindlimbs had almost fully disappeared like those in modern cetaceans and they had developed a full aquatic lifestyle (Jamieson and Miller, 2007, 111).

The Odontoceti, also known as the toothed whales, have as their name suggests teeth and this is the major aspect that sets them apart from the Mysticeti. Modern Odontoceti nearly all possess a homodont dentition (a set of teeth that possess the same tooth morphology), and are unlike the early Odontoceti species, like the Agorophiidae, Waipatiidae, Simocetidae, and Squalodontidae families which are primarily heterodont (Uhen, 2009b, 304).

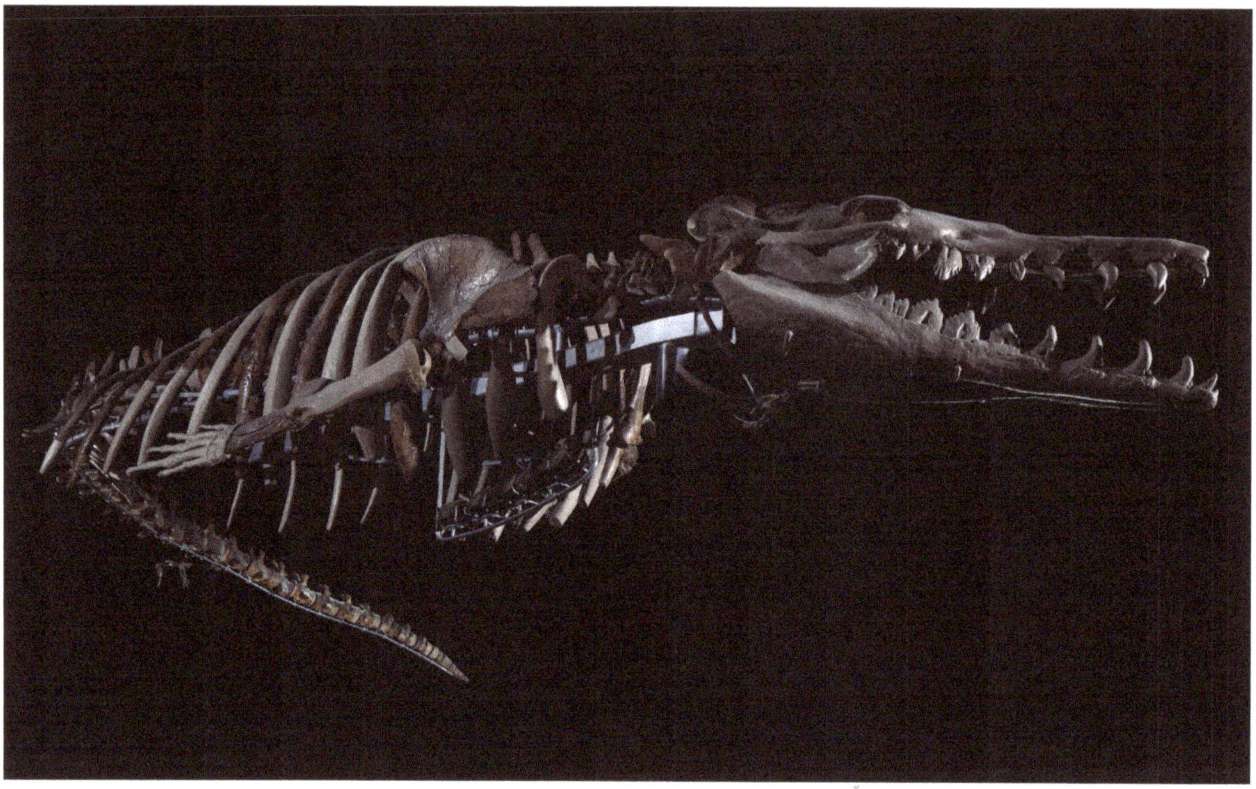

Figure 3. *Basilosaurus cetoides* skeleton. USNM Number V4675. Reproduced courtesy of the Smithsonian Institution. CC0 License. Available at: <https://www.si.edu/object/basilosaurus-cetoides-owen-1839:nmnhpaleobiology_3342386>.

Another aspect that sets the Odontoceti apart from the Mysticeti, is their ability to use echolocation. This ability developed around the time when the Mysticeti and Odontoceti split into two separate parvorders, approximately 34 million years ago (Lazarus, 2006, 15). This adaptation resulted in them no longer relying on their sense of sight anymore, but rather on their echolocation to hunt prey. This furthermore resulted in them being able to dive deeper to places with little light, opening new feeding grounds. The family of the Squalodontidae are considered to be the first cetacean family to have possessed the ability to use echolocation and lived around 33 to 14 million years ago (Whitmore and Sanders, 1977, 304-320).

The earliest Mysticeti whales had not developed baleen just yet and still had teeth. The toothed Mysticeti species evolved in the late Eocene and went extinct before the start of the Miocene (Berta and Deméré, 2009, 749). Since baleen rarely fossilizes, the origin of it is still poorly understood (Marx and Fordyce, 2015, 1-14). There appears however, to have been a gradual transition from Mysticeti whales with exclusively teeth (Llanocetidae, Janjucetidae, and Mammalodontidae), to an intermediate form with both teeth as well as baleen (Aetiocetidae), to a form with just baleen (Eomysticetidae, Cetotheriidae, and all the modern baleen whale families; Deméré *et al.*, 2008, 15-37; Uhen, 2009b, 304). Embryotic modern baleen whales still possess deciduous teeth which are resorbed prior to birth (Karlsen, 1962). Besides having teeth, early members of the Mysticeti order differed in more ways from their modern descendants. The degree of telescoping of the skull was less evolved and their blowhole was only positioned midway between the tip of the rostrum and the orbit (Berta and Deméré, 2009, 749-753).

Eventually, cetaceans developed into the most successful marine mammal group to exist, with numerous extant, as well as extinct, species. Their osteological features are perfectly developed to a marine environment, with a long vertebral column that aids their propulsion, front limbs that turned into pectoral fins, the almost complete reduction of hind limbs, as well as several adaptation to their skulls, which allows them to effectively hear underwater, and in the case of the Odontoceti, use echolocation in order to find prey.

1.2. The North Atlantic Ocean

Before going into the topic of cetaceans, first the geographic range this study is concerned with will be discussed. All the archaeological material discussed in this study was excavated from sites from Northern and Western Europe. However, cetaceans are unlike terrestrial mammals, not found in terrestrial settings, but in the bordering seas and oceans. Past communities whom exploited cetaceans must have gone onto sea by boats and head into the domain of the cetaceans. As there are numerous cetacean species, which all have their own habitat and diet. It is therefore important to discuss all the different parts of the Atlantic Ocean separately, as there is quite a variation in bathometry, water characteristics and ecosystems, which can support different cetacean species.

The Atlantic Ocean is the second largest ocean on earth. The total area of the ocean is 106,460,000 square kilometres. It is connected to the Arctic Ocean in the north, the Pacific Ocean in the southwest, the Southern Ocean in the south and the Indian Ocean in the southeast. This Atlantic Ocean is divided into the North Atlantic Ocean and the South Atlantic Ocean, which is divided by the Equatorial Counter Current, which is located eight degrees north of the equatorial plane.

Located within the North Atlantic Ocean is the North Atlantic Gyre. This is a circular system of ocean currents that stretches from Europe to North America. The gyre is composed of several currents, including the Gulf Stream and the North Atlantic Current. The Gulf Stream originates in the Gulf of Mexico and follows the eastern coastline of the United States up north to Newfoundland. At this point it crosses the North Atlantic and changes into the North Atlantic Current which runs to Europe. As a result, the waters around north-western and western Europe are much warmer. Furthermore, the current is believed to have a strong effect on the Western Europe's and Northern Europe's climate.

Each spring (and to a lesser extent in the fall as well) there is a huge bloom in phytoplankton in the North Atlantic. This is the base of a complex food-chain with at the top cetaceans (and humans) (Szabo, 2008, 76). The abundance of plankton in Atlantic waters, attracts numerous species of cetaceans. 35 species are frequently sighted in or are known to have inhabited the North Atlantic and will be discussed further on in this chapter.

The North Atlantic Ocean additionally has numerous gulfs, bays and seas. This study is concerned with a number of them, including the Baltic Sea, Denmark Strait, North Sea, Norwegian Sea, Barents Sea, Celtic Sea, Irish Sea, Bay of Biscay, and the English Channel, and all of these will be shortly discussed. The Northeast Atlantic geographic regions will be subdivided based on the divisions developed by the OSPAR Commission (2016). This is a commission that is concerned with protecting and conserving the North-East Atlantic and its resources. OSPAR has divided the North-East Atlantic into five areas: Arctic Waters, Greater North Sea, Celtic Seas, Bay of Biscay and Iberian Coast, and the Wider Atlantic. Additionally, a sixth area will be added to this study, called the "Baltic area".

1.2.1. Arctic Waters

The Arctic waters are the most northern waters this study is concerned with and include the Barents Sea, the Norwegian Sea, and the Denmark Strait. The area is characterised by its harsh climate and its ice coverage. The ecosystems in the region are however still rich (OSPAR, 2016).

The most northern sea in this region is the Barents Sea. This sea, named after Dutch navigator Willem Barentsz, is located north of Norway and the Kola peninsula (Russia), west of Novaya Zemlya, east of the Norwegian Sea and south of the archipelagos of Svalbard/Spitsbergen. Unlike all the other seas discussed in this chapter, the Barents Sea is actually a marginal sea of the Arctic Ocean. It is located in the upper north of the research area this study is concerned with. It is a shallow shelf sea, with an average depth of 230 metres (Loeng, 1991, 5-18).

The Barents Sea has a high biological production in comparison to other seas and oceans of similar latitudes. Close to the ice edge, the spring bloom of phytoplankton starts early, which feeds zooplankton. Zooplankton is subsequently consumed by fish and whales. Baleen whales were common in the area, especially the bowhead whale (*Balaena mysticetus*). The North Atlantic right whale (*Eubalaena glacialis*) is no longer present in the Barents Sea (Loeng, 1989, 327-365).

Another sea part of the Arctic Waters regions is the Norwegian Sea. It is a marginal sea, part of the North Atlantic Ocean and is located northwest of Norway, north of the Faroe Islands and Shetland Islands, and east of Iceland. It borders the Barents Sea to the northeast. The Norwegian Sea is not part of a continual shelf, unlike many other seas. As a result, it is much deeper with a depth of about two kilometres on average. The sea is relatively warm which is the result of the warm North Atlantic Current, which ensures that the Norwegian Sea is ice-free year-round (Bjelland *et al.*, 2011, 101-117).

Fish are abundant in the coastal regions of the Norwegian Sea, which use that area for spawning, especially cod (*Gadus morhua*) and herring (*Clupea harengus*; Bjelland *et al.*, 2011, 101-117). The Norwegian Sea is also rich in plankton, which attracts the large Mysticeti whales. Rorquals, are abundant in the Norwegian Sea, with especially high numbers of minke whales (*Balaenoptera acutorostrata*). Other include humpback whale (*Megaptera novaeangliae*), sei whale (*Balaenoptera borealis*), and fin whale (*Balaenoptera physallus*). The blue whale (*Balaenoptera musculus*) was also frequently sighted in the past, but since the twentieth century, whalers started to target this species and it is rarely seen in the region anymore. The North Atlantic right whale was once also abundant, but the population has been decimated by whaling practices. The bowhead whale is recovering from whaling practices and is more often seen in the Norwegian Sea. The white-beaked dolphin (*Lagenorhynchus albirostris*) and the sperm whale (*Physeter macrocephalus*) are also often sighted in the coastal regions of the Norwegian Sea (Reid, Evans and Northridge, 2003).

The Denmark Strait is also incorporated into the "Arctic Waters" region. This strait is located between Iceland and Greenland, with the island of Jan Mayen located northeast of it. It is a relatively narrow strait, reaching a maximum depth of about 200 metres (Swift, Aagaard and Malmberg, 1980, 29-42). Baleen whales are common in the area, North Atlantic right whales have however disappeared from the strait as well (Reid, Evans and Northridge, 2003).

1.2.2. Greater North Sea

The North Sea is a marginal sea of the Atlantic Ocean located between Scotland, England, France, Belgium, the Netherlands, Germany, Denmark, and Norway. It is a relatively shallow sea; the southern part is only 25-55 meters deep, in the north the sea is around 100 to 200 meters deep. The sea is 970 kilometres long and 580 kilometres wide and covers an area of approximately 750,000 square kilometres (MacGarvin, 1990, 14; Camphuijsen and Peet, 2006, 13-31).

Through the English Channel and the northern part of the North Sea, the Atlantic Ocean's currents reach the North Sea, bringing warm water into the (SeaOnScreen, 2013; OSPAR, 2016). In the North Sea itself an anti-clockwise current direction is present. (MacGarvin, 1990, 14-18; International Centre for the Environmental Management of Enclosed Coastal Seas, 2003; Camphuijsen and Peet, 2006, 13-31).

The Romans considered the North Sea to be one of the roughest in their world. Storms and squalls started without warning, making it a risk to cross it (MacGarvin, 1990, 26). The North Sea is home to hundreds of thousands of herrings and other fish (MacGarvin, 1990, 7; SeaOnScreen, 2013). Most cetaceans are not that abundant in the North Sea. The harbour porpoise is an exception to this, being present in high numbers, especially in the central and northern part of the North Sea. Minke whales, Atlantic white sided dolphins (*Lagenorhynchus acutus*), white beaked dolphins and common bottlenose dolphins (*Tursiops truncatus*) are also relatively frequently sighted (Camphuijsen and Peet, 2006, 47-87).

1.2.3. Celtic Seas

The Celtic Sea is the part of the Atlantic Ocean south of Ireland, west of the English Channel, and north of the Bay of Biscay. In the south and west the Celtic Sea is delimited by the continental shelf. In the northeast, the Celtic Sea has a depth around 100 metres, while in the southwest the sand ridges have a similar depth, but they are separated by troughs which are approximately 150 metres deep (OSPAR, 2016). The common bottlenose dolphin, short-beaked common dolphin (*Delphinus delphis*), harbour porpoise, killer whale (*Orcinus orca*), minke whale, long-finned pilot whale (*Globicephala melas*), Risso's dolphin (*Grampus griseus*), sperm whale (*Physeter macrocephalus*), striped dolphin (*Stenella coeruleoalba*), white beaked dolphin and Atlantic white sided dolphin are all frequently sighted in the Celtic Sea (Reid, Evans and Northridge, 2003).

Included in the "Celtic Seas" area is the Irish Sea. The Irish Sea is the body of water that separates the islands of Great Britain and Ireland. In classical times it was known as the *Oceanus Hibernicus*. It is connected to the Inner Seas off the West Coast of Scotland in the north and the Celtic Sea in the south. The sea is about 210 kilometres long and 240 kilometres wide. The sea is relatively shallow with its greatest depth being 175 metres. It supports a variety of fish, phytoplankton and zooplankton, that serve as food sources for larger creatures, including larger fish species, such as the basking shark (*Cetorhinus maximus*), common seal (*Phoca vitulina*), and grey seal (*Halichoerus grypus*). Cetaceans are frequently sighted in the Irish Sea, with the harbour porpoise and common bottlenose dolphin being the most frequently encountered species (OSPAR, 2016). Several of the rorqual species are also frequently encountered, including the common minke whale, sei whale, fin whale and the humpback whale. The sperm whale, northern bottlenose whale (*Hyperoodon ampullatus*), killer whale, long-finned pilot whale, Risso's dolphin, striped dolphin and the white-beaked dolphin are also common visitors (Reid, Evans and Northridge, 2003).

1.2.4. Wider Atlantic

This area is located between the abyssal plain and the Mid-Atlantic Ridge and is defined by deep waters. Several currents run in the upper layers from west to east. There are outflows to the northwest, which are important in maintaining the relatively mild climate of North-West Europe. The region also sustains a relatively high biomass. Pelagic species of fauna are twice as diverse in the southern part of the region than in the northern part. The biomass however shows the reverse (OSPAR, 2016).

Numerous cetacean species are present within the region. Rorquals, sperm whales, beaked whales, pygmy sperm whale (*Kogia breviceps*) and dwarf sperm whales (*Kogia sima*) are more frequently found in pelagic waters than closer to the European shores. Several Delphinidae species are also found in the deeper waters. This region is of a lesser importance to this study as the only landmasses located within its reach are the Azores, which are excluded from this study.

1.2.5. Bay of Biscay and Iberian Coast

The Bay of Biscay is a gulf of the northeast Atlantic Ocean, south of the Celtic Sea and it borders France to the east and Spain to the south. Parts of the bay are relatively shallow as the continental shelf extends far into the bay, while further off coast regions can be as deep as 4735 metres (Pingree and Le Cann, 1990, 857-860). There is a high diversity in fish in the bay, with the sardine, mackerel, anchovy and blue-whiting (*Micromesistius poutassou*) being abundant in the pelagic habitat. Hake, albacore tuna (*Thunnus alalunga*) and Bluefin tuna (*Thunnus thynnus*) are also present and feed on smaller pelagic fish (Poulard and Blanchard, 2005, 1436-1443). Zooplankton is abundant in the area as well, especially in April and August as a result by seasonal upwelling (Valdés *et al.*, 2007, 98-114).

Because of the high biodiversity, cetaceans are common. Fin whale, minke whale, harbour porpoise, short-beaked common dolphin, sperm whale, striped dolphin, northern bottlenose whale, Sowerby's beaked whale (*Mesoplodon bidens*), Cuvier's beaked whale (*Ziphius cavirostris*), Risso's dolphin, common bottlenose dolphin and long-finned pilot whale are all frequently sighted in the area (Kiszka *et al.,* 2007, 1033-1043). In the past, the North Atlantic right whale was also abundant in the region, probably using the region for feeding and calving. However excessive hunting by the Basques and others wiped them out in the region (Reid, Evans and Northridge, 2003, 12-13).

1.2.6. Baltic Sea

The Baltic Sea is enclosed by Sweden, Denmark, Finland, Russia, Estonia, Latvia, Lithuania, Poland, and Germany and includes the Gulf and Bay of Bothnia, the Gulf of Riga, the Gulf of Finland, and the Bay of Gdansk. The Baltic Sea is connected to the North Sea by way of the Skagerrak, Kattegat and the Danish Straits, though limited water exchange happens between the North Sea and the Baltic Sea, making it the largest brackish inland sea by area. Approximately two hundred freshwater rivers flow into the Baltic Sea and give it its brackish component. In the southeast the salinity is at its highest, while in the north the water is no longer salty and as a result many fresh water fish species live in that area of the sea, including perch (*Perca fluviatilis*), northern pike (*Esox luciusi),* and roach (*Rutilus ritilus*; Voipio, 1981).

Because it is a shallow sea, the only cetacean species that inhabits the Baltic Sea is the harbour porpoise. Other species are occasionally sighted, including the Atlantic white-sided dolphin, common bottlenose dolphin, common minke whale and killer whale. Other species are rarely sighted (Skóra, 1991, 67).

1.3. Cetacean Species in the Northeast Atlantic

After millions of years of evolution, modern cetaceans are perfectly adapted to an aquatic lifestyle and have diverged into at least 88 separate extant species. For some species, it is still not sure whether the defined species truly represents only one species or should be subdivided into two or more species. The killer whale (*Orcinus orca)* is an example of this and this species might be subdivided into two, three or even more separate species, raising the number of extant species significantly (Jefferson, Webber and Pitman, 2008, 158-163).

In this section, all 35 cetacean species that inhabit (or inhabited) the Northeast Atlantic Ocean will be described. The geographic setting, the species, sex, age of an individual, pod size, and size are all decisive factors determining the hunting methods, weaponry, gear, and organization for the taking. All these aspects are important when considering cetacean exploitation, and are therefore considered here (Lindquist, 1997). In the short descriptions, an actual abundance estimation in the North Atlantic, lifestyle (pelagic or coastal), the distribution, the length (also displayed in figure 5), the weight, the pod size, the main prey and the maximum speed will be provided. This information was extracted from Shirihai and Jarrett (2011). The IUCN status was extracted from IUCN (2015).

This data is primarily modern data, while many of these aspects are dynamic and subject to change over longer time periods. Many of these aspects were probably different during the Medieval Period. Especially the distribution might have been different. Several species have been hunted and many populations might have been reduced in size or even completely disappeared. As a result, prior to this period of industrial whaling, several species might have been more abundant in some areas of the eastern North Atlantic. Furthermore, even their morphology might have changed. Many of the large individuals of a species were hunted, reducing their genetic diversity, resulting in smaller individuals to be more likely to pass on their genes. As a result, several species might be smaller than they were in the past. Therefore, while this data is still of importance to this study, it might have been different in the past and should be treated with caution when considering past cetacean biology and exploitation.

The Northeast Atlantic distribution will be based on the geographic areas the OSPAR Commission developed (OSPAR commission, 2016). These six sections are the: Arctic Waters, Greater North Sea, Celtic Seas, Bay of Biscay and Iberian Coast, the Wider Atlantic and the Baltic Sea. All of these areas have a Roman numerical, which will be used in the actual descriptions of the separate species to define their Northeast Atlantic distribution (figure 4).

I. Arctic Waters. This includes the far north of the Northeast Atlantic. This area is located between Iceland, the Faroe Islands, northern Norway, the Russian oblasts of Arkhangelsk and Murmansk, the republic of Karelia and the Nenets Autonomous Okrug to the south, Svalbard and the Franz Josef Land to the very east, the north pole to the far north and the whole eastern half of Greenland to the west.

II. The Greater North Sea. This includes the area that borders the North Sea, (the Orkney and Shetland islands, north and east Scotland, east England, north France, Belgium, the Netherlands, northwest Germany, west Denmark), Wadden Sea, English Channel, the Kattegat and the Skagerrak.

III. The Celtic Seas. This area includes the actual Celtic Sea, the Irish Sea, the St. Georges Channel and the Firth of Clyde, including the western part shoreline of mainland Scotland, the Hebrides, western England, Ireland and Northern Ireland.

IV. The Bay of Biscay and Iberian Coast. This area includes the coastline of France south of Brittany, and the Atlantic coastline of Spain and Portugal with the most south-eastern point being Gibraltar.

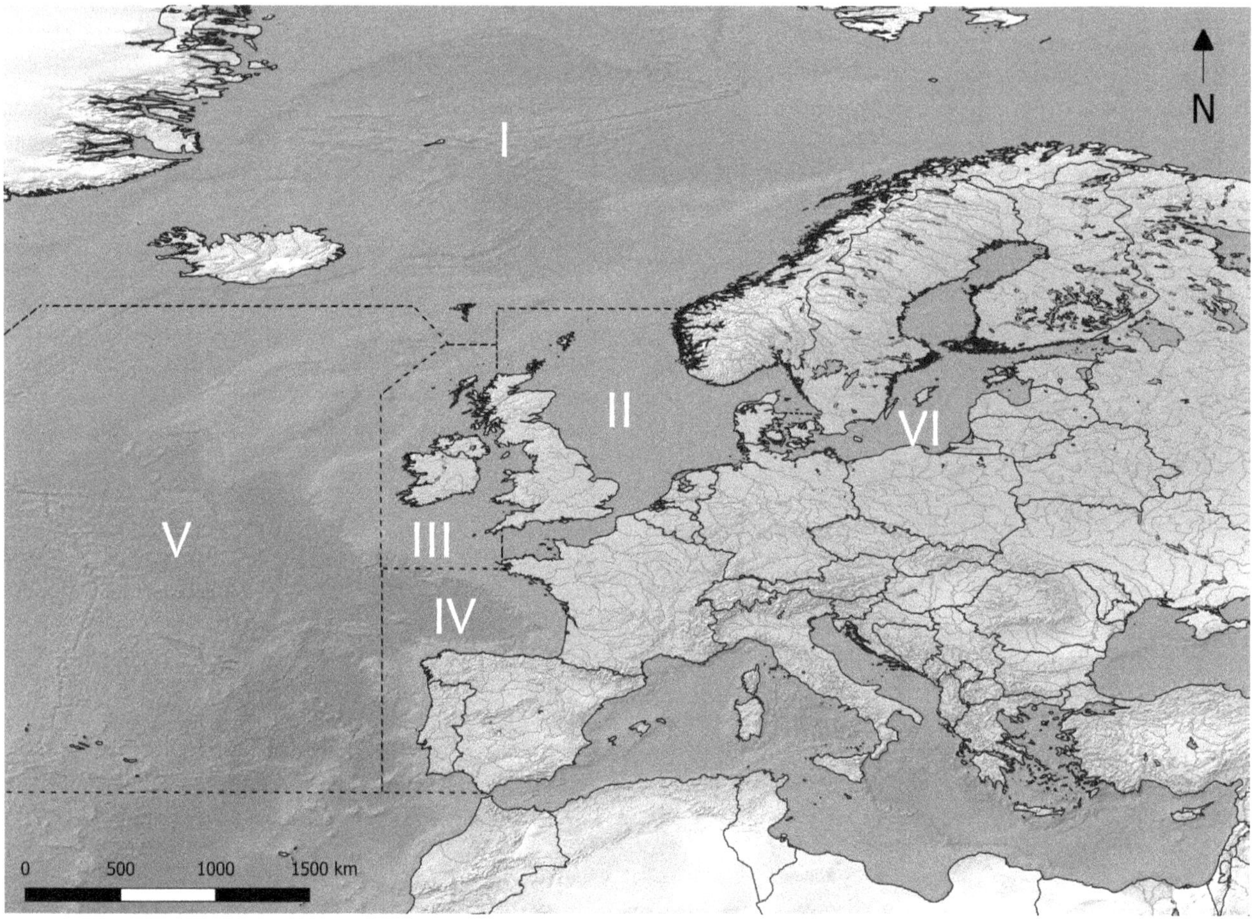

Figure 4. Geographic area of the Northeast Atlantic. I. Arctic Waters, II. Greater North Sea, III. Celtic Seas, IV. Bay of Biscay and Iberian Coast, V. Wider Atlantic, and VI. Baltic Sea. Created by author.

V. The Wider Atlantic region. This region is the offshore, deeper region of the Northeast Atlantic in which only the Azores are located.
VI. Baltic Sea. This region is not incorporated within OSPAR Commissions Areas, but in this study includes the actual Baltic Sea and the Danish Straits.

1.3.1. Parvorder Mysticeti

Fifteen species of the Mysticeti parvorder are currently recognized of which eight species inhabit the North Atlantic Ocean (Reid, Evans and Northridge, 2003, 13-23). The grey whale additionally inhabited the North Atlantic Ocean as well until the seventeenth- or eighteenth-century AD, but only inhabits the North Pacific Ocean now (Jefferson, Webber and Pitman, 2008, 70-73). The Mysticeti parvorder includes all the whales that use the baleen plates in their upper jaw to feed, and this technique, called filter-feeding, has proven to be extremely successful since all of the species can grow to an enormous size. Their primary target is krill and schooling fish species. Additionally, all whales of the Mysticeti parvorder have two blowholes, opposed to the members of the Odontoceti parvorders which only have one. Four families are part of this parvorder: Balaenopteridae, Balaenidae, Eschrichtiidae, and Neobalaenidae, of which members of the first three families can be found in the North Atlantic (Shirihai and Jarrett, 2011, 25-68; figure 6).

1.3.1.1. Family Balaenopteridae

Members of the Balaenopteridae family are known as rorquals and this is the largest family of baleen whales with nine extant species spread over two genera with several pygmy and dwarf subspecies. Of the nine recognized members, six can be found in the North Atlantic. In the past they were far more abundant, but because of commercial whaling in the twentieth century their numbers diminished strongly (Shirihai and Jarrett, 2011, 46).

Rorquals are streamlined and slender whales in comparison with their right whale relatives. All members of the family have longitudinal folds of skin running from the mouth to the navel, which allow the animals to open their mouth very wide when feeding, allowing them to take in more water than their own body size. Following this the animals close their mouth and then push out all the water through the baleen plates using their tongue, leaving only crustaceans such as krill and various fish species there which they subsequently consume. The blue whale (*Balaenoptera musculus*), believed to be the largest animal to have ever lived, is one of the members of this group (Shirihai and Jarrett, 2011, 46).

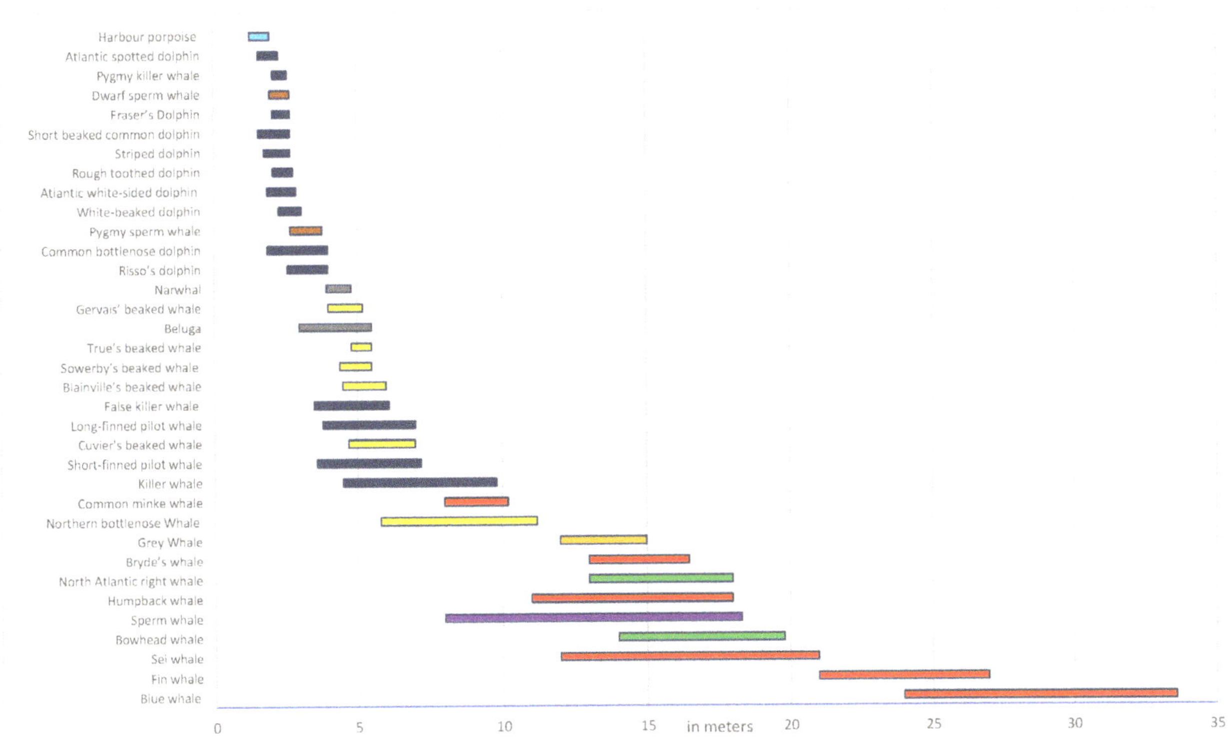

Figure 5. Size of all the North Atlantic cetacean species (adult individuals) from small to large. Cyan: Phocoenidae, Blue: Delphinidae, Brown: Kogiidae, Grey: Monodontidae, Yellow: Ziphiidae, Red: Balaeopteridae, Orange: Eschrichtiidae, Green: Balaenidae. Based on the data from Shirihai and Jarrett (2009), created by author.

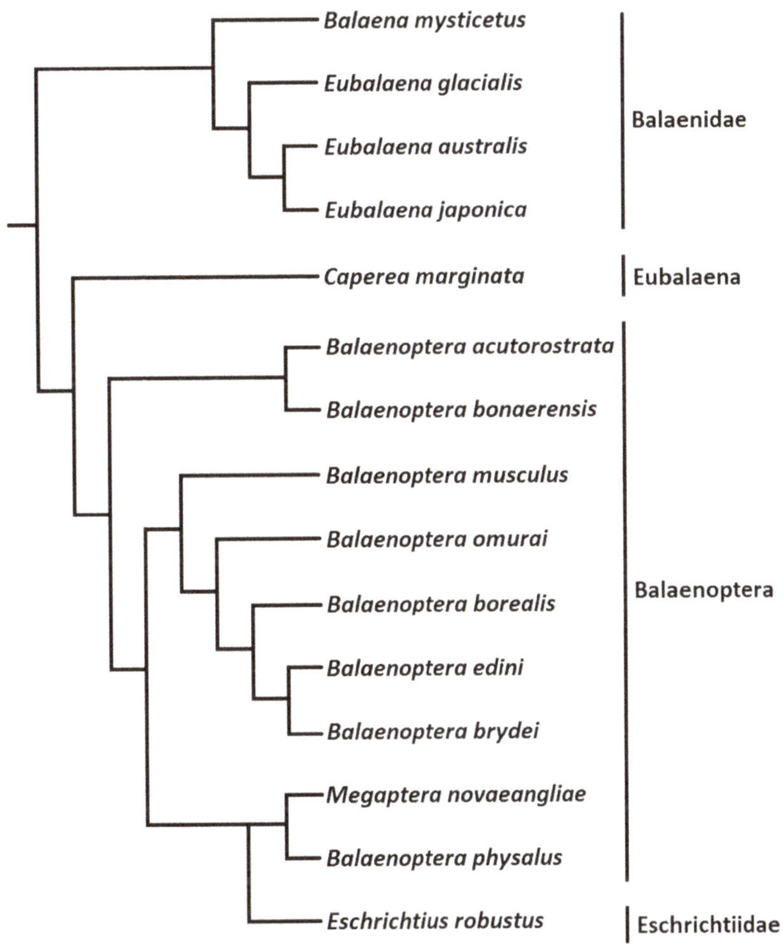

Figure 6. Phylogenetic classification of the Mysticeti family, based on Hooker (2009) and Deméré *et al.* (2008), created by author.

1.3.1.1.1. Blue whale (*Balaenoptera musculus*)

- Numbers estimation	<10,000 (600-1500 in the North Atlantic)
- Pelagic/coastal	Predominantly pelagic
- Distribution	Cosmopolitan
- North Atlantic distribution	I, IV, and V
- Length	24-33.6 m (females larger than males; figure 7)
- Weight	80-190 ton
- Group size	1-2 individuals
- Main prey	Euphausiids (krill)
- Maximum speed	50 km/h
- IUCN Status	Endangered

1.3.1.1.2. Fin whale (*Balaenoptera physalus*)

- Numbers estimation	119,000 (52,800 in the North Atlantic)
- Pelagic/coastal	Predominantly pelagic
- Distribution	Cosmopolitan
- North Atlantic distribution	I, IV, and V
- Length	21-27 m (females larger than males)
- Weight	30-80 ton
- Group size	3-7 individuals
- Main prey	Euphausiids, planktonic crustaceans, schooling fish and small squid
- Maximum speed	50 km/h
- IUCN Status	Endangered

1.3.1.1.3. Sei whale (*Balaenoptera borealis*)

- Numbers estimation	80,000 (10,000 in the North Atlantic)
- Pelagic/coastal	Predominantly pelagic
- Distribution	Cosmopolitan
- North Atlantic distribution	I, IV, and V
- Length	12-21 m (females larger than males)
- Weight	15.2-30.4 ton
- Group size	1-5 individuals
- Main prey	Copepods, euphausiids, shoals of fish and squid.
- Maximum speed	50 km/h
- IUCN Status	Endangered

1.3.1.1.4. Bryde's Whale (*Balaenoptera brydei*)

- Numbers estimation	90,000 (North Atlantic population not known)
- Pelagic/coastal	Predominantly pelagic
- Distribution	Cosmopolitan
- North Atlantic distribution	V
- Length	Max 15.6 m
- Weight	12-25 ton
- Group size	1-7 individuals
- Main prey	Pelagic schooling fish, euphausiids, copepods, cephalopods, and pelagic red crabs
- Maximum speed	25 km/h
- IUCN Status	Data deficient

1.3.1.1.5. Common Minke whale (*Balaenoptera acutorostrata*)

- Numbers estimation	185,000 in the North Atlantic with unknown numbers elsewhere
- Pelagic/coastal	Both pelagic and coastal
- Distribution	North Atlantic and North Pacific
- North Atlantic distribution	I, II, II, IV, and V
- Length	Max. 8.8 m (females larger than males)
- Weight	1-8 ton
- Group size	1-3 individuals
- Main prey	Schooling fish, euphausiids and copepods
- Maximum speed	30 km/h
- IUCN Status	Least Concern

1.3.1.1.6. Humpback whale (*Megaptera novaeangliae*)

- Numbers estimation	35,000 (12,000 in the North Atlantic)
- Pelagic/coastal	Both pelagic and coastal
- Distribution	Cosmopolitan
- North Atlantic distribution	I, II, III, IV, and V
- Length	11-18 m
- Weight	24-40 ton
- Group size	1-15 individuals
- Main prey	Euphausiids and schooling fish
- Maximum speed	25 km/h
- IUCN Status	Least Concern

1.3.1.2. Family Balaenidae

This family, known as the right whales, contains four extant species; the three species of right whales (genus *Eubalaena*) and the bowhead whale (*Balaena mysticetus*). The group is characterised by their large rotund bodies, large heads, being finless and broad flukes (Shirihai and Jarrett, 2011, 34). Because of their large size, with large quantities of meat, oil and baleen and the fact that they are relatively slow, these species were the first to be commercially hunted. Only the North Atlantic right whale (*Eubalaena glacialis*) and the bowhead whale inhabit the North Atlantic (Reid, Evans and Northridge, 2003, 12-13).

Figure 7. Blue whale skeleton hanging in the entrance hall of the Natural History Museum in London, UK (Photo by author).

1.3.1.2.1. North Atlantic Right whale (*Eubalaena glacialis*)

- Numbers estimation	300-500 (Western North Atlantic Ocean)
- Pelagic/coastal	Both pelagic and coastal
- Distribution	North Atlantic
- North Atlantic distribution	I, IV, and V
- Length	Up to 18.3 m (females larger than males)
- Weight	40-70 ton
- Group size	1-6 individuals
- Main prey	Copepods, euphausiids and pteropods
- Maximum speed	15 km/h
- Lifespan	50-70, possibly up to 100 years
- IUCN Status	Endangered

1.3.1.2.2. Bowhead whale (*Balaena mysticetus*)

- Numbers estimation	10,000
- Pelagic/coastal	Pelagic
- Distribution	Arctic or sub-Arctic waters
- North Atlantic distribution	I
- Length	14-19.8 m (females larger than males)
- Weight	75-100 ton
- Group size	1-6 individuals
- Main prey	Copepods, euphausiids and amphipods
- Maximum speed	10 km/h
- Lifespan	>200 years
- IUCN Status	Least Concern

1.3.1.3. Family Eschrichtiidae

The family Eschrichtiidae is one of the four baleen whale families, with only a single extant species, the grey whale (*Eschrichtius robustus*), though more species existed in the past. In the past there were two populations of grey whale, one in the North Pacific and one in the North Atlantic. The North Atlantic population was wiped out in the late 1600s or 1700s in the North Atlantic, while the populations in the Northern Pacific Ocean survive (Wolff, 2000, 209-217). Grey whales are relatively large and are slow moving and spend almost their entire lives in near shore areas. Because of these traits, grey whales were the ideal targets for early whalers and this probably led to their extinction in the North Atlantic area (Jefferson, Webber and Pitman, 2008, 70-73, Bryant, 1995, 860).

1.3.1.3.1. Grey Whale (*Eschrichtius robustus*)

- Numbers estimation	20,000-22,000 (North Atlantic population extinct)
- Pelagic/coastal	Predominantly coastal
- Distribution	North Pacific
- North Atlantic distribution	None
- Length	12-15 m (females slightly larger than males)
- Weight	15-35 ton
- Group size	1-3 individuals
- Main prey	Benthic amphipods and similar organisms
- Maximum speed	17.5 km/h
- Lifespan	75-80 years
- IUCN Status	Least Concern (North Atlantic population is extinct)

1.3.2. Parvorder Odontoceti

The parvorder Odontoceti contains all the 73 species of toothed cetaceans, which are divided into ten different families. Six of those families have representatives in the Northeast Atlantic, those being the: Physeteridae, Kogiidae, Ziphiidae, Delphinidae, Monodontidae, and the Phocoenidae families. The other four families (those being the Platanistidae, Iniidae, Lipotidae, and Pontoporiidae families) are not present in the Northeast Atlantic (Carwardine, 1995, 1-256). Besides having teeth, members of the Odontoceti parvorder are active hunters, have only one blowhole in opposition to the members of the Mysticeti parvorder which have two, and some of the members are believed to be highly intelligent.

1.3.2.1. Family Physeteridae

The only extant member of the Physeteridae family is the sperm whale (*Physeter macrocephalus*), though more members are known based on the fossil record.

1.3.2.1.1. Sperm Whale (*Physeter macrocephalus*)

- Numbers estimation	300,000
- Pelagic/coastal	Pelagic
- Distribution	Cosmopolitan (excluding the High Arctic)
- North Atlantic distribution	I, IV, and V
- Length	8-18.3 meters
- Weight	13.5-55.8 tons
- Group size	1-25 individuals
- Main prey	Principally large squid
- Maximum speed	34-43 km/h
- Lifespan	60-70
- IUCN Status	Vulnerable

1.3.2.2. Family Kogiidae

The Kogiidae family is closely related to the Physeteridae family, though their members are much smaller in size. Only two members are recognized, the pygmy sperm whale (*Kogia breviceps*) and the dwarf sperm whale (*Kogia sima*).

1.3.2.2.1. Pygmy Sperm Whale (*Kogia breviceps*)

- Numbers estimation	Unknown (at least 3000+ off California)
- Pelagic/coastal	Pelagic
- Distribution	Tropical and temperate waters
- North Atlantic distribution	IV and V
- Length	< 3.5 meters
- Weight	315-450 kg
- Group size	1-7 individuals
- Main prey	Mid- and deep-water cephalopods, fish and crustaceans
- Maximum speed	Unknown
- Lifespan	Up to 22 years
- IUCN Status	Data deficient

1.3.2.2.2. Dwarf Sperm Whale (*Kogia sima*)

- Numbers estimation	Unknown (at least 11,000 in tropical East Pacific)
- Pelagic/coastal	Pelagic and coastal
- Distribution	Tropical and temperate waters
- North Atlantic distribution	IV and V
- Length	<2.7 meters
- Weight	135-272 kg
- Group size	1-10 individuals
- Main prey	Cephalopods, fish and crustaceans
- Maximum speed	Unknown
- Lifespan	Unknown
- IUCN Status	Data deficient

1.3.2.3. Family Ziphiidae

The Ziphiidae family is the second largest cetacean family with 21 extant species. However, because of their pelagic lifestyle and the fact that they are deep-diving and almost never come near shore, they are also the most poorly understood family of cetaceans. Only males have teeth in general. The family can be divided into six genera: Mesoplodon, Hyperoodon, Indopacetus, Beradius, Tasmacetus, and Ziphius. Of these six genera, only members of the Mesoplodon, Hyperoodon, and Ziphius genera can be found in the Northeast Atlantic (Reid, Evans and Northridge, 2003, 30-33; Shirihai and Jarrett, 2011, 22).

1.3.2.3.1. Sowerby's beaked whale (*Mesoplodon bidens*)

- Numbers estimation	Unknown
- Pelagic/coastal	Pelagic
- Distribution	Cool and warm-temperate waters of North Atlantic
- North Atlantic distribution	I, IV, and V
- Length	4.4-5.5 meters
- Weight	1-1.3 tons
- Group size	8-10 individuals
- Main prey	Squid and pelagic fish
- Maximum speed	Unknown
- Lifespan	Unknown
- IUCN Status	Vulnerable

1.3.2.3.2. True's beaked whale (*Mesoplodon mirus*)

- Numbers estimation	Unknown (probably quite rare)
- Pelagic/coastal	Pelagic
- Distribution	Warm temperate waters in North Atlantic and southern Indian Ocean
- North Atlantic distribution	IV and V
- Length	4.8-5.5 meters
- Weight	0.89-1.5 tons
- Group size	1-6 individuals
- Main prey	Probably deep-water squid
- Maximum speed	Unknown
- Lifespan	Unknown
- IUCN Status	Data deficient

1.3.2.3.3. Blainville's beaked whale (*Mesoplodon densirostris*)

- Numbers estimation	Unknown
- Pelagic/coastal	Pelagic
- Distribution	Warm-temperate and tropical seas
- North Atlantic distribution	IV and V
- Length	4.5-6 meters
- Weight	0.7-1.03 tons
- Group size	3-12 individuals
- Main prey	Squid and deep-water fish
- Maximum speed	Unknown
- Lifespan	Unknown
- IUCN Status	Data deficient

1.3.2.3.4. Gervais' Beaked whale (*Mesoplodon europaeus*)

- Numbers estimation	No estimates
- Pelagic/coastal	Pelagic
- Distribution	Tropical, sub-tropical and temperate waters of the North Atlantic
- North Atlantic distribution	IV and V
- Length	4-5.2 meters (females overall larger than males)
- Weight	1-2.6 tons
- Group size	Probably small groups
- Main prey	Squid and fish in deep waters
- Maximum speed	Unknown
- Lifespan	Unknown
- IUCN Status	Data deficient

1.3.2.3.5. Cuvier's beaked whale (*Ziphius cavirostris*)

- Numbers estimation	No estimates
- Pelagic/coastal	Pelagic
- Distribution	Tropical, sub-tropical and temperate waters
- North Atlantic distribution	I, IV, and V
- Length	4.7-7 meters (no significant sexual dimorphism)
- Weight	2.03-3.4 tons
- Group size	1-12 individuals
- Main prey	Squid and fish in deep waters
- Maximum speed	Unknown
- Lifespan	Up to 60 years
- IUCN Status	Vulnerable

1.3.2.3.6. Northern bottlenose Whale (*Hyperoodon ampullatus*)

- Numbers estimation	>40,000
- Pelagic/coastal	Pelagic
- Distribution	Cold temperate waters of the North Atlantic
- North Atlantic distribution	I, IV, and V
- Length	5.8-9.8 meters, Males larger than females
- Weight	5.8-7.5 tons
- Group size	4-20 individuals
- Main prey	Primarily squid
- Maximum speed	Unknown
- Lifespan	At least 37 years
- IUCN Status	Data deficient

1.3.2.4. Family Delphinidae

Oceanic dolphins are members of the Delphinidae family. Almost 40 species belong to this family. They vary in size greatly and can be found in all oceans, most seas and even some river systems.

1.3.2.4.1. Killer whale (*Orcinus orca*)

- Numbers estimation	>50,000
- Pelagic/coastal	Coastal and Pelagic
- Distribution	Cosmopolitan (figure 8)
- North Atlantic distribution	I, II, II, IV, and V
- Length	4.5-9.8 meters (males larger than females)
- Weight	3.8-5.5 tons
- Group size	5-55 individuals
- Main prey	Catholic, bus usually specialises regionally
- Maximum speed	56 km/h
- Lifespan	Up to 90 years
- IUCN Status	Data deficient

1.3.2.4.2. False Killer Whale (*Pseudorca crassidens*)

- Numbers estimation	No global estimate - 50,000+ in the Pacific
- Pelagic/coastal	Principally pelagic
- Distribution	Cosmopolitan except for the Arctic waters
- North Atlantic distribution	IV and V
- Length	3.5-6.1 meters (males larger than females)
- Weight	1.01-2.03 tons
- Group size	10-50 individuals
- Main prey	Principally fish and cephalopods, but also other cetaceans
- Maximum speed	22 km/h
- Lifespan	Up to 63 years
- IUCN Status	Data deficient

1.3.2.4.3. Long-finned pilot whale (*Globicephala melas*)

- Numbers estimation	No global estimate – 800,000 in North Atlantic
- Pelagic/coastal	Pelagic and coastal
- Distribution	Sub-tropical and temperate waters of the Pacific and North Atlantic
- North Atlantic distribution	I, II, II, IV, and V
- Length	3.8-7.0 meters (males larger than females)
- Weight	2-3 tons
- Group size	20-100 individuals
- Main prey	Principally fish and squid
- Maximum speed	50 km/h
- Lifespan	Up to 60+ years
- IUCN Status	Data deficient

1.3.2.4.4. Short-finned pilot whale (*Globicephala macrorhynchus*)

- Numbers estimation	No overall estimate – 150,000 in tropical eastern Pacific
- Pelagic/coastal	Pelagic and coastal
- Distribution	Tropical, subtropical and warm-temperate waters
- North Atlantic distribution	IV and V
- Length	3.6-7.2 meters (males larger than females)
- Weight	1-4 tons
- Group size	20-100 individuals
- Main prey	Principally fish and squid
- Maximum speed	50 km/h
- Lifespan	Up to 63 years
- IUCN Status	Data deficient

Figure 8. Killer whale in Vancouver, Canada (photo by author).

1.3.2.4.5. Pygmy Killer Whale (*Feresa attenuata*)

- Numbers estimation	No overall estimate - 40,000 in the east tropical Pacific
- Pelagic/coastal	Principally pelagic
- Distribution	tropical and subtropical
- North Atlantic distribution	IV and V
- Length	2.1-2.6 meters
- Weight	110-170 kg
- Group size	15-25 individuals
- Main prey	Cephalopods and small fish, but also small cetaceans
- Maximum speed	Unknown
- Lifespan	Unknown
- IUCN Status	Data deficient

1.3.2.4.6. Risso's dolphin (*Grampus griseus*)

- Numbers estimation	No overall estimate – 175,000 in tropical east Pacific
- Pelagic/coastal	Pelagic and coastal
- Distribution	tropical, subtropical and temperate waters
- North Atlantic distribution	I, II, III, IV, and V
- Length	2.6-4.0 meters (males slightly larger than females
- Weight	300-500 kg
- Group size	10-30 individuals
- Main prey	Octopus, cuttlefish, squid, and krill
- Maximum speed	26 km/h
- Lifespan	At least 30 years

- IUCN Status	Least concern

1.3.2.4.7. Common bottlenose dolphin (*Tursiops truncates*)

- Numbers estimation	No overall estimate
- Pelagic/coastal	Pelagic and coastal
- Distribution	tropical, subtropical and temperate waters
- North Atlantic distribution	II, III, IV, and V
- Length	1.9-4.1 meters (males slightly larger than females)
- Weight	150-650 kg
- Group size	1-15 individuals
- Main prey	Fish, krill, and other crustaceans
- Maximum speed	35 km/h
- Lifespan	Up to 52 years
- IUCN Status	Least concern

1.3.2.4.8. White-beaked dolphins (*Lagenorhynchus albirostris*)

- Numbers estimation	Probably in the low 100,000s with most in NE Atlantic
- Pelagic/coastal	Predominantly pelagic
- Distribution	Temperate and cold shelf waters of North Atlantic
- North Atlantic distribution	I, II, III, and IV
- Length	2.3-3.1 meters (males larger than females)
- Weight	180-354 kg
- Group size	5-50 individuals
- Main prey	Fish, crustaceans, and octopus
- Maximum speed	47 km/h
- Lifespan	Unknown
- IUCN Status	Least concern

1.3.2.4.9. Atlantic white-sided dolphin (*Lagenorhynchus acutus*)

- Numbers estimation	No overall estimate - At least 42,000 off North America
- Pelagic/coastal	Predominantly pelagic
- Distribution	Temperate and cold waters of the North Atlantic
- North Atlantic distribution	I, II, III, IV, and V
- Length	1.9-2.9 (males larger than females)
- Weight	165-230 kg
- Group size	30-100 individuals
- Main prey	Fish and squid
- Maximum speed	47 km/h
- Lifespan	Up to 27 years
- IUCN Status	Least concern

1.3.2.4.10. Striped dolphin (*Stenella coeruleoalba*)

- Numbers estimation	No overall estimate - >3,000,000
- Pelagic/coastal	Predominantly pelagic
- Distribution	Tropical and temperate waters
- North Atlantic distribution	IV and V
- Length	1.8-2.7 meters (males larger than females)
- Weight	90-156 kg
- Group size	20-50 individuals
- Main prey	Schooling fish and cephalopods
- Maximum speed	60 km/h
- Lifespan	Up to 58 years
- Maximum speed	Least concern

1.3.2.4.11. Atlantic Spotted dolphin (*Stenella frontalis*)

- Numbers estimation	No overall estimate
- Pelagic/coastal	Predominantly pelagic
- Distribution	Warm-temperate and tropical waters
- North Atlantic distribution	IV and V
- Length	1.6-2.3 (males larger than females)
- Weight	100-143 kg
- Group size	1-50 individuals
- Main prey	Small fish, cephalopods, and invertebrates
- Maximum speed	30 km/h
- Lifespan	Unknown
- IUCN Status	Data deficient

1.3.2.4.12. Short Beaked common dolphin (*Delphinus delphis*)

- Numbers estimation	No overall estimate - >100,000 in Northeast Atlantic
- Pelagic/coastal	Both pelagic and coastal
- Distribution	Warm-temperate and tropical waters
- North Atlantic distribution	II, III, IV, and V
- Length	1.6-2.0 meters (males larger than females)
- Weight	70-235 kg
- Group size	10-30 individuals, but larger groups also common
- Main prey	Small fish and squid
- Maximum speed	30 km/h
- Lifespan	Up to 35 years
- IUCN Status	Least concern

1.3.2.4.13. Rough Toothed dolphin (*Steno bredanensis*)

- Numbers estimation	No global estimate
- Pelagic/coastal	Predominantly pelagic
- Distribution	Tropical to warm temperate waters worldwide

- North Atlantic distribution	IV and V
- Length	2.1-2.65 meters (males larger than females)
- Weight	90-160 kg
- Group size	10-20 individuals
- Main prey	Fish and cephalopods
- Maximum speed	24 km/h
- Lifespan	Up to 36 years
- IUCN Status	Least concern

1.3.2.4.14. Fraser's Dolphin (*Lagenodelphis hosei*)

- Numbers estimation	No global estimate
- Pelagic/coastal	Predominantly pelagic
- Distribution	Principally in deep tropical waters
- North Atlantic distribution	IV and V
- Length	2.1-2.7 meters (Males larger than females)
- Weight	160-210 kg
- Group size	100-1000 individuals
- Main prey	Mesopelagic fish, crustaceans, and cephalopods
- Maximum speed	28 km/h
- Lifespan	Unknown
- IUCN Status	Least concern

1.3.2.5. Family Monodontidae

This Monodontidae family consists of only two separate species: the beluga/white whale (*Delphinapterus leucas*) and the narwhal (*Monodon Monoceros*). Both species are medium-sized cetaceans and are highly distinctive from all the other species. They can be found in the coastal regions and pack ice region in the High Arctic region and the far north of both the Atlantic and the Pacific Oceans (Shirihai and Jarrett, 2011, 97).

1.3.2.5.1. Beluga/White Whale (*Delphinapterus leucas*)

- Numbers estimation	110,000+
- Pelagic/coastal	Coastal
- Distribution	High Arctic
- North Atlantic distribution	I
- Length	< 5.5 meters (males larger than females)
- Weight	400-1500 kg
- Group size	1-10 individuals (with aggregations of more than a 1000 in some areas)
- Main prey	Fish, cephalopods, crustaceans, and large zooplankton
- Maximum speed	22 km/h
- Lifespan	50+ years
- IUCN Status	Near Threatened

1.3.2.5.2. Narwhal (*Monodon monoceros*)

- Numbers estimation	50,000+
- Pelagic/coastal	Pelagic and coastal
- Distribution	High Arctic
- North Atlantic distribution	I
- Length	< 4.7 meters (excluding the tusk, males larger than females)
- Weight	400-1500 kg
- Group size	5-10 individuals (with aggregations of more than a 1000 in some areas)
- Main prey	Pelagic fish, squid, and shrimp
- Maximum speed	20 km/h
- Lifespan	25-50 years
- IUCN Status	Near Threatened

1.3.2.6. Family Phocoenidae

The word "porpoise" is commonly used to attribute any small dolphin, but it is only the correct name for the six members of the Phocoenidae family. Though porpoises are related to dolphins, they have a shorter beak and spade-shaped teeth, while dolphins have more conical shaped teeth (Carwardine, 1995, 236). Additionally, they tend to be much smaller than their Delphinidae relatives. The family includes the Vaquita (*Phocoena sinus*), a species of porpoise that lives off western Mexico and with a length of 1.4 meters is the smallest cetacean species (Shirihai and Jarrett, 2011, 246-259).

1.3.2.6.1. Harbour Porpoise (*Phocoena phocoena*)

- Numbers estimation	No overall estimate, but in 1994 circa 340,000 in Europe
- Pelagic/coastal	Coastal
- Distribution	Subarctic and temperate waters
- North Atlantic distribution	I, II, II, IV, and VI
- Length	1.3-2.0 meters (females larger than males)
- Weight	50-70 kg
- Group size	1-8 individuals
- Main prey	Schooling fish, cephalopods, and crustaceans
- Maximum speed	23 km/h
- Lifespan	Up to 24 years
- IUCN Status	Least concern

1.4. Cetaceans and Medieval Climate Variability

The impact of climate change on marine mammals remains poorly understood. This can largely be ascribed to the difficulty of obtaining substantive evidence. In temperate regions, including those in north western Europe, environmental changes are likely to influence prey abundance and distribution as a result of warmer sea temperatures, and can potentially lead to range shifts. Species with relatively narrow habitat requirements, like the shelf sea species such as the harbour porpoise, common minke whale, and white-beaked dolphin, may come under increased pressure as they will face a reduced available habitat if they are forced to move to more northern waters. Furthermore, if overall secondary production is reduced, this could have a devastating effect on some baleen whale species that feed upon zooplankton (Evans and Bjørge, 2013).

Climate change has been suggested to negatively affect marine ecosystems and biota. Shifts in distribution and reproductive success have been associated with climatic factors, while extreme weather events are thought to influence stranding events. A reduction in the extent of key habitants, changes in breeding success, and greater incidence of strandings are also likely to occur (Lambert *et al.*, 2011; Schumann *et al.*, 2013).

In the Pacific, the frequency of the strongest El Niño events is increasing as a result of climate change. This is affecting cetacean populations as well. In March 2015, at least 343 primarily sei whales washed up in Southern Chile. This has been connected to a harmful algal bloom, which is a direct result of the El Niño (Häussermann *et al.*, 2017). The occurrence of mass strandings of baleen whales relating to climate change and harmful algal blooms have been tracked to the Northeast Pacific coast; the Valdés Peninsula, Argentina; Australia; and Brazil as well (Wilson *et al.*, 2015; Häussermann *et al.*, 2017). Many species have been involved including grey whales, humpback whales, and southern right whales (*Eubalaena australis*). This clearly indicates that cetaceans are strongly influenced by climatic changes (Häussermann *et al.*, 2017).

Furthermore, analysis of 639 stranding events of 39 different taxa from southeast Australia, revealed a clear 11-13-year periodicity in the number of events through time. This data positively correlated with regional persistence of zonal (westerly) and meridional (southerly) winds. These winds result in colder and presumably nutrient-rich waters being driven closer to southern Australia, which results in increased biological activity in the water column during spring. This suggests that climatic events influence stranding events (Evans *et al.*, 2005).

Climate change is also thought to affect the composition and structure of local ecological communities. An investigation on the correlation between ocean warming around north-west Scotland since 1981 and its association to changes in local cetacean communities, revealed that strandings of white-beaked dolphins, a colder water species, declined, while strandings of common dolphins, a warmer water species, increased. Sightings surveys confirmed that the white-beaked dolphin population in the area was declining, while the common dolphin population instead increased. This indicates that cetaceans are sensitive to climatic changes and might adapt to it by moving to other areas (MacLeod *et al.*, 2005).

Modelling efforts to predict species distribution under future climates have increased with concern about climate change and its effect on species ecology. A study by Lambert *et al.* (2014) concerned with ten different cetacean species from the eastern North Atlantic, revealed that for only four species a good prediction regarding future distribution changes over time in response to changes in water temperature could be made. For only one of these species this meant a range expansion, while for the other three this meant a range contraction. The white-beaked dolphin is likely facing a range contraction of up to 80%. Model prediction provide important information on the ability to predict how individual species will respond to future climate change, as well as set in place viable conservation and management strategies allowing to protect these species (Lambert *et al.*, 2014).

Furthermore, it has been indicated that grey whales are delaying their southbound migration in the Pacific, with some individuals even staying in Arctic waters year-round (Moore, 2008). It is predicted that moderate to extreme selection pressure on marine mammals will take place within the next century (Harwood, 2001). Because marine mammals integrate and reflect ecological variation, they are prime sentinels of marine ecosystem change. This can be used to guide human stewardship activities, demonstrate ecosystem vulnerabilities and health, and reflect the ocean's role in climate interaction across regions (Moore, 2008).

This concept of "ecosystem sentinels" additionally has implications for zooarchaeological and palaeontological research as well, as species identification (through aDNA, ZooMS, osteometry, morphological comparison, etc.) of palaeontological and zooarchaeological specimens can be reached, allowing to potentially reconstruct the past distribution of species and the associated environment.

While modern climate anomalies are increasingly associated with shifts in the distribution and population of cetaceans, other historic climate anomalies are less frequently the topic of extensive research on past cetaceans. One exception is the study concerned with the harbour porpoise using Bayesian coalescent modelling of microsatellite variation to track the population demographic history of the species. The combination of genetic inferences with palaeo-oceanographic and historical records has indicated that the populations of the harbour porpoises have clearly responded to recent climate-driven reorganization in the eastern North Atlantic food web. The population in Iberian waters has become isolated from populations further north approximately 300 years ago, contemporaneous with the warming trend since the Little Ice Age period. Additionally, the local extinction or exodus of the harbour porpoise from the Mediterranean has been associated with the Mid-Holocene optimum (approximately 5000 BP (Fontaine *et al.*, 2010)).

For the Medieval Period, two substantial changes in climate took place for northern and western Europe: the Medieval Warm Period (MWP) and the Little Ice Age (LIA). A study by Mann *et al.* (2009) based on a global climate proxy network used to reconstruct surface temperature patterns over the past 1500 years, revealed that the Medieval Period is found to display warmth that matches or even exceeds that of the first decennium of the twenty-first century for some regions. However, on a global scale the temperatures fall well below those levels. La Niña-like conditions occurred in the tropical Pacific. Recent modelling studies suggested that solar irradiance (potentially leading to changes in large-scale circulation patterns associated with the Arctic Oscillation), high levels of explosive volcanism, and widespread hydrological anomalies, might have led to local climatic anomalies, associated with the Medieval Warm Period (AD 950-1250; Bradley, Hughes and Diaz, 2003).

The Little Ice Age (AD 1400-1700) showed the greatest cooling over the extratropical Northern Hemisphere continents. The patters of temperature changes during the Medieval Period, suggest dynamical responses of climate to radiative forcing changes involving El Niño and the North Atlantic Oscillation-Arctic Oscillation (Mann *et al.*, 2009).

Though research on medieval climate variability and its impact on cetaceans has not been conducted, a study by Stige *et al.* (2006) revealed that modern cod recruitment in the North Sea is clearly affected by the North Atlantic Oscillation. This could have wider implications for other species, including cetaceans. Changes in the North Atlantic Oscillation, like those happened in the Medieval Period, could therefore potentially have even wider implications. This was also confirmed by a study by Drinkwater *et al.* (2003), indicating that changes in the NAO have a strong effect on marine ecosystems, including the survival rate of North Atlantic right whales. It might additionally also be that climatic variability influences strandings.

1.5. Strandings

Cetaceans that die at sea may be brought to shore by wind and wave action and wash up on a beach. To live individuals this might happen as well, and this is called a "stranding". Some species have the tendency to sink when they die (rorquals for example) and these species are therefore less likely to strand than species that float (the North Atlantic right whale for example). Peltier *et al.* (2012) performed a study regarding modern strandings in France. They fitted tags to 86 common dolphin and 14 harbour porpoise carcasses that were by-caught by fishermen. The tagged carcasses were released back in the sea and in only 8% of the cases these carcasses were recovered along the French coastline. Four of these were common dolphins and four harbour porpoises. The distance between the location the carcasses were released at and the stranding place was between 3 and 77.7 kilometres and the duration of the drift varied between 9 and 25 days. This indicates that carcasses can drift for a considerable amount of time and travel a significant distance.

It should be noted that as part of the study conducted by Peltier *et al.* (2012), only small cetaceans were incorporated. Larger whales, especially the members of the Balaenidae family, that have a large layer of blubber can potentially drift for a longer amount of time and a longer distance.

Occasionally, live individuals strand, sometimes even whole pods (this is called a "mass-stranding"; Perrin and Geraci, 2009, 1118-1122). It is often difficult to determine the cause of a stranding, but live strandings can have several natural as well as unnatural causes. Disease and malnutrition are two of the main reasons that cause an animal to strand, but parasites, biotoxins, collisions with ships, damage done by sonar from military vessels, entanglement in fishing gear (figure 9), contaminants, oil spills, injuries sustained by other animals or humans and environmental conditions (such as climate change) are all suggested causes as well. Navigational errors can also be a reason cetaceans strand. Some areas with strong currents, broad tidal flats, or extreme tidal changes can be dangerous to cetaceans (Perrin and Geraci, 2009, 1118-1122).

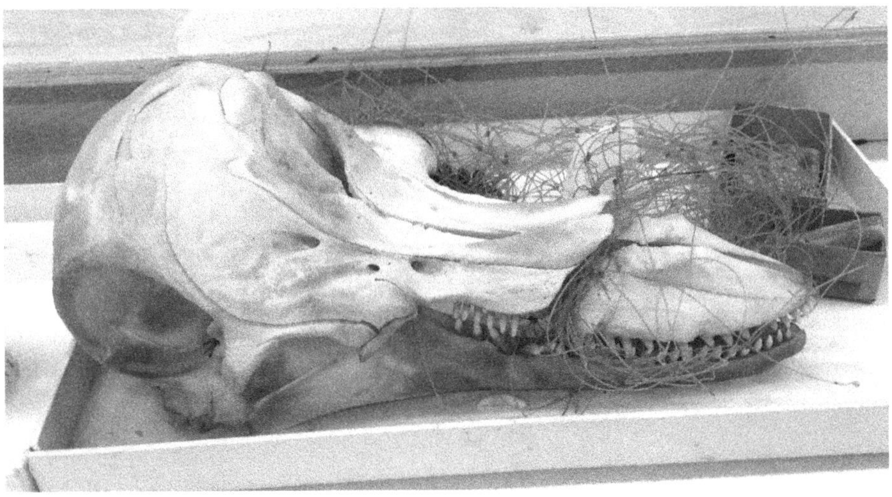

Figure 9. Cranium of a common bottlenose dolphin with a net cut deep inside its beak. Part of the Smithsonian Institution (photo by author).

Furthermore, some of the toothed whales live in pods with strong social structures and these species on occasions are involved in mass-strandings (Lazarus, 2006, 128-129). On those occasions one sick or injured member of the pod strands and is followed by the entire pod. Species that are known to mass-strand are: sperm whales, pygmy sperm whales, killer whales, long-finned pilot whales, false killer whales, striped dolphins, short beaked common dolphins, Atlantic white sided dolphin, Sowerby's beaked whales, Blainville's beaked whales, Cuvier's beaked whales and common bottlenose dolphins (Shirihai and Jarrett, 2006, 21-222).

Mass-strandings are poorly understood but are thought to be the result of panic (Gardiner, 1997, 175-176). However other explanations have been provided as well. Quite frequently whole pods of sperm whales strand along the shores of the southern North Sea area. It has been recently suggested that solar storms are the cause for these events (Vanselow *et al.,* 2017). These solar storms disrupt the Earth's magnetic field, which sperm whales use for navigation. Young bulls travelling from the Norwegian Sea south normally avoid the North Sea, which is known as the "North Sea sperm whale trap", but the disruptions in the magnetic field might affect their navigation and cause them to swim into the relatively shallow North Sea (Smeenk, 1997). The gradual slope of the shore in the southern part of the North Sea effects their sonar and they subsequently strand along the southern shores of the North Sea (Vanselow *et al.,* 2017). This happened to 29 young sperm whale bulls in early 2016 (figure 10), but has happened on other occasions as well, in for example January 1995 and April 1993 (both in the Netherlands). Mass strandings of young sperm whale bulls are also recorded in historical documents, for example 3 individuals stranded in November 1572 in Skallingen, Denmark, another 3 individuals stranded in November 1577 in Ter Heijde, the Netherlands and 18 individuals stranded in December 1723 in Neuwerk, Germany (Smeenk, 1997; Vanselow *et al.,* 2017). Smeenk (1997) has been able to accumulate data from historical and recent sources and has identified 89 sperm whale stranding events (ranging from 1 to 18 individuals) between 1560-1995 for the North Sea area.

Cetacean strandings are quite frequently described in historical records and on some occasions the carcasses were exploited. Cetacean stranding events were probably widely welcomed by past coastal communities and the carcasses were stripped of their nutritious meat and fat, sinews bones, baleen and other resources (Childe, 1931, 97; Mulville, 2002b, 34-37). This was for example registered by Martin (1716) in his description of the Scottish Western Isles where he mentioned that on one occasion around 160 little whales stranded themselves on the island of Tiree and that the natives consumed them all.

Egils saga Skalla-Grímssonar, written in the thirteenth century Iceland also mentions that a farmstead was built on Alftanes in the close vicinity of an area where whales

Figure 10. Sperm whale stranding in the Netherlands in 2016. (photo courtesy of Pepijn Kamminga and Rebecca Reurslag of the Naturalis Biodiversity Centre, Leiden, the Netherlands).

stranded in great numbers. This indicates that during the Medieval Period, people were already aware that cetacean often strand in the same areas. This is often in areas were the seafloor gradually inclines into shallow waters were the cetaceans' echo location does not work. It appears that during the Medieval Period, people took advantage of this (Szabo, 2008, 94-95).

Many of the zooarchaeological reports incorporated in this study suggest that the cetacean bone remains found as numerous archaeological sites were acquired by the past communities from stranded individuals and not by active whaling practices. This is for example suggested for the various Iron Age broch sites in Northern Scotland, from which numerous cetacean bones derive (Childe, 1931, 97).

It could well be that most of the medieval cetacean bone coming from archaeological sites were derived from stranded individuals. Indeed, cetaceans strand quite frequently along the coastline of western and northern Europe. Many of the countries in the region have an organisation that keeps a record of all the cetacean strandings. For the Netherlands, this is undertaken by Walvisstrandingen.nl (2017). It includes modern strandings, but also has incorporated stranding information extracted from historical sources. For Belgium, strandings have been recorded since 2001 by the Department VI of the Royal Belgian Institute of Natural Sciences (2017). Additionally, for Denmark a record is kept as well since 2001, by Kinze (Personal communication Kinze, 2014) as part of the "Saeler og Hvaler I Denmark". For the UK, this is undertaken since 1913 by the Natural History Museum and since 1990 as part of the UK Cetacean Strandings Investigations Programme (2017). The Centre de Recherche sur les Mammifères marins & LIENSs Laboratory (2014) has been recording cetacean strandings for France. Strandings in the German state of Schleswig-Holstein have been recorded by U. Siebert at the Institut für Terrestrische und Aquatische Wildtierforschung at the Hochschule Hannover (Personal communication Siebert, 2014). Norway does not have an organisation that collects cetacean strandings information (Personal communication Frafjord, 2014).

The information that can be extracted from these cetacean strandings organisations is of great value to science and provides information regarding population size, health of the animals and for some of the beaked whales which are extremely rarely recorded on sea, strandings are the only source of information that can be used to study the creatures. It is almost impossible however, to compare the modern situation of cetacean strandings with the past situation. Many of the causes of modern strandings are anthropogenic aspects, e.g. ingestion of plastic, ship strikes, oil spills, fishing bycatch or military sonar (Mulville, 2002b, 36).

Furthermore, modern whaling undertaken in the nineteenth and twentieth century has depleted numerous cetacean population stocks, diminishing their numbers, while their numbers were far higher during the Medieval Period. The North Atlantic right whale is the species that was most strongly affected by whaling. Its population is below 500 now and the species has completely disappeared from the European side of the North Atlantic. Strandings were probably far more regular during the Medieval Period, but no such an event has been reported since 1913 (Mulville, 2002b, 36).

However, several archaeological and palaeontological studies have recently addressed the correlation between certain past environmental conditions and climate variation and cetacean strandings. In Cerro Bellena in the Atacama Region of Chile, a massive "whale graveyard" dating to the Late Miocene was found, preserving bones from over 40 marine mammals, including rorquals (n=31; probably all from the same species), sperm whales and *Odobenocetops* (a walrus-like whale species), but also aquatic sloths and seals (Pyenson *et al.*, 2014). They were found in four discrete horizons, indicating at least four different events that led to the death of multiple marine mammals in a short period of time.

It has been suggested that these events were the result of harmful algal blooms (HABs), as these cause organ failure in marine mammals. Modern examples of this are also known, as discussed in the previous sub-chapter. The stranding of fourteen Humpback whales over a short amount of time around Cape Cod, MA, USA in 1987-1988 is another example of this (Pyenson *et al.*, 2014). It could well be that those events also took place in Europe in the Medieval Period. If this happened, it would not have gone unnoticed and people might have taken advantage of the situation and stripping the carcasses of their raw materials.

Another, archaeological site where cetacean strandings are linked to climate change is the site of Grotta dell'Uzzo in North West Sicily, Italy. At this site 224 specimens were identified belonging to a member of the Cetacea order (Mannino *et al.*, 2015). Most of the remains were fragmented and therefore could not be identified to the species level, however long finned pilot whale, Risso's dolphin and short beaked common dolphin were identified amongst the remains (Mannino *et al.*, 2015). Additionally, several of the fragments belong to large member of the cetacean order, e.g. large baleen whales or sperm whale, indicating that a wide variety of species was exploited.

Their remains were found in the Mesolithic-Neolithic transition layer, while cetacean remains are absent for preceding or following periods. Stable isotope analysis on collagen from bones of humans as well as on the remains of a red fox (*Vulpus vulpus*) from the Mesolithic-Neolithic transition layer, indicate that they acquired at least one third of their protein from marine fauna consumption. Foxes are generalist predators using resources opportunistically according to the availability, though in the Mediterranean Basin and in the rest of Europe, they rarely consume fish. Based on this it has been suggested that the high $\delta^{13}C$ could only be explained by the availability of plentiful stranded cetaceans. This suggests that numerous cetacean carcasses were available on a regular basis for at least a decade (Mannino *et al.*, 2015).

The Mesolithic-Neolithic transition layer coincides with the drainage of the Laurentide lakes in the Atlantic Ocean, which happened around 8.47± 0.3 kyr cal. After this, temperatures plummeted for various regions, including the Mediterranean Basin. It has been demonstrated that cetacean strandings are linked to large-scale climate variability, since it affects the oceanic circulation as well as prey distribution. This suggests that cetaceans might have stranded in the area regularly during the Mesolithic-Neolithic transition time frame, making them an easy meal for resident humans (and foxes) who eagerly took advantage of the situation (Mannino *et al.*, 2015).

Another study concerned with modern strandings also linked climate change to cetacean strandings. This study was undertaken by Bradshaw, Evens and Hindell (2006) and was concerned with modern strandings in Tasmania. Mass-stranding events happen quite frequently in Tasmania and Bradshaw. Evens and Hindell (2006) showed a correlation between an increase in zonal and meridional winds resulting in colder (and presumably nutrient-rich) waters reaching the southern part of Australia and Tasmania and an increase in cetacean strandings (Evans *et al.*, 2005).

These case studies imply that during periods of climate variability cetacean populations are suspect to (mass-) strandings (Mannino *et al.*, 2015). This might have happened for different areas and different time frames as well, for example during the Medieval Warm Period and the Little Ice Age, two periods of climate change that fall within the time frame this study is concerned with. Based on numerous historical sources (which will be discussed later in this study), it is clear that past communities were aware of cetacean strandings and that these occasions were frequently exploited. An example of this is the already discussed *Egils saga Skalla-Grimssonar* (Szabo, 2008, 95).

1.6. Exploitable products

Numerous resources can be extracted from cetacean carcasses. These resources are different for each species but include: meat, organs, baleen, bone, teeth, ivory, blubber, oil, sinews, entrails, spermaceti, and ambergris. Through various point in history all these resources were prized and exploited (Lindquist, 1997, 11). Not all these resources survive in the archaeological record, but historical sources can be a useful tool in filling those gaps.

In order to acquire these products, cetaceans had to be butchered. However, when cetaceans die, especially for large species, their internal body heat rises with decay, which eventually leads to a build-up of internal gas pressure. People unaware of this trying to butcher a large whale might set off an explosion (Foote, 1996). From historical sources it appears that people were already aware of this (Szabo, 2008, 87-88).

From a fresh carcass, a vast quantity of resources could be extracted from one single cetacean individual, providing food for a whole community for a considerable amount of time. In this section the various products will be discussed and their (zoo)archaeological potential will be analysed.

1.6.1. Meat

The most obvious product that can be obtained from cetaceans is their meat and since several species can reach enormous sizes, a lot of meat can be harvested from just one individual. Whale meat has a deep, red colour and looks like the flesh of other mammals (figure 11). It indeed has a very high nutritional value though it has fewer calories than for example beef or pork (Bloch, Dam and Hanuardóttir, 2003; Suzuki, 1993).

Moreover, it is an excellent source of protein. Protein is broken down into amino acids inside the human body. There are eight amino acids that the human body cannot produce but are still required. These eight amino acids are essential amino acids, and these are well represented both in quantity and balance in whale meat (Suzuki, 1993).

Furthermore, whale meat has a high content of many macro- and microminerals. Especially iron and selenium are abundant in whale meat. Furthermore, studies have indicated that meat from narwhals is a good source of vitamins A, B1, and B2 (Bloch, Dam and Hanuardóttir, 2003). Whale meat is also rich in eicosapentaenoic acids, which inhibits the coagulation of blood (Suzuki, 1993). Additionally, whale meat has a low-fat content and is relatively low in cholesterol (Bloch, Dam and Hanuardóttir, 2003).

Geraci and Smith (1979) analysed the diet of the Inuit hunters from Holman, Northwest Territories and analysed the vitamin C intake. The Inuit relied for a major part of their vitamin C intake on beluga exploitation. Their analysis indicated that the thin epidermis (part of the skin) contains 38mg/100g of vitamin C, making it one of the highest sources of vitamin in the Inuit diet.

During the Christian period in Europe, the meat of whales was however perceived as fish and was classified as "cold food", allowing it to be consumed on Fridays and holy days. This *Kings Mirror* attested this by stating "...whale flesh may be eaten on fast days like other fish food...". Something similar is stated for medieval Iceland in the Grágas laws "...eat this when he fasts: fish of all kinds and whales other than walrus and seal – these may only be eaten when meat may be eaten..." (Szabo, 2008, 84).

Because of the geometry of surface-to-volume ratios, the percentage of meat is larger for bigger species. For example, the harbour porpoise weighs on average 41 kg of

Figure 11. Dolphin autopsy at the Statens Naturhistorisk Museum Copenhagen, Denmark showing the red meat (photo by author).

which 45 percent is blubber and 25 percent meat. The fin whale however, weighs up to 45 tons, of which 25 percent is blubber and 45 percent meat (Smith and Kinahan, 1984, 95). One large whale would supply a whole community with enough meat to last months.

However, with a large quantity of meat, preservation must have been a major concern. Caught cetaceans might have been consumed raw without any problems, however in order to make the meat last longer preservation techniques are required. Cooking or boiling the meat might have been an option and was certainly practiced on dead stranded cetaceans in order to eliminate the oils or excreted materials which made the material inedible.

The development of salting preservation made it possible to transport cetacean meat, as well as fish further inland. In many regions salt was already available by the Early Medieval Period, though during the High Medieval Period it became more common (Fagan, 2000, 71).

Salt making in England dates back to at least the Late Bronze Age. At that point seawater was boiled in clay dishes supported by clay firebars. From the Roman period onwards, brine was used instead of sea water. Between two successive high tides, saltwater impregnated the ground and would eventually dry out by natural evaporation, after which the surface contained dried salt. This surface sand was scraped off the beach and subsequently washed with fresh water in order to produce brine, which was then filtered into sunken receptacles (Greenwood, 2011). Salt making was undertaken especially in Sussex from the pre-Roman to medieval times and was an important industry in the Adur Valley (Ridgeway, 2000). Many salt processing sites have been identified in England. These sites were valuable, and salt appears to have been accessible to the wealthiest only (Hammond, 1998, 110; Banham, 2004, 40).

In the Netherlands, peat was used as fuel, but due to the high salinity, it could also be burned for the creation of salt. However, after drainage took place, and peat digging and marine erosion of peat deposits took place, Eel-grass (*Zostera marina*) expanded. This was also used for salt-making, probably after many peat deposits had become difficult to obtain (Van Geel and Borger, 2005).

In the Mediterranean, after the tenth century, salting became more common and was transported along the Atlantic coast. The Basques are known to have acquired salt through the evaporation of seawater from estuaries in Portugal and Spain in order to preserve whale meat (Fagan, 2000, 71; Fagan, 2006, 50-57). The Norse were also known to both boil and salt whale meat (Tinker, 1988, 53). This clearly indicates that in many parts of Europe, salt preservation would have allowed the transportation of cetacean meat further inland, from the Early Medieval Period onwards already.

Not every type of whale appears to have been preserved in a similar fashion and there might have been difference between people and their taste. Olaus Magnus noted that sixteenth-century Norwegians preserved whale meat from stranded and captured cetaceans using salting techniques (Olaus Magnus, 2010, 1094). However, when Olaus came to Britain in 1532 he witnessed how a whale (probably dead stranded) was only partially butchered and limited resources were carted away, as the animal was already decomposing and stank (Olaus Magnus, 2010, 1095). Szabo (2008, 86) additionally noted that it appears that not all whales were treated in a similar way. From the *King's Mirror* (a Norse text dating to around AD 1250) and Albertus Magnus' *De Animalibus* it appears that some were deemed too oily or stank.

British and Scandinavian historical sources suggest that butchering a large whale took between five and six days (Linquist, 1995b, 28). However historical sources do not provide any information regarding butchery techniques or methods (Szabo, 2008, 87). It should be noted that it is far easier to remove flesh of marine mammals than it is for terrestrial mammals. Ethnoarchaeological investigations by Savelle and Friesen (1996, 717-719) of modern Inuit in the Arctic, showed that the scalp of belugas and narwhals can easily be separated from the carcass. However, since the scalp is not associated with any skeletal elements it is impossible to trace the removal of it based on zooarchaeological specimens. It is possible to butcher a whale, without removing any of the bones (Mulville, 2002b, 40). Therefore, butchery practices can be invisible to zooarchaeologists.

Savelle and Friesen (1996, 713-721) have performed a section on a sub-adult harbour porpoise to create a meat utility index and in this way find out what parts of the harbour porpoise contained the most flesh weight. The scalp (the skin plus the subdermal blubber) turned out to be largest anatomical unit and this was similar for previous research conducted by Omura *et al.* (1969) on a North Pacific Right Whale. From these studies it appears that especially the lumbar, caudal and thoracic vertebrae regions contain a lot of meat, while the mandible (apart from the tongue, which can be easily removed from the carcass) and the pectoral fins do not possess a high amount of meat (Savelle and Friesen, 1996).

1.6.2. Baleen

Baleen is made of keratin (just like fingernails, hair, horn and hoofs) and is often referred to as whalebone (figure

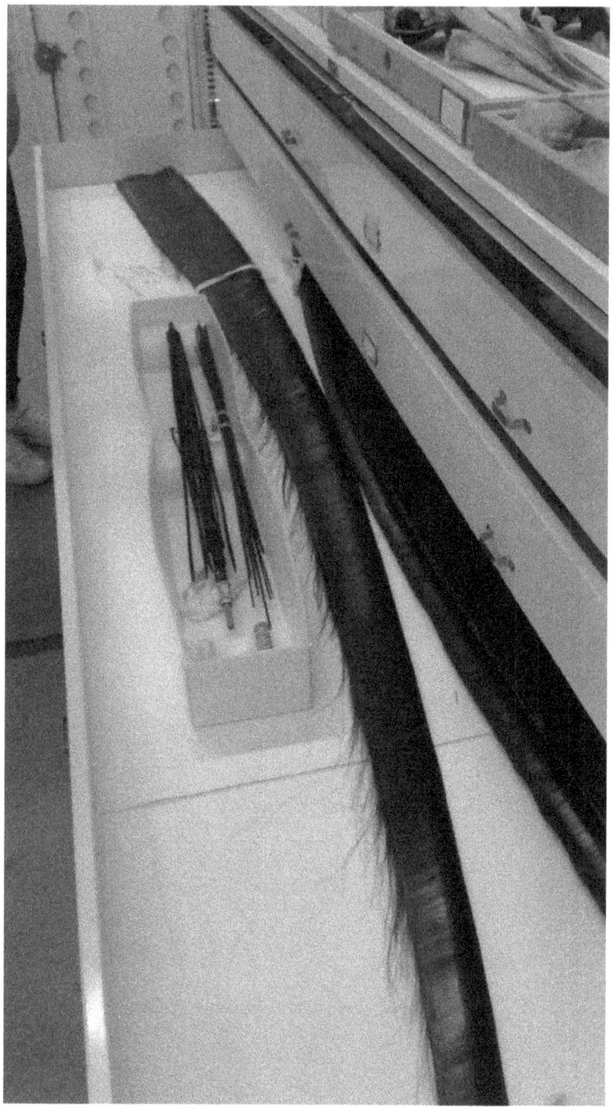

Figure 12. Whale baleen in the collection of the Smithsonian Institution (photo by author)

12). It consists of a homogeneous cortical layer around three to four layers of horny tubes. Since baleen is hollow, it allows for inner movement, making the material more elastic. Baleen is exposed to the movement of water and the tongue and is worn off. Therefore, baleen Is continuously being replenished at its root, so it retains the proper length (Young *et al.*, 2015).

Baleen plates descend in two rows from the upper jaw. The plates are located under one centimetre from another. At the top the baleen plates are broader and at the bottom the plates are pointy. There is quite a difference between species (table 2). The baleen of right whales is long and narrow and can reach a length of 2.5 meters. The baleen plates of the bowhead whale are even longer, reaching a length of almost 4,5 meters. To handle the high numbers of baleen plates within the jaw and the length of the individual plates, the right and bowhead whales have an arched rostrum and a slim lower jaw with lips of about 1.5 meters for sealing the sides of the mouth. When the jaws of a whale are shut, the baleen plates are

Table 2. Number and length of baleen plates per species

Family and Species	Baleen length (in meters)	Laminae count on either side of jaw	References
Balaenidae			
North Atlantic Right Whale	2.5	200-390	Rice, 2009d, 78-80; Young *et al.*, 2015
Bowhead Whale	4.5	250-350	Rice, 2009d, 78-80; Young *et al.*, 2015
Eschrichtiidae			
Grey Whale	0.4	130-180	Leatherwood *et al.*, 1988, 79; Young *et al.*, 2015
Balaenopteridae			
Blue Whale	1	250-400	Young *et al.*, 2015
Fin Whale	0.7	262-473	Leatherwood *et al.*, 1988, 79; Young *et al.*, 2015
Sei Whale	0.6	219-402	Young *et al.*, 2015
Bryde's Whale	0.4	285-350	Young *et al.*, 2015
Common Minke Whale	0.2	230-395	Leatherwood *et al.*, 1988, 79; Young *et al.*, 2015
Humpback Whale	0.75	270-400	Leatherwood *et al.*, 1988, 79; Young *et al.*, 2015

bent inwards. Rorquals and the grey whale have much shorter baleen plates and as a result their rostrum is not arched, and they do not have a large lower lip (Rice, 2009d, 78-80).

Baleen is a raw material that can easily be worked with knives and saws. From the thirteenth century onwards, baleen was the preferred material for the production of composite crossbow staves. In the early fourteenth century, baleen was also used for tourney equipment in which the gauntlets were often made of baleen and also sleeves were sometimes made of baleen (Solazzo *et al.*, 2017). In early modern times it was used for various other purposes as well, such as for the creation of umbrellas and parasol spars, whips, fans, and corsets (Lazarus, 2006, 25-26). Ambroise Paré additionally mentions that baleen was dried and polished and served to make busks for women, little staves, and to stiffen garments (Frank, 1993).

Though baleen is a relatively strong material, it is rarely retrieved from archaeological contexts because of its susceptibility to fungal and bacterial attacks. Additionally, alkaline and reducing environments will break down the material as well. Szabo (2008, 149) has also pointed out that literary references of the use of baleen are also rare, obscuring our knowledge of baleen use even more. Moreover, the fact that in the English language baleen is often called whalebone, makes it difficult to tell the differences between baleen and actual bones of whale.

1.6.3. Bone

Since whales live in water and their bodies do not require to be supported by their bones, their bones make up a smaller proportion in comparison with their muscles, than terrestrial mammals. Osteological remains of animals are frequently used for the creation of various artefacts and tools. This was also the case for the Medieval Period and in many cases antler and bone from terrestrial mammals appear to have been the preferred raw material.

Terrestrial bone, as well as antler has unaligned, heterogeneous cavities of various sizes (Hennius *et al.*, 2018). This is different for example for antler, for which the spongy structure varies for the different deer species and even within one single piece of antler. The orientation of the cavities of the spongy part is heterogenous. However, this is problematic for the creation of artefacts and tools as there is little workable volume. As a result, most tools and artefacts made of antler are small, though some larger pieces can be made of moose antler. Bone of terrestrial mammals also has a spongy core like antler, with unaligned cavities, but the outside layer is much thicker, making them relatively heavier (Hennius *et al.*, 2018).

Cetacean bone is unlike terrestrial mammalian bone material. It has a relatively thin outside layer of compact bone, while the interior consists of spongy bone, with thread-like, elongated cavities and is furthermore coarse and homogenous. As bones of whales are often sizeable, this leaves a large workable volume for the creation of various tools and artefacts. Especially mandible pieces of large baleen whales are suitable for this (Hennius *et al.*, 2018).

The cavities in the spongy part of the bone is filled with bone marrow. In young animals this is red bone marrow in which red and white blood corpuscles are created. When an animal gets older however, this red bone marrow gets replaced by yellow bone marrow. For the vertebrae this starts at the cervical and caudal vertebrae and after some time only the thoracic vertebrae contain red bone marrow (and some parts of the ribs). This yellow bone marrow consists entirely of fatty tissue, while red bone marrow also contains some fatty tissue, though to a lesser extent. This fatty tissue also contains oil. The bones of a cetacean individual contain about a third of the total oil yield, with the skeleton having a fat content of 51 percent.

The head contains 84 percent, the vertebrae containing yellow marrow between 32-68 percent and the vertebrae with red bone marrow 24 percent (Slijper, 1962, 109-110). The vestigial femur of cetaceans does not have medullary cavity unlike bovid or avian femurs (Higgs, Little and Clover, 2011, 9-17).

Because of the oil present within the bones, bone became an important source for oil extraction as well. By drilling a hole in the bone, a large amount of oil could be extracted. Monks (2005, 138-153) has created an oil utility index for whale bones, which are based on the dry weight of humpback whale bones. The head contains 80 to 84 per cent of all the skeletal oil (Monks, 2005, 140). Based on the index, Monks (2005) argues that by looking at the presence of the skeletal elements at a particular site it is possible to see whether oil was frequently extracted from the bones. However, it should be noted that the skeletal elements are from different sizes and it is much easier to transport a small caudal vertebra to a settlement than a whole cranium. Therefore, the index should be applied with some caution.

Next to oil extraction, cetacean bone material has also been used for various other purposes. Bone has been worked to create combs, chopping blocks, scrapers, blades, pivots, whale bone plaques (figure 13), hide working tools, cups/vessels, handles, keys, gaming pieces, sword pommels, fishing weights and various other tools and artefacts. In areas where wood is a limited resource, cetacean bone is frequently used, also for architectural features, with sometimes whole tents made of whale bone (Savelle, 1997). Lestocquoy (1948) furthermore states that vertebrae of large cetaceans could be used as chairs.

Figure 13. The Scar plaque. Created from whale bone (photo by author).

1.6.4. Teeth and Ivory

The teeth of the Odontoceti have occasionally been used for the creation of various artefacts. For most Odontoceti emphasis was placed on the extraction of other product. However, the teeth of three different species were especially sought after. These three species are the narwhal, sperm whale, and killer whale.

The tusks of elephants (and mammoths) have been traditionally called "ivory", however the teeth and tusks of all mammals have the same chemical structure. Therefore, Espinoza and Mann (1991) defined ivory, as "any mammalian tooth or tusk of commercial interest which is large enough to be carved or scimshawed" (figure 14). The teeth of the sperm whale and the killer whale, as well as the tusk of the narwhal are therefore frequently defined as "ivory" (Espinoza and Mann, 1999).

Teeth and tusks consist out of four regions. The innermost region is the pulp cavity, an empty space within that tooth

Figure 14. Scrimshaw on sperm whale tooth from the mid nineteenth century. Record ID chndm_a-a-813. Reproduced courtesy of the Smithsonian Institution. CC0 copyright. Available at https://www.si.edu/object/scrimshaw:chndm_a-a-813.

with odonotoblastic cells that produce dentine. Dentine is a combination of an organic matrix of collagenous proteins and inorganic dahllite that forms a layer around the pulp cavity. At the microscopic scale, dentine consists of small tubules with micro-canals. These canals have different configurations in different species and are therefore helpful in determining the species. The third layer is the cementum layer. The primary function of cementum, that surrounds the dentine of both the root and tooth, is to adhere the tooth to the mandible or maxilla. Enamel is the fourth layer and covers the surface of the tooth and tusk, which is most subject to wear (e.g. the crown). Ameloblasts create enamel and exhibit a prismatic structure. These prism patters are also useful for species identification (Espinoza and Mann, 1999).

Odontoceti teeth display annual dentinal growth layer groups (GLGs). These GLGs can be used to determine the age of an individual. Age determination was for example undertaken on free-ranging bottlenose dolphins from Florida (Hohn *et al.,* 1989). To the author's knowledge, age determination based on GLGs has not been undertaken yet on archaeological material.

From archaeological contexts, teeth of dolphins are occasionally found and in some sites the teeth have been scrimshawed. At the Salme boat grave, Estonia, four dice made out of dolphin teeth were recovered, as well as 328 gaming pieces of which most are assumed to have been crafted from whale bone (personal communication, Jüri Peets, 2015). Killer whale teeth have been recovered from the Danish site of Borrebjerg dating to the Iron Age or Viking age period (Degerböl, 1933, 396). Sperm whale teeth have been recovered from the terp site in Eenumerhoogte, the Netherlands, and eight Risso's dolphin teeth have been at the site of Burrain, Orkney Islands, (van Giffen, 1914; MacGregor 1974).

Ivory has been used for the creation of various objects, e.g. jewellery, piano keys and small statuaries. Sperm whale and killer whale teeth, as well as narwhal tusks (figure 15) have also been superficially carved, which is known as "scrimshawing". Sperm whale teeth have an average height of approximately twenty centimetres and are especially useful for the scrimshawing. Four warders and two pawns, part of the Lewis chess pieces, are actually made out of sperm whale teeth (Stratford, 1997; Tate, Reiche and Pinzari, 2014, 11-28). Killer whale teeth are smaller in comparison the sperm whale teeth but can also still be used for scrimshawing. The tusk of the narwhal is actually a modified (usually left) upper incisor. The tusk is spirally twisted with a counter-clockwise direction and has a length between two and seven meters (Espinoza and Mann, 1999).

These archaeological sites show that dolphin and sperm whale teeth were occasionally used during the Medieval Period and prior to it, for the creation of various artefacts. Sperm whale teeth probably derived from stranded individuals, as sperm whales were most likely only commercially exploited from circa 1712 onwards, when the Americans started undertaking this (Frank, 1993)

Narwhal tusks are surprisingly missing for the archaeological record of medieval Europe (Arneborg, 2000). Narwhals are rarely sighted in European waters, though in 2016 a dead narwhal was found on the banks of the river Scheldt, in Bornem, Belgium (Haelters, 2018). However, this is only one of eleven known cases of narwhal sightings in the North Sea area between 1588 and 2018. For the Medieval Period, narwhal tusks were probably acquired through trade with northern regions of Europe and Greenland. At Norse sites on Greenland narwhal tusks are occasionally found and historical sources indicate that narwhal tusks were transported to Northern Europe in which they were appropriated as horns of unicorns (Pluskowski, 2004).

Charles the Bold, Duke of Burgundy, had four unicorn horns (narwhal tusks) in the four corners of the sideboard placed during his wedding with Margaret of York in 1468. Ornaments were often placed in the corner of sideboard in order for them to be on display. Unicorn horn was thought to be a recognized ward against poisoning. An inventory of the duke also records a unicorn horn carved with the image of Christ being held by the Virgin. The Imperial treasury in Vienna holds a narwhal tusk and a "unicorn sword" with the pommel, hilt and scabbard being made of narwhal tusk, that originally belonged to Charles, but came into the possession of Emperor Maximilian I after he married Mary of Burgundy, the daughter of Charles (Pluskowski, 2004).

Medieval sources also mention that ivory, including narwhal tusk, could be a seigneurial or royal gift to churches. Processional staffs are also often created out

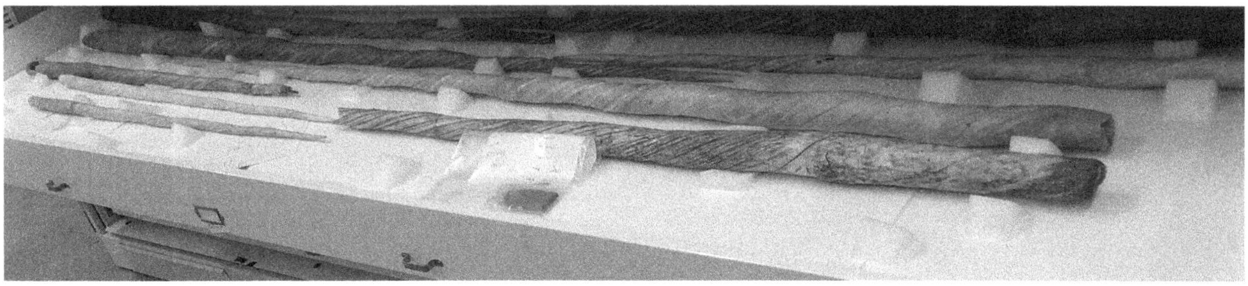

Figure 15. Narwhal tusks at the Natural History Museum, Smithsonian Institution, Washington DC, USA. Photo by author.

of ivory. The processional staff from the treasury of Salzburg Cathedral, dated to the late twelfth century, is an example of this. Another two components of processional staffs originate to twelfth century England are elaborately scrimshawed. Historical sources also indicate that narwhal tusks were present at the sixteenth century church of St. Denis, France and the fifteenth and sixteenth-century basilica of St Mark, Venice, Italy (Pluskowski, 2004).

Based on these historical sources, Pluskowski (2004), suggested that narwhal tusks, perceived as unicorn horns, were often displayed on ceremonial occasions and were only available to the social elite. Due to its supernatural origin (unicorn) and its curative qualities, the narwhal tusk was a highly valuable commodity from the twelfth century onwards till at least the sixteenth century (Christen and Christen, 2011). The lack of narwhal tusks being uncovered from medieval archaeological contexts might suggest that the tusks were indeed highly sought after, not often discarded and therefore missing from the archaeological record.

1.6.5. Blubber and Oil

Like humans, the skin of cetaceans consists out of three layers. The epidermis and dermis are the two outer layers of the skin and are thin and have no industrial value. The third layer, the hypodermis, is the blubber layer and was a highly sought-after product (figure 16; Slijper, 1962, 296-297). Blubber is a subcutaneous layer of fat beneath the skin present in all cetaceans, pinnipeds and sirenians, and might make up 50% of the total body mass of several species (Iverson and Koopman, 2018, 107-110). Not only is it fat and thus energy-storage for the cetaceans, but it also serves as an insulator and is of vital use to the thermoregulation of the animals, keeping them warm in their cold environment. They have to maintain a body temperate of about 37°C in an environment that is often between -1 to 25 °C. Since heat is lost 25 times more rapidly to water than to air, keeping warm is essential to marine mammals (Lazarus, 2006, 24). Furthermore, blubber serves as a body streamliner for the cetaceans allowing them to move more swiftly through the water and it affects their buoyancy as well.

The outer layer of blubber tends to be the coolest near the surface of the skin and warmer deeper inside the body near the muscles. Blubber consists of fat cells or oil called adipocytes held in place by a mesh of structural collagen fibres, giving it a tough and firm character. The adipocytes of many new born mammals are often empty of lipid but are filled up quickly after birth by weening. Blubber

Figure 16. Dolphin autopsy at the Statens Naturhistorisk Museum Copenhagen, Denmark showing the white blubber layer (photo by author).

contains not only blood vessels, but so called arterio-venous anastomoses (AVAs) as well, which allows more blood to flow more swiftly through the blubber than by blood vessels alone. During periods of food shortage or fasting, the lipid is mobilized from the adipocytes and thus serves as a reserve (Iverson, 2009, 115-120).

The structure, thickness and biochemical composition of blubber can vary strongly over the body as well as between separate species. Larger species tend to have a thicker layer of blubber than smaller species. 27 per cent of the blue whale's weight is made up of blubber. For the fin whale this is 23 per cent, the sei whale 21 per cent, sperm whale 32 per cent and for the right whale this is even somewhere between 36 and 45 per cent. The harbour porpoise body contains also a lot of blubber and in one case this was 60 per cent (Slijper, 1962, 296-297). In order to sustain their blubber layer and its insulative capacity, smaller species have to almost continuously feed in. Large whales can fast and mobilize reserves of blubber and go without feeding for weeks or months (Iverson and Koopman, 2018, 107-110).

Blubber floats in the water so the layers cannot be too thick since that would make the whale too buoyant to dive (Lazarus, 2006, 24-25). Because of the large amount of blubber, as well as the large amount of fat in the bones, the specific gravity of most whales is approximately one, meaning that they neither rise nor sink, but float in the water. This is however, a generalisation as there is quite a difference between species. Right whales, sperm whales and humpback whales have the tendency to float after they have been killed. This can be ascribed to the fact that they have thick layers of blubber. Because of this, these species were most frequently exploited by early whalers. The other rorquals sink more frequently, however at the end of the whaling season, when the blubber layer is at its thickest, rorquals also frequently float after being killed. Additionally, the quantity of air in the lungs also plays a role in this. Live dolphins can float thanks to air in their lungs (Slijper, 1962, 109-110).

Whale oil had many applications and has been used for oil lamps, soap, lubricant, and margarine. The medieval writer Olaus Magnus states that whale oil is a rich substance that can be used for oil lamps to supply light for altars and other places but can also be used for domestic purposes in the wintertime. Furthermore, he states that blubber is smeared on the outside of ships to prevent freezing of the boats and sinking them (Olaus Magnus, 2010, 1103-1104). Furthermore, it is used for greasing carriage whales, and the Arabs presumably used it on their camels to keep away gadflies. Additionally, it is stated that whale blubber was used for medicinal purposes and when mixed with honey, it sharpens the vision. The Norse additionally believed that the sperm of some whale species was an effective remedy for headache, leprosy, eye troubles, and any other illness (Olaus Magnus, 2010, 1104). This was stated in the *King's Mirror* (dating to AD), a Norse educational text.

Olaus Magnus furthermore registers that blubber was both salted and preserved for consumption as well as rendered potentially to produce oil, by sixteenth century Norwegians (Olaus Mangus, 2010, 1104). A stranded whale often stinks and is no longer suitable for consumption. Blubber however, can still be extracted from the carcass for oil production.

From the fifteenth to the eighteenth-century whale oil was also used in woollen textile manufacturing in England, the Netherlands, Normandy and Castile. Wool had to be washed before it could by dyed. Whale oil, as well as olive or linseed oil, was used to "regrease" wool in order to make it flow smoothly though the weaving and spinning machines (Loewen, 2009).

1.6.6. Spermaceti

Spermaceti is the liquid wax present in the head of the sperm whale. The word spermaceti comes from the Latin word *sperma* meaning "sperm" and *ceti* meaning "whale" or "sea monster". It was given this name because of its resemblances to actual sperm and it was thought that the sperm whale stored its sperm in its head. The spermaceti is actually stored in the spermaceti organ and the junk, both present within the head (Clarke, 1978, 1-17). The spermaceti organ consists of spongy tissue saturated with spermaceti, while the junk is located underneath it, which is more solid than the spermaceti organ and is divided into segments and is also saturated with spermaceti (Rice, 2009b, 1098-1099).

Spermaceti consist of a mixture of triglyceridic oils and wax esters, though the proportions of these vary between age and sex groups as well as the position within the body. The wax esters of sperm whale, together with that of the Amazon dolphin (*Inia geoffrensis*), differs from that of other toothed whales in that it consists predominantly of a relatively long chain of fatty acids (C10-C22). Because of this heterogeneous lipid composition and the internal temperature gradients, the spermaceti organ can focus sound waves effectively, making it possible for the animal to channel acoustic emission (Lazarus, 2006, 38; Rice, 2009b, 1098-1099).

Albertus Magnus saw the locals of Friesland butchering a sperm whale, piercing the animals eye resulting in the spermaceti flowing out of the hole. The locals filled eleven large flagons with it and also stripped the blubber from the animal (Albert the Great, 1987, 338). This is the only medieval source to make a specific mention of spermaceti, though based on this it can be argued that during the Medieval Period, spermaceti was a valued product.

1.6.7. Ambergris

Ambergris, also known as "floating gold", is an odoriferous fatty substance that is formed in the stomach and the intestines of sperm whales (figure 17). It is thought that ambergris is formed as a result of the sperm whales' diet of

Figure 17. Example of a piece of ambergris. Reproduced courtesy of Arthur Oosterbaan and Pam Lindeboom of the Ecomare Museum, Texel, The Netherlands.

cephalopods. The beaks of these animals are indigestible and are normally regurgitated. However, if this does not happen and enters the intestines, ambergris might be formed that may way up to 10 kg, though bigger pieces have been found as well. Ambergris consists of a mixture of unsaturated, waxy, high molecular–weight alcohols (Rice, 2009a, 28-29). The material will eventually pass through the intestines and will be defecated. Only one out of a hundred sperm whales produce ambergris, making it a rare resource (Clarke, 2006, 7).

While fresh ambergris possesses a rather unpleasant smell of sperm whale faeces, aged ambergris has a pleasant musky odour. It is both a rare and valuable substance and since the sperm whale is a cosmopolitan species, it can be found worldwide. The word ambergris comes from the old French *amber gris*, which means grey amber (Rice, 2009a, 28-29).

In the Medieval Period, ambergris was used for several purposes, e.g. as an aphrodisiac, as a laxative source, or as a spice and for candles (Brito, Jordão and Pierce, 2015, 1-12). The main purpose of ambergris was however its use in medicine and the perfume industry. It was believed that ambergris cured heart diseases, sore throat, cough, paralysis, cardiac diseases and hysteria (Romero, 2006, 6). It was introduced to western Europe by the Arabs and it remained an important natural resource until the nineteenth century (Rice, 2009a, 28-29).

Though it was a sought-after product, the origin of ambergris remained unknown during the Medieval Period. Marco Polo travelled through the Arabian Sea in the thirteenth century and mentions that the local islands caught whales and whilst butchering the animals they extracted ambergris from their stomachs. Many other authors thought ambergris was a product of excess whale sperm or originated as a result of reproduction (Szabo, 2008, 91). In the herbaria; *Hortus Sanitatis* by Jacob Meydenback dated to 1497, ambergris had its own entry. Several theories regarding the origin of the material were stated in it, including it being the fruit of a tree that solely grows in the sea, it growing on the seafloor, it being sea foam or being produced by fish (Brito, Jordão and Pierce, 2015, 1-12). By 1574 the Flemish botanist Carolus Clusius had suggested that ambergris is a product that is formed in the intestines of whales.

Since ambergris is less dense than water, it floats and can be found washed up on beaches, though most of it is directly recovered from caught or stranded sperm whale individuals (Rice, 2009a, 28-29). Ambergris has even been recovered from a 1.75-million-year-old Pleistocene palaeontological context in Western Umbria, Italy, though these coprolite remains are the only Pleistocene examples (Baldanza *et al.*, 2013, 1075-1078). No archaeological contexts are known from which ambergris has been recovered.

1.7. Pre-whaling population size

Knowing cetacean population sizes prior to the start of whaling is vital to our understanding of cetaceans and the impact humans have made on their populations during the period of whaling. These numbers allow the prediction of their future recovery under modern protection levels. This will aid the recovery of the species to their original population sizes. However, predicting pre-whaling population has proven to be difficult.

The (relative or absolute) presence of zooarchaeological material cannot be used in providing an estimate of historical abundance. There are many anthropogenic factors that can give a bias in the zooarchaeological record. Some species might have been more frequently exploited (either actively caught or opportunistically scavenging of a stranded whale), giving the false assumption that those species were more abundant than other. Additionally, zooarchaeologists still face identification problems, resulting in some species being more easily identified than other, again creating a bias in species representation. Zooarchaeological remains should therefore not be used as an indication of past abundance.

Genetic analysis of zooarchaeological cetacean remains can however provide information regarding pre-whaling abundance. Genetic variation decreased as a result of whaling, resulting in the loss of numerous genetic lineages. This genetic information is however still stored in zooarchaeological samples. Genetic research on zooarchaeological (and palaeontological) samples can therefore provide a wealth of information regarding past genetic variation, as well as past population sizes.

Another source of information, especially valuable for the reconstruction of pre-whaling population seizes of the species that were exploited during the twentieth century (most notably the rorquals, e.g. blue whale, fin whale, sei whale, and minke whale) are whaling logbooks. Whalers often recorded their catches and based on this, estimates of populations can be made. Whaling logbooks can however be incomplete or can underreport catches (Roman and Palumbi, 2003). Furthermore, those logbooks are not available for early periods. During these early periods the North Atlantic right whale and the bowhead whale were most frequently exploited. As a result, uncovering the pre-whaling population size of these two species is impossible based on whaling logbooks.

Other techniques have also been utilized to estimate pre-whaling population sizes. Roman and Palumbi (2003) performed mitochondrial DNA (mtDNA) analysis on North Atlantic fin whale, humpback whale, and common minke whale populations. The levels of neutral genetic variation can be a useful tool in estimating population sizes. The genetic diversity suggests that pre-whaling populations sizes for the species numbers 360,000 for fin whale, 240,000 for humpback whale, and 265,000 for common minke whale. Present-day population estimates number 56,000; 10,000; and 149,000 respectively, indicating that especially fin whale and humpback whale numbers drastically declined as a result of commercial whaling.

Another species that was heavily exploited is the North Atlantic right whale. Nowadays the population numbers somewhere in between 300 and 500 individuals. Waldick *et al.* (2002) performed DNA analysis on the pre-whaling population size and their research indicated that the species numbered around 12,000 individuals in the eleventh century AD. Their research showed that the North Atlantic right whale did not lose a significant amount on genetic diversity within the past 200 years. Genetic variation is however, in comparison to other large cetaceans, relatively low. This suggests that prior to eighteenth century AD, the North Atlantic right whale had already lost a significant amount of genetic diversity (Waldick *et al.*, 2002). This might be the result of centuries of erosion of the genetic variability, which probably started with the onset of a more commercialised form of whaling, potentially initiated by the Basques during the eleventh century AD.

Outside of the North Atlantic similar studies have been undertaken. Tulloch *et al.* (2018) conducted research on the pre-whaling population size of the large baleen whale species present in the southern Hemisphere. They found that especially the populations of the Antarctic blue whale (*Balaenoptera musculus intermedia*), the fin whale, and the southern right whale (*Eubalaena australis*) were affected by whaling and show a slow recovery rate. These species are thought to not even reach half of the pre-whaling population size by 2100, indicating that protection of these species is still important. The humpback whale shows better resilience and is currently at approximately a third of its pre-whaling population size and is estimated to have fully recovered by 2050. Additionally, Antarctic minke whale (*Balaenoptera bonaerensis*), also show an increase in numbers, but various variables, including recent Japanese exploitation of this species, make it hard to track the recovery rate of this species.

Alter, Rynes and Palumbi (2007) also conducted DNA analysis, but on the Pacific grey whale. Their research indicated that DNA variability of the grey whale would be typical for a population of around 76,000-118,000 individuals, while the current population only numbers 22,000. The Pacific grey whale was thought to have returned to its pre-whaling population size, as mortality had increased the years prior to 2007 which some researchers interpreted as the species had reached its carrying capacity. Alter, Rynes and Palumbi (2007) however showed that the current population resembles at most 27-56% of its historical population size. Alter, Newsome and Palumbi (2012) performed DNA analysis on zooarchaeological and palaeontological grey whale remains and confirmed that the population declined due to the historical exploitation of the species, decimating it from roughly 96,000 to just 22,000 in the modern Pacific Ocean.

Jackson *et al.* (2016) preformed a similar research on the southern right whale population of New Zealand. With limited historical whaling logbooks available, their study combined whaling logbook data with genetic data and individual sightings histories. They concluded that the New Zealand pre-whaling population numbered between 28,800 and 47,100 individuals. During the nineteenth century the population was almost hunted to extinction with just 30-40 mature females surviving between 1914 and 1926. Currently, the population stands at 12% of its pre-whaling size (Jackson *et al.*, 2016).

These studies show the potential DNA research can have on our understanding of modern and past cetacean populations. Estimates of the pre-whaling abundance of cetaceans continue to be an important baseline for judging the modern recovery rate and providing a target for ecological restoration.

For this study pre-whaling population size data are also of importance as it provides rough numbers of species abundance during the Medieval Period. Based on this and the general lifestyle and habitat of the different species, assumptions can be made on which species were most frequently encountered or stranded in medieval Northern and Western Europe, as well as which species would be most accessible to medieval whalers.

2

Cetaceans in a Medieval Historical Context

During the Medieval Period, perceptions of cetaceans were extremely varied. These conceptions were frequently based on both fear and fascination people had for them. This leads to the question: How were cetaceans perceived during the Medieval Period?

While Europe has always been a very dynamic region with diverse people and cultures, this is a hard question to answer. The perceptions of cetaceans varied for all these people as well as through time. This perception first of all seems to have been influenced by writers from the Antiquity period, preceding the Medieval Period. Writers such as Pliny, Strabo of Amassya, and Arian, wrote on cetaceans and the legends in which they played a role. Biblical texts also appear to have influenced medieval writers on cetaceans, in which they were often described as fearsome monsters (Szabo, 2008, 32).

During the Medieval Period, cetaceans appeared in numerous legends, myths and folktales, but they are also described in a more naturalistic setting, with various authors describing different species as well as their behaviour. However medieval sources often do not give any detail regarding the species exploited or consumed. Frequently cetaceans are merely described by the Latin *crassus piscis*, *craspesius*, *cetus*, *balaena*, *mereswyn*, *marsvin*, *parcus maris*, or *delphinus* (these all describe whales, dolphins, and porpoises, but appear to have been used interchangeably) providing little detail in regard to species exploitation (Gardiner, 1997).

In general, the perception of ferocious animals that could easily sink ships lived on during the Medieval Period. Indeed, numerous sources make it clear that while cetaceans were generally feared by medieval people, the quantities of products including meat and oil that could be extracted from one single carcass were more than rewarding. For this reason, several people along the northern and western European coastline decided to actively hunt cetaceans.

People like the Basques and Norse are probably the most well-known examples of this, but historical sources suggest that several other people also conducted active whaling from time to time. This however, probably to a lesser extent than the Basques, which have often been described as the first European industrial whalers.

2.1. Methods

In this section first an overview of the classical and biblical texts will be provided which heavily influenced the medieval mindset in regard to cetaceans. Then an overview of the most predominant whaling people along the northern and western medieval European coastline will be given. Special attention will be given to active whaling and opportunistic scavenging methods applied by these people. By doing this, the symbolic as well as socio-economic significance of cetaceans during the Medieval Period will become clear, creating a solid historical background to which the zooarchaeological analysis (that will also be a part of this study) can be compared to.

Some studies have been performed on the historical sources for some cultures, while for other limited analysis has been conducted just yet. These cultures include the: Norse (Schnall, 1993; Lindquist, 1995b, 19-22; Lindquist, 1997; Szabo, 2012, 76), Norse-Gaels/Hiberno-Norse (Laist, 2017, 95), Normans (Musset, 1964; Proulx, 1986, 10; Guizard, 2011), Danes (Schnall, 1993; Hybel and Poulsen, 2007, 55), Sámi (King Alfred the Great, 1855, 9-11; Bately, 2007, 46), Basques (Aguilar, 1981; 1986; Kurlansky, 1991), Portuguese (Brito, 2011; Brito and Sousa, 2011), Anglo-Saxons and the English (Swanton, 1975, 110-111; Gardiner, 1997), Flemish (De Smet, 1981 Chevallier, 2014), Frisians (Albert the Great, 1987, 338), Dutch (Coenen, 1585), and the Polish (Ropelewski, 1957).

For most of these cultures secondary sources were used in order to assess their past whaling or whale exploitation activities. Though some important sources (especially those concerned with England and the Netherlands) translations of the original sources were reviewed as well. These sources include editions of Olaus Magnus (Olaus Magnus, 2010, 1098), Albertus Magnus (Albert the Great, 1987), Ohthere/Ottar (King Alfred the Great, 1855, 9-11), Aelfric's Colloquy (Swanton, 1975, 110-111), and Adriaen Coenen (Coenen, 1585). This allowed for a critical and original approach by combining these sources with zooarchaeological evidence. Additionally, a critical evaluation of the sources was necessary as some of these are merely purely literary, while others provide fantastic representations of cetaceans.

Furthermore, a subchapter is concerned with stranding events being mentioned in historical sources. The strandings of especially large species did not frequently go unnoticed and are especially for the High and Late Medieval Period, frequently recorded.

In addition, shortly after AD 1000 historical sources suggest that European society was subdivided based on the philosophical feudalism concept of three orders. These three orders were called "*orant*" ("those who pray"), "*pugnant*" ("those who fight"), and "*laborant*" ("those who work"). Special attention was given to their dietary practices and how cetaceans might be incorporated in this.

It seems that for at least some periods and regions the consumption of cetaceans was restricted to a select few. Moreover, it appears that marine resource exploitation was not stable throughout the Medieval Period and the Fish Event Horizon launched a gradual increase in interest in the exploitation of fish. This might have affected cetacean exploitation as well, since many medieval societies believed cetaceans were fish as well.

2.2. Pre-Medieval and Medieval perception of cetaceans

Strabo of Amassya, Pliny, and Arrian are all writers from classical antiquity that wrote about cetaceans and the many legends in which they played a role in. They were especially fascinated by their ability to spout water from their blowholes. An important work, the *Physiologus* ("He who talks about nature"), is a didactic, Christian text written in Greek between the second and fourth centuries AD in Alexandria, Egypt. It comprises a description of animals, birds, and mythical creatures. Each animal is described and moral and symbolic qualities for that specific animals are given as well. It was translated into Latin around AD 700 and during the Medieval Period it remained a popular work. The characteristics, qualities and illuminations of each animals remained generally the same (Szabo, 2008, 46-50).

Cetaceans were initially categorized as *balena*, however later it became part of the *aspidochelone*, which also included sea turtles. The *aspidochelone* had the ability to release a seductive odour from its mouth which attracted fish. When the fish entered its mouth, the creature would close it, trap them within and subsequently swallow them. *Aspidochelone* were additionally often associated with the Devil and were so big, that humans often mistook them for islands. Sometimes their backs were even covered with trees, rocks, dirt or whole valleys. Being located in the middle of the sea, this attracted sailors who set up camp on its back. Only when they started cooking and created a cooking fire (figure 18), the creature would suddenly dive into the depth and drown the sailors on its back. One of these events is described by Saint Ambrose in the *Hexameron* and there are many medieval depictions showing similar events (Szabo, 2008, 46-50).

Whales were thus perceived as the embodiment of evil and observations of actual whales became enmeshed with mythology. The descriptions given in the *Physiologus* had a huge impact on the later bestiaries (Szabo, 2008, 46-50). The Voyage of St Brendan (Navigatio) is another tale influenced by the *Physiologus.* This manuscript first appeared in the tenth century AD, but was probably composed during the ninth century AD. It tells the story of the monk Brendan and some fellow monks whom spend the night on an island in the Irish Sea. The following morning, they celebrated mass and started to cook when the island suddenly started to heave, and the monks quickly left the island. Brendan then tells the other monks that God informed him that it was the sea creature Jasconius they spent the night on. In contrast to the *Physiologus* the whale is not evil and does not drown the monks on its back (Szabo, 2008, 51-53; Guizard, 2011).

Another tale about an encounter with a whale is the Life of St Columba, written by Adomnán in the late seventh century AD. This story involves two monks wanting to cross the open see between Iona and Tiree. They are however warned by Columba of a monster on this sea route. The first monk, Berach, ignored Columba's warning and he came across a whale the size of a mountain, showing its huge teeth. He only just managed to avoid the wash created by the motion of the whale. The second monk, Baithéne, listened to the counsel of Columba and when they encountered the whale, Baithéne raised his hands and blessed the sea and the whale. Immediately the whale retreated into the sea and was not seen again (Szabo, 2008, 51-53). Though there are fabulous elements to this story, and some accurate parts as well. Monks travelling between the islands had to do this in small vessels and even a moderate sized cetacean could accidentally or purposefully engulf a vessel like that (Szabo, 2008, 51-53).

Many legends about cetaceans were passed from antiquity to Christian Europe where they often played the role of creatures that challenged sinners. The Books of Job, Jonah, Psalms, and Revelations are all examples of biblical works that "informed" the medieval audience about the creatures that patrolled the seas. It is also in this way that the authors from late antiquity differed from the Roman authors. Roman authors were primarily concerned with observations and describing the natural world, while the authors of the Christian late antiquity perceived the natural world as a symbolic one which could illuminate God's will.

With the spread of Christianity during the Late Roman period and the subsequent Early Medieval Period, whales are occasionally the topic of various stories. Though most of these creatures were actual observations of the natural world, their impact on the medieval perception of marine creatures was enormous. Many medieval Christian authors used whales to teach humility, obedience and faith to their readers and showed that even the spiritually pure were not safe. It is likely that these tales also formed a source of entertainment. Additionally, it is shown that whales were not merely viewed as symbols but were also seen as resources since in several tales, people set out to hunt and kill whales, trying to acquire the vast quantities of resources that could be exploited from their carcasses. But they were always met with caution, knowing that their powerful bodies could sink the ships pursuing them without warning (Szabo, 2008, 54-65).

Probably the most famous whale encounters in western literature would be the whale (or fish) that swallowed Jonah, as written in the book of Jonah in the Old Testament (figure 19). In this book Jonah is commanded by God to visit the city of Nineveh to warn them of their great wickedness. Jonah however, does not listen to this

Figure 18. A Fisherman on the Back of a Whale. Object Number 83.MR.174.94v. Digital Image courtesy of the Getty's Open Content Program. Available at :http://www.getty.edu/art/collection/objects/4881/unknown-maker-a-fisherman-on-the-back-of-a-whale-franco-flemish-1277-or-after/.

command and tries to flee by boat. A storm arises, and the sailors find out it is Jonah whom is at fault and they throw him overboard. Subsequently, Jonah is swallowed whole by a *piscus grandis* (large whale or fish). Immediately the sea gets calm again and the sailors offer sacrifices to God. After three nights in the whale's belly, Jonah prays to God and vows that he will follow his commands, after which Jonah is spit out by the whale. Jonah then follows Gods commands and visits the city of Nineveh and warns the inhabitants that the city shall be overthrown. The king and the people believe him and make a proclamation which decrees fasting, sackcloth, prayer, and repentance. God sees this and spares the city (Szabo, 2008, 43-44).

In this story the mouth of the whale can be seen as the gate to hell and the whale as hell itself or the devil. However, while being in the whale for three days, Jonah is humbled before God. St. Ambrose argues in his *Hexameron* that the whale in his perspective was a giver of life, making Jonah humble and giving the story a positive turn. The story about

Figure 19. Jonah spit out by the Whale by Johann Sadeler. Digital image courtesy of the Rijksmuseum Open Data Policy. Available at: <http://hdl.handle.net/10934/RM0001.collect.506784>.

Jonah and the whale had a huge impact on Christians and during the early Christian period many sarcophagi feature the Jonah theme and other sea monsters which probably represent a fear of maritime powers (Szabo, 2008, 43-44).

The tenth century written legend of Within, written by Letaldus which is heavily inspired by the tale of Jonah, tells about an Englishman Within whom encountered a whale whilst in his boat. Within was subsequently swallowed by the whale, but instead of staying inside, he killed the whale from within. The carcass with Within still in it, drifted in the sea for some time, but eventually stranded along the English shore near Rochester. When the local people discovered the whale, they set out to butcher it, but were horrified when they found that something was still alive of the carcass. Within spoke from within the whale and the locals and their bishop assume the whale to be demonic or possessed. The bishop then exorcised the whale with holy water and Within is freed from the whale. Once freed however, he discovered that his hair had fallen out, he lost his vision and the skin surrounding his nails had been eaten away, though his looks return to normal after some time and he returned to his normal life.

Another tale of a person being swallowed by a whale is recorded in the *Gesta Romanorum* (a collection of Latin stories of moral import and entertainment). According to one tale, the king of Ampluy sent out his virginal daughter to marry the son of Emperor Anselm of Rome. On her way however, she suffered a shipwreck and was swallowed by a great whale. Whilst in the whale, she used her knife and wounded the whale on several places. The whale felt this and started to swim to land. The whale was spotted from the land by a noble earl named Pirris and once upon spotting the creature he gathered a group of people and set out to kill it. Upon doing this, they heard the voice of the girl in the whale, whom proclaimed she needed help and was a true virgin daughter of the King. Pirris opened the side of the whale and took the princess out (Szabo, 2008, 52-54).

2.3. Whaling People

Historical sources have been used to analyse cetacean exploitation and various cultures have been associated with it (figure 20 and 21). Some of these cultures like the Basques and the Norse are closely associated with cetacean exploitation and appears to make up an important part of their cultural history. Other cultures appear to have infrequently undertaken cetacean exploitation or undertook it on a small scale.

This section will give an overview of cultures associated with cetacean exploitation and will primarily focus on historical sources that argue for the existence of these practices. This will give a solid historical background to which the archaeological material can be set out. This interdisciplinary combination of both historical sources and zooarchaeological sources will provide an idea of in which regions and time periods cetacean exploitation was practiced and whether it was an important aspect of local or regional cultural practices and subsistence economies.

However, it should be noted that the presence of one does not rule out the second. Historical sources regarding cetacean exploitation undertaken by the Basques are for example abundant, but zooarchaeological sources are not thoroughly analysed. On the other hand, cetacean material might be commonly found at an archaeological site, but this does not necessarily imply that whale hunting was undertaken but can merely indicate that stranded individuals were scavenged.

Figure 20 gives an overview of which cultures practiced whaling. This figure is based on historical sources suggesting that active whaling was undertaken. For some cultures this might mean the exploitation of large whales, while for other merely the harbour porpoise was exploited. Moreover, for some cultures numerous historical sources mention whaling activities, while for others only a handful suggest this. The various sub-chapters will discuss the primary sources used to create this figure. Additionally, the Sámi will be discussed in the sub-chapters as well, as it has been suggested that they also performed whaling, though historical sources are not clear about this, and as a result they are not incorporated into the figure.

2.3.1. Norse

The Norse are a people originally from Scandinavia. This Norse culture experienced its height during the Viking Age, which is dated from the ninth to the eleventh century AD, though Norse culture lasted until at least the fourteenth century. The Norse expansion was led by the development of the sail in the seventh or early eighth century and was probably initiated by an increase of economic contact with the continent or population pressure. Contact between Scandinavia and continental trading centres might have started in the eighth century AD already. Eventually they were able to colonise, the Shetland islands and the Orkney Islands (800 AD), the Faroe Islands (around 825 AD), the Scottish Western Islands, Sutherland, Caithness, Iceland (870 AD), Greenland (986 AD) and even Newfoundland (1000 AD; Fitzhugh, 2000, 19; Laist, 2017, 88).

During these periods of colonization, the settlers relied more heavily on scavenged and hunted resources, before they could establish self-providing settlements. Zooarchaeological analysis from several early Icelandic sites show a higher percentage of wild resources being exploited, including marine mammals and birds (Fitzhugh, 2000, 19).

The reference often cited as the oldest reference to Norse whaling is *The Voyage of Ohthere*. Ohthere was a Norwegian navigator and traveller, who visited the court of King Alfred of Wessex at the end of the ninth century AD. He reported that there were good walrus hunting places in northern Norway (close to modern-day Tromsø). It has often been interpreted that Ohthere was talking about

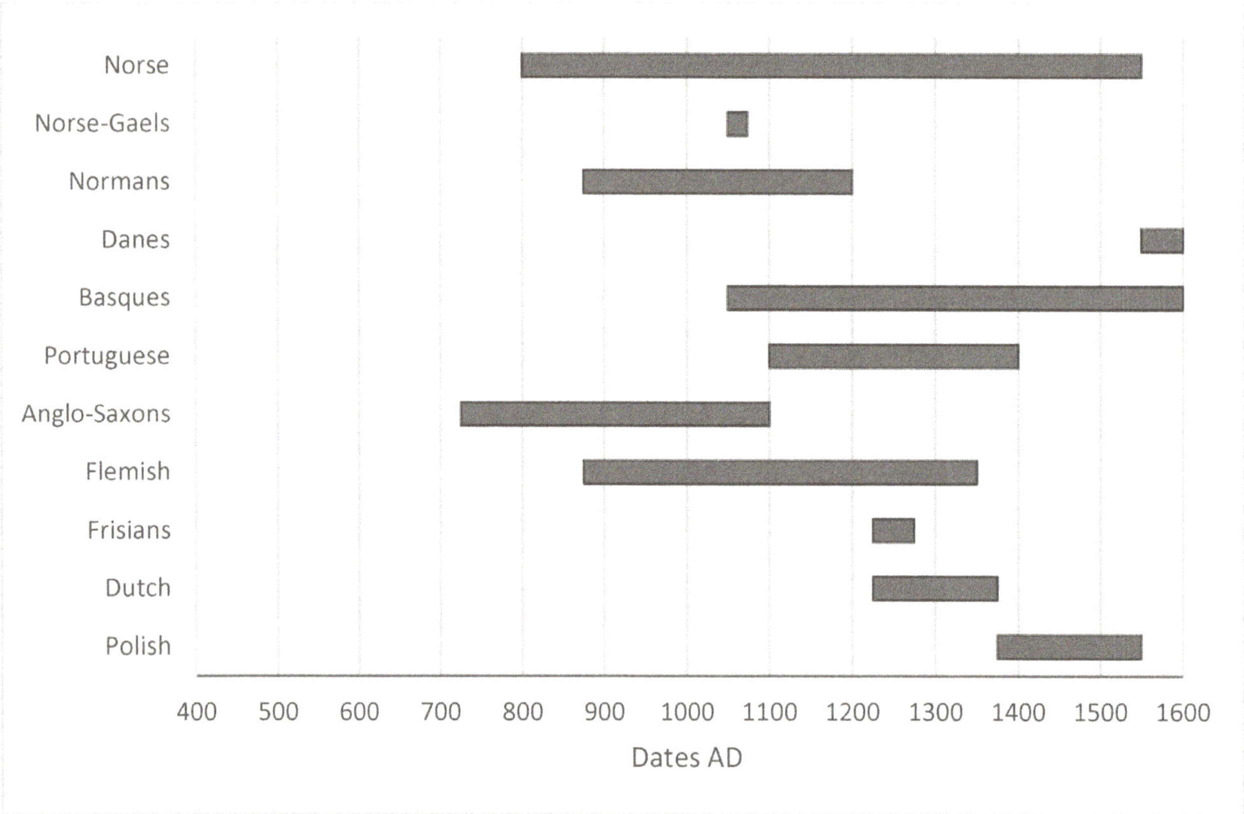

Figure 20. Overview of whaling activities based on historical sources. Not all the cultures targeted the same species. The Polish for example probably only targeted the harbour porpoise, while the Basques primarily targeted the North Atlantic right whale.

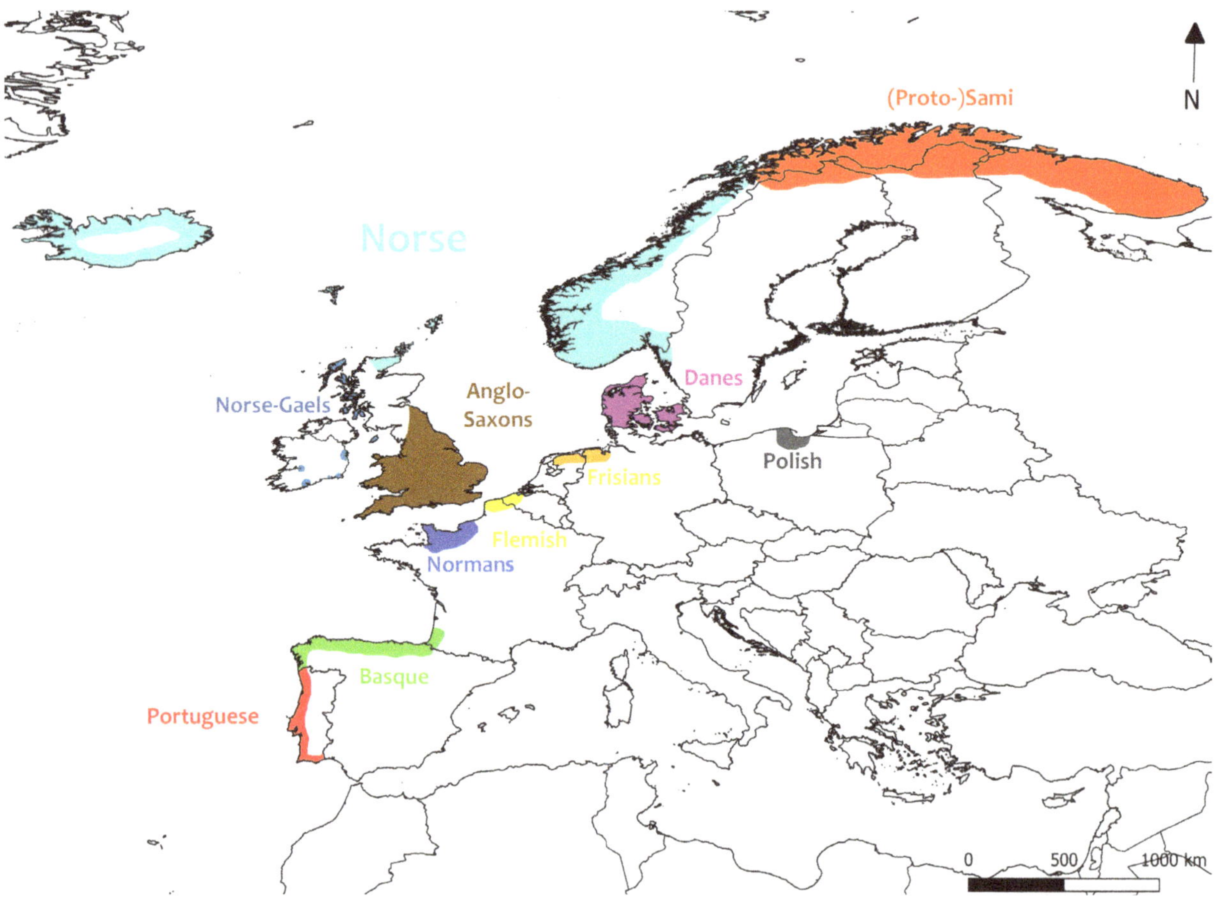

Figure 21. Map with geographical distribution of medieval people and cultures associated with whaling. Created by author.

whale hunting, stating that that the best "whale hunting" is in his own country, where the "whales" are forty-eight to fifty ells long. He states that he was able to kill sixty of these whales in merely two days (King Alfred the Great, 1855, 9-11). It seems impossible that Ohthere was able to kill sixty whales in just two days, but it has been suggested that instead of whales, walruses are meant.

Though *The Voyage of Ohthere*, is probably concerned with walrus hunting, it is clear that cetaceans were a valuable source to the Norse as well. Icelandic and Norwegian laws also suggest this. The *Gulaþing* (Norway's oldest legal code, dating to the mid-eleventh century AD, which travelled with Norse settlers to the Orkney Islands) declared that if a whaler was able to take a whale on open sea it was his property, however if the wounded whale beached, the whale had to be divided between the whaler and owner of the land the whale beached upon (Laist, 2017, 89).

Furthermore, a man of a higher rank was allowed to claim a stranded whale under eighteen ells in length, while any other man can only claim one that is half as long. Moreover, if after butchering the whale products are taken from the foreshore to the land and are transported over land of someone else, a portion has to be given to the landowner as well. Furthermore, if no witnesses are present, the whale's backbone, fin and head must be left in place to testify that the whale was below the maximum size allowed to be exploited (Szabo, 2005).

The laws regarding the procurement of whales became even more elaborate during the mid-thirteenth century in both Iceland and Norway (Fiskeripublikasjoner, n.d.). For Iceland, the person who first struck the whale, whether the animal died at sea or after stranding, the size of the animal, the owner of the land the whale stranded upon, and even the shore profile of that particular piece of land, all mattered for the division of the carcass. On top of that the King also regularly claimed a portion (Schnall, 1993). This was all recorded in the *Jónsbók*, the Icelandic law enforced in 1280. All Norse littoral and inshore legal regimes were allodial. As part of these regimes, the rights of coastal estates encompassed the resources present within the littoral and inshore waters and therefore included cetaceans (Lindquist, 1997, 11).

The *Saga of Grettir*, a purely literary source, notes that stranded whales were highly valued and even worth fighting over. A large whale washed up on the Island of Rifsker, off eastern Iceland. It was soon discovered, and two rivalling groups started fighting over ownership of the carcass. The fight was fought with axes, knives and cleavers, intended for butchering the carcass, and even whale bone and rotting blubber were used as weapons. At least two people were killed in the dispute and shows that people were willing to go to extreme measures in order to secure a stranded whale (Schnall, 1993; Laist, 2017, 87). Even though this is a purely literary source, stranded whales were probably highly valued.

Another case that took place in tenth century Iceland, was when a large whale was driven ashore in the Icefjord. One side of the fjord belonged to Thornbjørn and the other to Håvard. Everyone agreed the whale stranded on Håvard's side, but when Thornbjørn threatened the judge, he determined it was Thornbjørn's (Schnall, 1993).

The Icelandic sagas, as those described above, probably took part during the ninth or tenth century AD but were only written down during the thirteenth century. Often these sagas indicate that whales were acquired through natural strandings and not through active whaling. This indicates that the Norse were merely opportunistic whalers and that whaling was not an industry (Szabo, 2012, 79). However, several Icelandic sagas describe the sale of whale meat, including *Ljósvetninga saga* and *Eyrbyggja saga*, suggesting at least some form of commercialisation of whale meat.

The *King's Mirror* furthermore indicates that the Norse preferred the meat of rorquals (either blue, fin, sei, Bryde's, or minke whale). Szabo (2012) argued that probably the blue or fin whales were described in the text. The text describes the whales as quiet and peaceful and the whales are stated to not sink ships. It additionally is stated that their meat tastes and smells better than any of the other whales (Szabo, 2012, 75-76). Several of the Icelandic sagas display the pre-eminence of the rorqual, including the *Grœnlendinga saga*, *Saga of Grim Shaggy-Cheek*, and the *Hávarðar saga Ísfirðings*, which display the social conflicts that erupt when a large rorqual stranded on contested ground (Szabo, 2012, 77).

It has often been deemed unlikely that the Norse targeted these large species, as they are fast swimmers and can grow very big, but a recent zooarchaeological analysis by Szabo indeed suggests that the rorquals were the favoured species (personal communications Vicki Szabo, 2018). They might have been acquired through the described usage of marked spears and exploited the carcasses of those animals that died after being struck.

A thirteenth century record, the *Grágás,* indicates that wild resources (including stranded whale carcasses) could be scavenged even when other work was not prohibited. "A drifting or stranded whale may be moved and secured or cut up if it cannot be secured. If fish 'come ashore,' then men may catch them. Nets and hooks may not be used…" (Szabo, 2008, 72). The *Grágás*, furthermore states that the Icelanders indeed used personal marks on their weapons to hunt whales (Szabo, 2012, 80-81). These sources seem to suggest that whaling was performed on an opportunistic basis in a major part of Norse territory.

After colonizing Iceland and Greenland, the Norse spread to North America and at the archaeological site of L'Anse aux Meadows their remains have been uncovered. The remains of the site indicate that cetaceans were infrequently exploited, but the people did not rely on whaling. The Norse diaspora in the westernmost part of their territory appear to not have ventured into the regions to hunt whales (Szabo, 2012, 67).

The *King's Mirror* clearly indicates that the Norse were very familiar with whales (Szabo, 2012, 76). Szabo (2012) has argued that though the Norse had set up laws in several regions in regard to controlling the exploitation of whales. For whaling purposes, the Norse used spears. The Norse appear to not have used harpoons (a spear with a line attached to it) until at least the fifteenth century (Lindquist, 1995b, 19-22; Lindquist, 1997).

Whalers marked their spears with a personal sign, so when a whale was killed, credit for the kill could be based on the spears inside the body of the whale. The hunter could therefore claim the "shooter's share". A detailed text about the Icelandic nobleman Björn Einarsson, also discusses this, as the nobleman got stranded on Greenland and came across a large stranded whale (probably a rorqual). By harvesting the meat of the whale, the nobleman and his group could survive, but they came across the spear of one of the people from his district in Iceland. The whale was apparently killed by this spear, and upon returning to Iceland, Einarsson was obligated to reimburse the whaler. This account indicates that spear-whaling was an inefficient technique, with many of the whales escaping, and those whales that died could strand thousands of kilometres away. A large dose of luck was therefore needed to acquire the killed animals (Laist, 2017, 90-94).

From the Jónsbók it is also clear that various other forms of whaling were undertaken. Fjord- or bay-whaling, where a pod of whales or dolphins is driven into a fjord or bay where the animals were subsequently killed using spears, lances or poles, was one of them. The common minke whale seems to have been the primary target. It was targeted in the areas around the archipelago region of Bergen, especially in fjords with narrow inlets, such as Skogsvaag, Tellevaag, Østfjordenspollen, and Florevaag.

In some areas with narrow inlets the Norse are known to have ebb-trapped whales behind reefs. They simply had to be patient and wait for a whale to venture into the inlet and wait for the tide to fall, blocking the whale's way out of the inlet. This was for example the case in Steingrímsfjörður, Strandasýsla in Iceland. Here, large rorquals were ebb-trapped behind a reef and attacked by killer whales and chased ashore (Knag, 1938, 23). The Norse might have simply relied on the killer whales for killing the animal and might have exploited the carcass (Lindquist, 1997, 29). Another possibility is that the Norse copied the killer whale's tactics (Lindquist, 1997, 29). Cetacean exploitation like this might have happened at various localities in Norse territory.

Another whaling method frequently associated with the Norse was the "poisoning" of whales (Lindquist, 1995a). Lindquist (1995a, 488) has however argued that "poison", seems not to have been used in Norse whaling activities. Norse whaling however, seems to have relied on clostridium infection accompanied by spear, lance, and arrows whaling, provided the whale's thick blubber layer was penetrated.

Another technique was the driving ashore of a large pod. This technique was practiced on especially pilot whales, but other dolphin species are also known to have been targeted. It is still practiced in the Faroe Islands, during the *Grind*, in which pilot whales are targeted (Fielding, 2018, 92). This form of whaling is known to have been practiced in Iceland, the Shetlands, the Orkneys, the Hebrides, and Ireland during the seventeenth to early nineteenth centuries, but might have been practiced during the Norse period as well (Lindquist, 1997, 29). Drive seining, using specially made nets, has been recorded to have been practiced from at least the mid-eighteenth century onwards (Lindquist, 1997, 29).

Several whale traps existed in Norway as well and several are mentioned from the early fourteenth century onwards. Place names containing "hvalvágr" and "hvalvík", have been interpreted to have been suitable whale beaching grounds. Lindquist (1997) has argued that these traps must have required a socio-economic organisation in order to maintain them and their existence relied on the efforts of several farms and powerful chieftains with jurisdiction over the grounds.

In a letter dating to the 1300s the Bishop of Bergen appears to have claimed the whales caught in Skogsvaag (Collett, 1912, 564). The letter confirms that the clergy of Norway took interest in cetacean exploitation from at least the twelfth century onwards (Collett, 1912, 564).

The Crown also had an interest in whale exploitation. In Magnus Lagabøte's national law of 1274, cetacean exploitation is treated in various ways depending on the procurement method (stranding or hunting) as well as the size of the animal. Stranded animals were divided between the king, landowner and the finder. However actively caught animals were merely divided between the hunter and the landowner (Fiskeripublikasjoner, n.d.). Furthermore, it was believed that large whales drove herring towards fishermen, aiding them in their fishery activities. These whales were called "fish drivers". As a result, whaling during herring fishery periods was prohibited and those people killing a whale during a herring fishery event were fined (Lindquist, 1997, 25).

Moreover, it has been suggested that whaling might have been undertaken at the royal manor of Fitjar. The manor belonged to the Crown as early as the eleventh century and remained their possession until the seventeenth century. Indeed, from at least the twelfth/thirteenth century a partial royal prerogative in cetaceans was established on allodial ground (Lindquist, 1997). Several of the geographic locations in the Fitjar area have names linked to marine resource exploitation such as "*Hvalbeinvik*" (Whalebonebay). The crown granted Fitjar as a fief to Niels Gjordson in 1529, however this was with the exception of "the Crown's delights and hunting", suggesting that whaling might have been restricted to the Crown (Iversen, 2013).

Outside of Hordaland, Norway, whaling in this manner disappeared during the Late Medieval Period. In Iceland, whaling endeavours appear to have ceased at the end of sixteenth century AD, as an account by the Icelandic bishop Oddur Einarsson, dating to 1588 or 1589, states that Icelanders abstained from hunting whales except when it stranded along the shore. It is however not clear whether just a particular kind of whale species was meant, or all. It might be that Icelanders still hunted other species during the end of the sixteenth century, though it has been interpreted that merely the "fish drivers" were meant. These "fish drivers" were probably the rorquals which were believed to aid fishermen by driving fish towards them and the coast (Lindquist, 1997, 26).

2.3.2. Norse-Gaels/Hiberno-Norse

The Norse-Gaels were of a mixed Norse and Gaelic ancestry. After the Norse settled in Scotland, the Scottish Isles and Ireland, they gradually adopted Gaelic culture and intermarried with Gaels, giving rise to the Norse-Gaels. From the ninth to the twelfth century they dominated much of the Irish Sea and Scottish Sea region and founded several kingdoms.

Around 1058 AD the Muwallad geographer al-'Udhri visited the Norse-Gaelic region and described that during the months of October to January, the people performed whaling there. He recorded that especially calves were targeted and were struck by iron-tipped spears with large rings attached to which ropes were tied. Once killed the carcass was dragged ashore and the meat was preserved with salt. It has been argued that this is the first citing of the use of harpoons with attached lines (Laist, 2017, 95).

This targeting of the calves was probably also undertaken by the Basques and can be decimating to a slow reproducing animal like a large whale. How long the Norse-Gaels kept performing whaling and how frequently it was practiced remains unclear.

2.3.3. Normans

Prior to the arrival of the Normans, large parts of Normandy (modern France) were part of western Gaul. According to ancient rights, whatever stranded along the coast belonged to the local lord. In case of cetacean strandings, the animal was butchered by the subjects of the local lord. From this source it appears no active whaling was undertaken by the Gauls in the region and they relied on natural strandings (Guizard, 2011). This changed with the arrival of the Normans.

The Normans were a Norse people who settled the region of Normandy in the early tenth century AD. It has been suggested that they brought whaling with them from Scandinavia. However, historical sources indicate that cetacean exploitation was present already before the Normans settled in Normandy. The Parisian abbey of Saint Denis had a fishery on the Cotentin Peninsula already by AD 832 for *crassus piscis* (which might be either porpoises, dolphins, or whales; Tardiff, 1866, 85). This clearly demonstrates that cetacean exploitation predates the Norman Conquest.

In *The Life of the Abbot Philibert* (608-684), the ninth century Ermantarius, wrote that the abbot Philibert prayed for oil and not much later a dead fish (likely to be a whale) washed up, from which thirty *modii* of oil could be extracted. It is additionally mentioned that fifty feet long fish (again most probably whales) were caught for food and oil for illumination, indicating that active whaling was already practiced during the seventh century AD (Musset, 1964; Chevallier, 2014).

Historical texts also mention whaling after the formation of the Duchy of Normandy in 911. A charter of the priory of Héauville describes that restrictions on whaling were set in place at the port of Cape La Hague (Musset, 1964). Another charter describes that a whalers' society was present at Dives-sur-Mer (Proulx, 1986, 10). This charter registers the agreement between the abbots of Caen and Fécamp on the division of sturgeons and Craspois (probably cetaceans) taken at Dives. Both abbots contributed ships to the society of whalers. It has been suggested that drive hunting was undertaken in this area by the whalers, in which they hunted pods of dolphins, potentially pilot whales, and used nets and spears to drive them towards the shore where they were subsequently killed (Musset, 1964). Furthermore, the *Miracles of Saint Arnoul de Soissons* describes that large whales were captured as well, but that ships often sank as the whales attacked the ships (Guizard, 2011).

Whaling societies were present in the Saire, La Hougue, Saint-Vaast, Lestre, Quineville, and Saint-Marcouf. All these whaling societies were under the control of abbeys, indicating that ecclesiastical institutions had a lot of interest in cetacean exploitation (Musset, 1964). Further to the west in Brittany, according to a charter dating to 1181, all wrecks and stranded cetaceans belonged to the Archbishop of Dol (Musset, 1964).

Furthermore, Raoul Tortaire, a poet and monk, around the year 1115, took part in a whaling expedition on the Bessin coast. The hunt was undertaken in shallow waters in wintertime and nets and whaling boats were used to surround the whales. They used noise to pursue them and struck them with three pronged spears. Norman law furthermore made a distinction between actively caught cetaceans and opportunistically hunted or the exploitation of stranded individuals. When actively caught, a symbolic portion of the whale was gifted to the duke, but in the other cases the entire catch had to be handed over to the duke. This law was inspired from Scandinavia (Musset, 1964; Proulx, 1986, 10).

Vincent de Beauvais, a Dominican friar, in the early thirteenth century wrote about whaling being undertaken in Northern France. He wrote that the whalers used music to lure the whale and subsequently used harpoons to wound the animal and then the whalers moved away with great haste waiting for the animal to resurface and if necessary, impale the animal with more harpoons and lances in order to kill it (Chevallier, 2014).

In Normandy, cetaceans were also frequently claimed by ecclesiastical institutions, royalty, and nobility. This already started before the Normans. Charles the Bald, King of West Francia (843-877), granted the abbey of Saint Ouen in Rouen, all fish (including whales) and other things cast up by the sea (Giry and Prou, 1952, 410). This continued into the Norman period. In 1066, William the Conqueror, soon to be King of England, granted St. Trinity of Caën a tenth of the tolls on whales. In 1145 Pope Eugene II granted a church in Coutances a tenth of any whales (more specifically a tenth of the tongue) taken at Merry (Chevallier, 2014). The Abbey of Jumièges even ordered whaling to be undertaken near the Seine estuary. The hunted cetaceans were probably harbour porpoises (De Smet, 1981).

In the thirteenth century Henry II granted the monks of Montebourgh Abbey to keep the right flipper (pectoral fin) from all whales caught or beached near Coutances. Additionally, in Genêts a whale stranded in 1283 and belonged to the abbey of Mount Saint-Michel (Musset, 1964).

Furthermore, in the law-code known as "IV Æthelred", that has been attributed to Æthelred II and the late tenth century AD, merchants from Rouen were taxed in order to sell *craspois* or 'fat fish' (i.e. whale meat) in London, indicating that cetaceans were also commercially interesting products (Middleton, 2005). However, a recent re-evaluation of the law code indicated that the portion concerned with the *craspois* most likely dated to the aftermath of the Norman Conquest (Naismith, 2019). During the twelfth century Henry II, Duke of Normandy confirmed the privileges of merchants from Rouen, though he reserved his rights on whales (Middleton, 2005).

Rouen was also known for its business activities involving the creation of ornaments from baleen plates, from at least the beginning of the fifteenth century onwards. Furthermore, at the village of Saint-Vaast-la-Hougue whale bones were also used for a variety of purposes (De Smet, 1981).

These sources indicate that a taste for cetacean meat by the social elite was apparent in Normandy from at least the mid/late ninth century up until at least the twelfth century. Eventually they enforced laws to ensure they would receive a portion of the entire whale caught or stranded. This indicates a relatively longstanding tradition of cetacean exploitation. It was especially frequently undertaken in the estuaries of the Saire, Dive, and to a lesser degree, the Seine and Bresle, as well as on the eastern coast of Cotentin and the Bessin coast. From the thirteenth century onwards, active whaling seems to have disappeared and only records dealing with strandings are identified (Musset, 1964).

2.3.4. Danes

For Jutland, Denmark the "Den Jyske Lov" (Law for Jutland) was enforced in AD 1241. This law contained one paragraph about stranded whales. All stranded whales, dolphins, sturgeons and other large fish that a man could not carry, belonged to the King. The finder of a stranded whale, was however allowed to take a portion of whale, depending on whether he arrived at the scene by foot, horse, or boat. Though the finder was obliged to first tell the "ombudsman" of his finding (Schnall, 1993; Hybel and Poulsen, 2007, 55). Similar laws were enforced in Scania and Zealand (Tamm and Vogt, 2016, 84; 229; 291).

Inventories of the castles of Dragsholm and Helsinborg dating to 1536 and 1537 respectively, both listed numerous fishes, but also listed harbour porpoise meat and whale blubber (Hybel and Poulsen, 2007, 55-57). From these sources it appears that active whaling was not frequently undertaken, though at least one cetacean species was actively hunted: the harbour porpoise.

The harbour porpoise is the most common cetacean species in Danish waters and appears to have been frequently exploited. In Middelfart, on the western side of the Funen island, a harbour porpoise guild was set in place during the Early Medieval Period. As part of the guild, 10 boats with a total of 30 men hunted on the harbour porpoise during the winter season by driving them into the Gamborg Fjord and slaughtering them there (Middelfart museum, 2017).

2.3.5. Sámi

The Sámi people are a Finno-Ugric people inhabiting northern Scandinavia. During the Medieval Period the Sami were hunter-fishermen-pastoralists (Lindquist, 1997, 8). They are thought to have undertaken whaling as well, though historical documents regarding the Sámi are rare. Ohthere states that he collected taxes from them which included animal' skins, birds' feathers and whale's bone (though it is more likely that walrus tusks were meant) (King Alfred the Great, 1855, 9-11; Bately, 2007, 46).

Further to the south of Scandinavia, numerous gaming pieces of whale bone dating to the Vendel period and the Viking age (AD 550-1050) have been recovered. Cetaceans are rarely encountered in the Baltic Sea and therefore it was not known where the material originated from. Hennius *et al.* (2018) suggested that the gaming pieces were created by the Sámi in northern Scandinavia and were part of a trade network to central and southern Scandinavia. Indeed, at several Sámi sites in northern Scandinavia have whale bone material been unearthed (Nilsen, 2015), including several gaming pieces (Skomsvoll, 2012).

Hansen and Olsen (2004, 64) pointed out that after the collapse of trade the east around 0 C.E. the Sámi might have struggled obtaining metal products. Around this time, the Sámi started to exploit cetaceans and seals. Archaeological evidence for this besides the bone remains, are flagstone pits for the production of whale oil. The Sámi might have traded the oil, as well as the gaming pieces with Germanic agricultural groups to the south in return for metal products.

What species they might have exploited remains unclear, but Hennius *et al.* (2018) performed ZooMS on several of the gaming pieces from Sweden and determined them to be from the North Atlantic right whale, though they might have derived from bowhead whales as well, as ZooMS cannot differentiate between those two species.

In Meskfjord/Varangerfjord large baleen whales were trapped behind reefs during ebb and were attacked by killer whales (Knag, 1938, 23). This was recorded in the late seventeenth to early nineteenth centuries. It might be that the Sami made opportunistic use of these instances and scavenged the carcasses left behind by the killer whales. It might also be that the Sami copied the tactics of the killer whale and waited for an animal to be ebb-trapped and went in for the kill (Lindquist, 1997, 29). It is not known whether the Sami copied this from the Norse, whether it was the other way around, or that both developed the practice separately.

2.3.6. Basques

The Basques are considered to be the first medieval whalers. However, their whaling enterprises are little understood, obscured by limited archaeological as well as historical research (Huxley Barkham, 1984; Szabo, 2012, 69). It is not known when the Basques started their whaling endeavours. They are thought to have started whale exploitation by scavenging stranded individuals, followed by drive hunting the animals ashore, and subsequently pelagic pursuits (Szabo, 2012, 70). The earliest historical source that suggests Basque whaling dates to AD 670 when 40 *moyos* (casks of whale oil) were sold from Bayonne to the Abbey of Jumieges (near the city of Le Havre). From this source it is not clear whether this whale oil derived from an actively caught or a stranded whale. From this source however, it can be argued that the Basques already had the skills to render oil from whale blubber and had a trading network to sell it to other regions as well (Laist, 2017, 102-103).

In their initial stage the Basques practiced shore-based whaling. Whales were spotted from the shore. Eventually watchtowers were erected for spotting the whales as well. Watchtowers have been dated to the seventh century AD already, suggesting that whaling might have been practiced already from this point onwards (Kurlansky, 1999, 43). When a whale was sighted from the watchtowers, smoke signals or drum pounding was used to spread the news and the whalers took their small rowing boats, known as *txalupa* or *chalupa*, and pursued the animals and harpooned them with their harpooning irons and wounded them with lances (Aguilar; 1981; Laist, 2017, 102-103). The North Atlantic right whale was the primary target and especially calves were frequently taken, which was disastrous for the population. Additionally, the sperm whale and the grey whale were potentially occasionally taken (Aguilar, 1986).

The first historical records concerned with Basque shore whaling dates back to at least AD 1059. This record originates from Bayonne, the French Basque region, and is a regulatory measure taken that ensures that whale meat was concentrated at the market of Bayonne and a fee had to be paid on whales caught near the mouth of the Adour River (Aguilar, 1986; Proulx, 1986, 15).

It has been argued that the Basques acquired the techniques to hunt whales from the Norse whom were present along the European coast between 800 and 1000 AD, however this theory appears to have been based on unsubstantiated conjecture and is not generally supported anymore (Laist, 2017, 102). Whether the Basques acquired the techniques from other cultures or developed them themselves remains unclear. It is however clear that shortly after the first mentioning of whaling in 1059 in the French Basque region (where it especially frequently undertaken by the cities of Biarritz, Anglet, and Bayonne), whaling spread to the Spanish Basque region, and even later on further west. The first sources mentioning whaling date to 1150 for the Spanish Basque region, 1190 for Santander, 1232 for Asturias, and 1371 for Galicia, indeed indicating a westward expansion of whaling (Aguilar, 1981; 1986; Goyheneche Farnie, 1984).

Forty-seven whaling ports are known from the Basque region and appear to have been undertaking whaling for centuries, reaching a peak in the sixteenth century. The majority of the Basque whaling operations in the region were shore based whaling activities that were dependant on the whale watchtowers. However, whaling was also practiced using galleons, from which the small *chalupas* were launched as well. This was especially practiced in Galicia (Aguilar, 1986).

Aguilar (1986) has argued that one whale was caught in every port every one or two years. This made the fishery profitable and indicate relatively modest catches. The importance of whaling for the Basques can also be seen in the seals of several towns. These from the late thirteenth century onwards often depicted a whale. Examples of these are the towns of Bermeo, Fuenterrabía, Biarritz, Motrico, Guetaria, Lequeitio, Ondarraoa, Castro-Urdialles, Laredo, and Hendaye (Goyheneche Farnie, 1984; Kurlansky, 1991).

From the mid-twelfth century AD whalers in the Basque region were taxed a tribute. Up until the mid-twelfth century, whalers gifted part of what they caught to their lords or to the church. In 1150, the King of Navarre, Sancho the Wise granted rights to store whale bone to the town of Saint Sebastian. He even demanded a piece of the meat from each catch, as well as the tongue. Alphonse VIII of Castille and Ferdinand III extended those rights to the towns of Fontarabie (1203), Guetaria and Motrico (1209), and Zaraux (1237; Proulx, 1986, 16).

The whaling season started in September and lasted almost all winter. This was the period the North Atlantic right whale migrated south to calve, and subsequently migrated back north again (Goyheneche Farnie, 1984). Whaling was of such an importance during these months, that Edouard I forbad the bailiff of Labourd to force the men of Biaritz to appear before his court, as they would miss out on whaling activities (Goyheneche Farnie, 1984).

The Basques utilized all products that could be extracted from the whales. The meat was consumed, with especially the tongue being perceived a delicacy, the oil was used for lighting and to caulk ships and was exported to England and the Netherlands. The bones were used for the construction of buildings, fences, and even chairs. The first source mentioning ambergris (that can be extracted from sperm whales) dates to the sixteenth century, indicating that sperm whales were not targeted in the Medieval Period (Goyheneche Farnie, 1984).

During the period of English domination under Henry II, King of England (1154-1189), numerous taxes and fees had been recorded in Basque archives (Proulx, 1986, 16). In 1199, King John of England (1199-1216) granted the authority to take 50 Angevian pounds annually from the sale of the first two whales caught by whale hunters from Biarritz, to the governor of Bayonne. Additionally, the seneschal (a steward or governor) of Biarritz had to be paid for each whale caught. Whalers from Guetaria presented the first whale caught each season to the King, who frequently returned half of it to the whalers (Proulx, 1986, 16).

In 1257, William Lavielle provided to the bishop of Bayonne a tithe of the whales caught in Biarritz. This tithe was paid until at least 1498. Furthermore, in 1261, the Abbey of Honce granted permission to pay a tithe on whales landed at Bayonne (Jenkins, 1921, 61).

These records clearly indicate that amongst the twelfth and thirteenth century Basques, the clergy and nobility also developed a taste for cetacean meat and enforced laws in order to make sure they received a portion. However, the exploitation of whales seemed to have declined and Alfonso XI of Castille, in 1334, in response to the declining fishing population of Lequeitio, declared that instead of one in fifteen whales, taxes were to be paid for one in eighteen whales. This might indicate the first signs of a collapse of the whale population in the Basque region. Whalers were also, not always content with the taxes and in 1498 the whalers of Labourd refused to hand over the tongues as a tribute to the cathedral of Bayonne (Kurlansky, 1991).

After the North Atlantic right whale population declined in the Bay of Biscay, the Basques set out in search of new whaling grounds. From 1353 to 1561 whaling was performed off Ireland, and in 1412 a whaling station was constructed at Grundarfjördhur, western Iceland, were the eastern population of the bowhead whale was hunted (Azpiazu 2000; Loewen, 2009).

Beginning around 1530 the Basques started to exploit the bowhead whale population in the Strait of Belle Isle, Canada and transport the valuable oil back to Europe in a 35- to 40-day journey, where it was sold especially in Castile, Normandy, England, and the Netherlands. The Basques set up several whaling stations in Quebec and Labrador, where shore-based whaling was practiced (Loewen, 2009). Both the French Basques and the Spanish Basques performed whaling in the region and it sometimes led to confrontations. Basque whaling peaked in the 1560s and 1570s. During this period anywhere between 6000 to 9000 barrels of whale oil was sent to Europe (including Bristol, London, and Flanders) from Red Bay and another 8000 or 9000 barrels from St. Modeste (Huxley Barkham, 1984).

After 1579, when England closed its ports to Spanish whale oil. Whaling was still practiced in the region, but the population was over-exploited (Huxley Barkham, 1984), and the whalers moved to the Gulf of St. Lawrence in the early 1580s. In the 1630s whaling stations were constructed along the coast of western and southern Newfoundland and Eastern Quebec. Basque whaling eventually disappeared in the 1730s, after which many of the whale populations had disappeared (Loewen, 2009; Laist, 2018, 120-126).

In each region the Basques encountered new stocks and populations and historical sources seem to indicate that the Basques were able to tell the difference between these as well. They were able to tell the difference between the bowhead whale, North Atlantic right whale, and grey whale, and even had two separate names for the western and eastern populations of the North Atlantic right whale. It appears the Basques were aware of the migration routes of the various populations and adapted their strategies based on these differences (Loewen, 2009).

This overview shows that the Basques were very successful at whaling and that is has been a long-standing part of their culture (figure 22). Based on these sources it can be stated that the Basques were the first industrial whalers, undertaking whaling on a scale previously unheard of. Their techniques proved to be highly successful and even gained the interest of other foreign nations that sought their expertise.

2.3.7. Portuguese

Another group that was potentially undertaking whaling were the Portuguese. While Portugal is not renowned for being a medieval whaling nation, Brito (2001) analysed a large number of historical documents and suggested that from at least the twelfth century onwards whaling was

Fig. 194.—Whale-Fishing.—Fac-simile of a Woodcut in the "Cosmographie Universelle" of Thevet, in folio: Paris, 1574.

Figure 22. Whale-Fishing Woodcut in the "Cosmopgraphie Universelle" or Thevet, in folio: Paris, 1574. Digital Image courtesy of the Gutenberg Project. Available at: http://www.gutenberg.org/files/10940/10940-h/10940-h.htm.

practiced by the Portuguese. Brito was able to identify 38 historical documents concerned with whale exploitation or whaling dating between 1201 and 1728, with the majority of them dating to the thirteenth and fourteenth century (Brito, 2011; Brito and Sousa, 2011).

It was noted that these whaling activities were contemporary to those undertaken by Basques in the French and Spanish Basque countries. There appears however to be no connection between the Portuguese and Basque whaling activities, nor seems there to be a geographical cluster were whaling was regularly undertaken, as the localities discussed in the historical documents are scattered along the entire Portuguese coastline (Brito, 2011).

The Portuguese sources indicate that the "black whales" were targeted, which is likely to have been the North Atlantic right whale. However, it is not clear what methods the whalers equipped to hunt the whales and how many whales were taken. Just as for many other regions in north western Europe, tithes were levied as part of the feudal system that was in place in thirteenth century Portugal, indicating that whales were a prized resource and were associated with the social elite (Brito, 2011).

2.3.8. Anglo-Saxons and the English

Another group that potentially carried out whaling are the Anglo-Saxons during the Early Medieval Period in Great Britain. *Beowulf*, an Anglo-Saxon epic story written somewhere between the eighth and eleventh century AD, describes a swimming competition between Beowulf and Breca. During this swimming competition they were attacked by *hron-fixas* (whale-beasts) which they fought off with swords they were carrying during the swim. In the story, whales are not spiritually threating and contain no ominous biblical allusions but indicates that cetaceans were perceived as large ferocious monsters (Szabo, 2008, 55).

Other sources seem to indicate that active whaling was already undertaken during the Anglo-Saxon period. In Bede's *Historia Ecclesiastica* (AD 731) it is mentioned that seals, dolphins, and sometimes even whales were caught in Britain (Gardiner, 1997). Additionally, another source that has been used to argue that whaling was already practiced during the period is *Ælfric's Colloquy*. This document, written by the abbot Ælfric of Eynsham (955-1010), was a text used to teach Latin vocabulary and grammar to pupils, but provides valuable information regarding medieval practices. One of the sections is about a fisherman who has a dialogue with a master about his work as a fisherman. He informs the master that he often goes fishing in the river and catches pike, eels, burbot, trout, minnows, and lampreys. Additionally, he tells him that on occasion he fishes in the sea and catches salmon, herring, sturgeon, plaice, flounders, small crustaceans, molluscs, and even porpoises. When asked whether he would ever be considering catching whales, the fisherman argues that "…it is a risky business catching whales. It's safer for me to go on the river with my boat, than to go hunting whales with many boats….I prefer to catch a fish that I can kill, rather than a fish that can sink or kill not only me but also my companions with a single blow." He however continues saying that many do catch whales and make great profit by it (Swanton, 1975, 110-111).

In the tale, porpoises are listed with the other fish and it appears that the fisherman was not able to distinguish the difference between porpoises and real fish. It is furthermore not clear how frequent whaling was undertaken, in exactly what area, which species were targeted, and whom undertook it. It is however suggested that it was a profitable undertaking and that many boats were needed to hunt the animals. This might suggest that drive hunting was undertaken in which several boats line up and chase the animals to a shallow bay where they can subsequently be caught and killed (Swanton, 1975, 110-111). This method is practiced on dolphins that live in pods, including for example the pilot whale, though they do not often attack boats when in danger.

Another possibility is that several boats speared or harpooned large species, for example the North Atlantic right whale or potentially the grey whale, in a similar fashion as the Basques undertook whaling. Whales are known to have been more abundant in the English Channel during the Anglo-Saxon period, as De Smet's research (1981) noted that historical documents recorded that in 1004 in the English Channel, several ships had encounters with whales and sank. Gardiner (1997) and Szabo (2012, 74) noted that although cetacean remains infrequently are recovered from archaeological sites, they probably did not undertake active whaling frequently.

Following and just prior to the Norman Conquest a different pattern appears to emerge. From at least the eleventh century stranded cetaceans and other large fish were claimed by the king as "royal fish" and between 1116 and 1118 the king made sure that "wreck by sea and things cast up by the sea" were his right, as recorded in the *Leges Henrici Primi* (Gardiner, 1997). The Crown's interest in whales also spread to Ireland, which was ruled by the King of England and under control by the loyal Anglo-Norman lords between 1177 and 1542. A record in the Calendar of the Justiciary Rolls of Ireland dating to July 1295, specifies that Robert de Clohulle was charged with appropriating a whale to his own use "in prejudice of the Crown". Robert however refuted the charge and stated that by ancient custom in Ireland whales were reported "wreck of the sea", a right he inherited from his father (Went, 1968). Moreover, in September 1295, William Macronan was reported to having made a fine for a large whale of two cows and ten shillings (Went, 1968). This clearly indicates that similar rights were installed in Ireland as well and additionally these records display the importance of stranded whales.

Furthermore, the IV Æthelred law-code, probably dating to the aftermath of the Norman Conquest, indicate that cetacean meat was also imported from Rouen (Middleton, 2005; Naismith, 2019). Hunting on porpoises was also undertaken already in the eleventh century by fisheries in Gloucestershire and Tidenham (Gardiner, 1997).

The royal right to take "wreck of sea" was like many other royal rights, granted to others with coastal manors, such as members of the nobility, but also ecclesiastical institutions (e.g. Battle Abbey, St. Pauls Cathedral, Chichester Cathedral, Rochester Cathedral, and the Abbey of Faversham). The ecclesiastical interest in cetaceans was for example recorded in *The life of St Godric* dating to the twelfth century AD, in which Godric came across a stranded *delphines* in the Wash area and cut it up and carried it home (Gardiner, 1997).

Just like game consumption, cetacean consumption was associated with a high-status diet (Gardiner, 1997). Therefore, the illegal consumption of game and cetacean meat infringed the social order and a breach of seigneurial rights. It was however difficult for the king, and to a lesser extent for the local elite, to enforce the "wreck of sea" claims, as it was too easy for beachcombers to conceal stranded goods. As a result, later royal claims merely claimed the head (including the tongue which was perceived as a delicacy) and tail, leaving a large portion for the finders (Gardiner, 1997).

Historical records indicate that "wreck of sea" rights were often confusing as to not being clear to which stranded cetaceans belonged to. The local elite or the King? Furthermore, coastal communities often illegally exploited cetaceans. Many commissions of "oyer and terminer" were granted to local lords to examine the theft of stranded cetaceans by peasants, but by the time these commissioners started their work the whales were probably already consumed or sold by the exploiters. The purpose of these commissions was to assert seigneurial rights.

The interest by the nobility and the clergy in cetacean meat can be linked to the Christian fasting practices. At the beginning of Lent in 1246, the sheriff on London bought 100 cutlets of whale meat and sent those to the King's larder at Westminster. By the late thirteenth century, the royal claims to wreck and cetaceans was largely lost by the royals and were restricted to the demesne manors. However, the interest in cetacean meat remained, especially for porpoise meat. Henry IV in 1399 and the bishop of

Lincoln in 1430 are known to have bought porpoise meat for feasts (Gardiner, 1997).

Many administrative rolls from England record cases dealing with rights to stranded cetaceans, but very few record actively caught specimens. An exception to this is a hearing in front of the King in 1255, in which the Bishop of Norwich claimed a "great monstrous fish" landed in his lordship. The fish was however caught by six boats taken at sea and the King stated that since this was the case it could not be treated as "wreck of sea". The case was however adjourned, and the final decision is not known. However, the confusion this case presents seems to indicate that active whaling was not regularly undertaken (Gardiner, 1997).

Furthermore, between Recliffe near Seaford in Sussex and Shoe Beacon, Maplin Light in Essex stranded and actively caught cetaceans came within the jurisdiction of the admiralty courts of Cinque Ports. Fifteenth century records indicate that especially porpoises stranded relatively frequently in the area, and additionally suggest that occasionally they were actively caught as well. However, these might have been taken accidentally and it appears that this happened infrequently (Gardiner, 1997).

Porpoises appear to have been the most frequently caught cetacean species and the historical sources suggest that active whaling (of large species), was only infrequently undertaken during the Medieval Period. This in contrast to the situation on the other side of the English Channel. Indeed, several records indicate that cetacean meat was imported from France and the inhabitants of Britain might have relied on foreign whalers in order to get access to cetacean meat.

2.3.9. Flemish

Northeast to Normandy, in Flanders whaling also seems to have been practiced early on. In *The Life of St. Vedastus*, dating to around 875, a group of fishermen from a monastery had a contest with another group to hunt a whale. The story indicates that the hunt was communally organized and that the participants paid a fee into a "contubernium" (a co-operative society) and agreed on sharing the catch. Eventually, the group that prayed to the St. Vaast eventually caught the whale (Chevallier, 2014).

Praying to ecclesiastical figures was often undertaken in order to catch whales in Flanders. In the tenth century, a Flemish fisherman prayed to St. Bavon (of Ghent) and thrusted his "iron" (probably a harpoon) into a whale and shortly after being hit, the animal resurfaced (Chevallier, 2014). Additionally, in *The life of St. Arnulf*, dating to the twelfth century, fishermen from Flanders surrounded a large whale and attacked it with spears, harpoons and lances (Carnandet, 1876, 256). However, the whale fought back and only after praying to St. Arnulf and vowing a portion of the whale in return for his aid, the fishermen succeeded in taking the animal (Chevallier, 2014).

Indeed, several sources record that cetacean meat was gifted to various ecclesiastical institutions. In 1121, the Count of Flanders gifted "pinam de cetam" (tail of a whale) to the Abbey of Sint-Winoksbergen (Bergen; Steevens, 2014). But the nobility tried to get access to cetacean meat as well. In 1178, Count Philip of Alsace was gifted a monstrous beast that was hunted by whalers from Bruges (De Smet, 1981). In 1300 the Count of Artois bought 33 pieces of whale meat from the market of Calais (De Smet, 1981). In 1371, the Flemish Count Louis of Male sent whale meat to his daughter Margareta at the Burgundian Court. Whale meat appears to be prized at the Burgundian Court, as Charles the Bold, Duke of Burgundy, also served whale meat at his wedding with Margareta, Countess of Flanders in 1468 (De Smet, 1981).

From the twelfth century onwards, active whaling is also more frequently mentioned. Historical sources describe that four whale hunting ships had their homeport in Blankenberge in 1147 (Charlier, 2004). Other sources indicate that whaling was a specific activity which required the permission of nobility. In 1163, several towns were granted the rights to hunt cetaceans in the Charter of Newport (Charlier, 2004). Additionally, in 1340 Wenduine was granted the right to hunt cetaceans, more specifically the harbour porpoise. A picture of a harpooned porpoise was also present at Wenduine's coat of arms (Charlier, 2004). Even the Duke of Burgundy, Count of Flanders had a ship undertaking whaling in the North Sea in 1456 (De Smet, 1981).

South of Flanders, in the County of Boulogne fishers (which was from the eleventh century onwards perceived as the centre of whaling) were required to give a part of their whaling endeavours to the Abbey of Saint-Wulmar (De Gryse, 1940-1945; Steevens, 2014). Moreover, the Duke of Boulogne is known to have had baleen plates on his helmet, as was recorded by the twelfth century chronicler Vilhelmus Brito (De Smet, 1981).

However, it appears that cetacean meat was not exclusively restricted to the nobility and clergy of Flanders. Sources indicate that in 1024 taxes had to be paid for every hundredth part of whale meat at the city of Arras. Taxes also had to be paid at Newport in 1163 (Steevens, 2014). Other cities, such as Boulogne, Calais and Damme are also known to have sold whale meat at the local markets between the eleventh and twelfth century (De Smet, 1981). This indicates that cetacean meat was widely available at medieval Flemish markets.

Cetacean strandings are also frequently recorded in historical documents. Eight whales were recorded stranded in Oostduinkerke in 1403 (Charlier, 2004). On the 2nd of June 1577 a sperm whale stranded and was killed in the Scheldt nearby Antwerp and two other sperm whales were found dead near Bieselinge (Coenen, 1585, 10). These might have been exploited by the medieval Flemish as well.

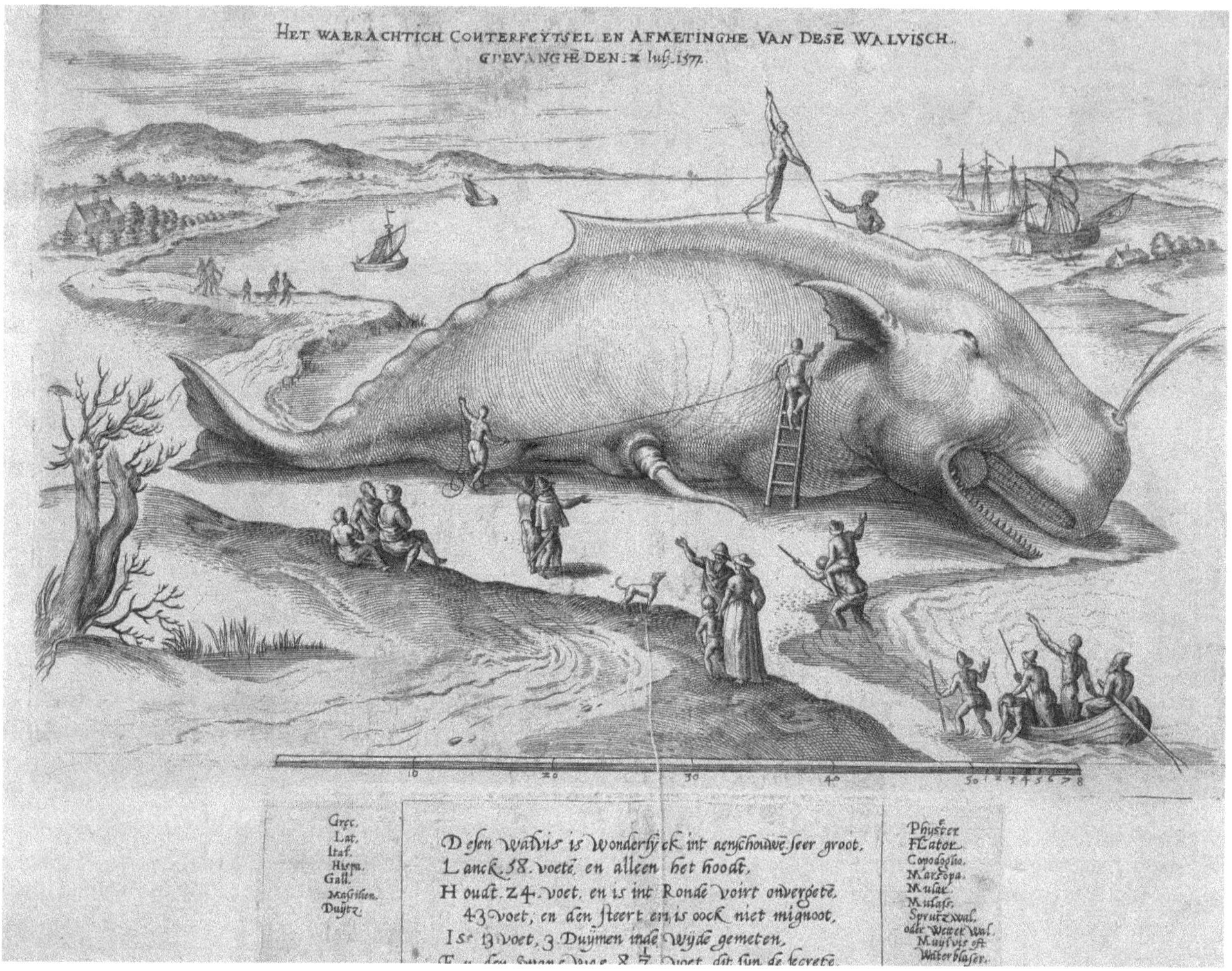

Figure 23. Stranded whale by Antwerp, 1577. Objectnumber: RP-P-OB-79.685. Digital image courtesy of the Rijksmuseum Open Data Policy. Available at <http://hdl.handle.net/10934/RM0001.COLLECT.446297>.

It appears that the Flemish conducted active whaling just like their neighbours in Boulogne and Normandy. Just like in the case of the Normans, most the historical sources date to the eleventh and twelfth century AD, though several of the sources date to the thirteenth and fourteenth century AD as well. This suggests that only from the High Medieval Period onwards the Flemish performed whaling more frequently and though this taste lived on during the Late Medieval Period, it became less apparent for this period. It remains unclear to what extent active whaling was practiced (figure 23).

2.3.10. Frisians

Though historical sources mentioning the Frisians are rare, their whaling activities were recorded by Albertus Magnus (Albert the Great, 1987). Albertus Magnus, in his time known as *doctor universalis* or *doctor expertus*, was a German Dominican friar and at Catholic bishop who lived from 1193 to 1280. He was perceived as one of the most prominent scientific minds of the Medieval Period and one of his works was entitled *De Animalibus*. Book twenty-four of this work was concerned with aquatic animals, including whales, dolphins and a variety of sea monsters. The book provides us with Albertus' own observations of sea creatures, as well as information derived from other classical and medieval works.

Several of his own observations took place in Frisia, and the Wadden Sea islands, in the current province of Friesland in the Netherlands, as well as one in Utrecht. He observed the local fishermen beach and flense the carcasses of several whales caught in the North Sea. He stated that they conserved the oil, rendered the whale blubber and retrieved the whale bone and baleen. He even registered the volume of bones and meat being recovered from each carcass (Albert the Great, 1987, 338). The Frisians are also known to have exploited stranded individuals as recorded by Albertus and used ropes to prevent the tide from washing away the stranded animal (Szabo, 2002, 73-74; Camphuijsen and Peet, 2006, 27).

Though Albertus had observed cetaceans on several occasions himself, his work still relied heavily on that of other authors, including Pliny and Thomas of Cantimpré and many of the sea creature categories he describes were obtained from their works. His work is however, different from that of previous writers since he focussed more on the scientific and economic aspects of whales and the quantities of resources that could be exploited from them.

At one event he noted that three hundred wagons were needed to transport all the exploited resources from merely one carcass. This shows the fast quantity of resources the Frisian people could exploit and the huge effort they put into transporting all the resources. These resources do not only include meat and oil, but also bone. Furthermore, Albertus notes that whales from varies sizes were being exploited and noted that the big one that required three hundred wagons were only rarely exploited (Szabo, 2008, 61-65).

Another aspect Albertus sheds light on, which other writers mostly ignored, were on the methods and technologies of how whaling was practiced by the Frisians. The technique Albertus describes include several small boats with a team of three hunters located on each (Albert the Great, 1987, 339.) Sometimes music or noise was created using kettledrums or other instruments to drive the whales in a direction (Proulx, 1986, 12). The thirteenth century chronicler Vincent de Beauvais also mentioned that the inhabitants of the German coast (potentially the Frisians) hunted whales by making noise using kettledrums (De Smet, 1981).

Harpoons were used to injure the animals and ropes were attached to the boats to prevent the animal from swimming away. However, when the animal attempted to swim away, the ropes were cut, and the animal was pursued, and whalers kept wounding it with harpoons, attaching all the lines to the boats. The harpoons appear to be typical barbed hooks, but the use of a "powerful ballista" is also mentioned by Albertus (Albert the Great, 1987, 339-341).

Cetacean remains are frequently found in the northern part of the Netherlands and these finds are normally ascribed to people scavenging stranded individuals. The sources by Albertus Magnus seem to indicate that the Frisians performed active whaling themselves. On what scale whaling was performed remains unclear however.

2.3.11. Dutch

Adriaen Coenen (1515-1587) was a fishmonger and an amateur ichthyologist based in Scheveningen, the Netherlands. He held an interest in sea creatures and in 1584-85 he wrote his *Walvisboek* a book concerned with porpoise, dolphins, whales, and various other (fictional) sea creatures (figure 24). His work is a valuable resource that aids the reconstruction between humans and sea creatures in the sixteenth century AD. He provides an account of whalers that utilized noise and music to lure the creatures to their ships. They then throw a harpoon with a line attached to a drogue that tires the whale and prevents its escape. The whalers then tow the carcass to shore where the animal is butchered. This activity is clearly described and depicted in his book (Coenen, 1585).

It however appears that Coenen did not witness this himself. Coenen frequently states that he witnessed particular events himself, but he probably did not in this case. It is therefore not clear whether this activity was practiced by the Dutch or by other people elsewhere (Coenen, 1585, 42).

Coenen (1585) furthermore describes that whales were also opportunistically exploited through strandings. He describes a live stranding of a large whale. The fishermen in the area gather and attach several lines to the creature to prevent it from floating away when the tide comes in again. When this is undertaken successfully all the people who helped get their share (Coenen, 1585, 72). Again, Coenen actually extracted this from a text from Olaus Magnus' *A Description of the Northern Peoples*. Therefore, it remains unclear whether this was also undertaken in the Netherlands (Olaus Magnus, 2010, 1098).

Coenen (1585) also describes the migration of whales past the coast of the Netherlands. He describes numerous whales migrating southwards. This happens once, twice or sometimes even more frequently every year. He describes the breaching of the animals and states that so many animals migrate southwards that their numbers cannot be counted. He however states that a northwards migration never occurs.

it is interesting that Coenen stated that animals frequently display breaching behaviour. Almost all of the large whales display this kind of behaviour, but it is more common for the North Atlantic right whale, sperm whale, and the humpback whale. Based on the depictions, Coenen probably did witness the migration of sperm whales. The fishermen referred to them as "potshoofden", which resembles the modern Dutch word "potvis" for sperm whale. Males sperm whales migrate southwards from the Arctic region normally do not enter the North Sea, but when they do and reach the shallow southern part of the North Sea they frequently get disoriented and get stranded (IJsseldijk *et al.*, 2018). This happens infrequently, but as modern whaling has greatly diminished the Atlantic sperm whale population, the animals might have entered the North Sea more frequently in the past. However, there were only twelve strandings or sperm whales for the period of 1519-1617 (prior to modern whaling activities that targeted the sperm whale). As Coenen said large numbers of these animals migrated past the Dutch shore, this low number of strandings for this period is curious.

Another possibility is that Coenen did not describe sperm whale migration, but North Atlantic right whale migration. This species is also known to migrate south during winter time and its eastern North Atlantic population has disappeared. This species is also known to breach frequently which is in line with the description provided by Coenen. This species was already heavily exploited by other medieval whaling people, most predominantly the Basques. Though it is unknown whether the North Atlantic right whale population had already been severely affected by whaling activities, it might be that Coenen (1585) described their migration.

Figure 24. Sperm whale in Adriaen Coenen Walvisboek. Coenen, Adriaen, Dat eerste boock van menich derleij walvischen ende ander selseme groote wonderlijke visschen, 707256, Collectie Stad Antwerpen, Erfgoedbibliotheek Hendrik Conscience, Bruikleen Koninklijke Maatschappij voor Dierkunde Antwerpen (KMDA). Available at: <https://dams.antwerpen.be/asset/jUZJkHhUZFeQeePAieiY0tyw/gDR5RTVYUgRDVYMNpUCjf2xN>.

Though cetaceans are occasionally mentioned in other medieval historical sources, none of these explicitly talk about active whaling activities organized in the Netherlands. This suggests that opportunistic scavenging appears to have been the main source of procurement and that only in the seventeenth century after the discovery of Spitsbergen/Svalbard the Dutch became a well-established whaling people.

2.3.12. Polish

Cetaceans are not commonly found in Polish waters. The harbour porpoise is the only species that is regularly seen. On the 17th of August 1378, Winricha von Kniprode confirmed the city rights of Helu (Hel) and recorded the taxes the fishermen from Hel had to pay to the head of the fishermen who had his headquarters in Pucku (Puck). Ships that caught dolphins (probably porpoises) had to pay a yearly fee of 2 mark. They made up an important component of the fishing industry (Ropelewski, 1957).

Furthermore, in the *Kronike Szymona Grunau* a record indicates that porpoises were caught along the Polish shore in 1526 and that the fishermen association was allowed to sell porpoises outside of strictly regulated market rules in 1538 (Ropelewski, 1957).

This suggests that the fishermen in Poland were well organized and that porpoises were regularly targeted in Polish waters. It seems unlikely that any other cetacean species were regularly exploited by the medieval Polish people.

2.4. Stranding Events in Historical Sources

Strandings of large whales are rare and receive a lot of media coverage. Since ancient history, people have been fascinated by cetacean strandings and questions have arisen on why these events happened. These events have often led to myths and legends regarding cetaceans. Native Americans in the New England area for example believed that the cultural hero Maushop caught whales and deposited them at the shore, for them to exploit (Russell, 2011, 575).

A large portion of the historical sources concerned with cetaceans describe strandings. For the Medieval Period, these records are rare, and there seems to be an increase in historical records concerned with strandings during the early modern period (figure 25). The records however indicate the inquiring nature as well as interest in cetacean strandings by medieval people (Barthelmess and Svanberg, 2006). This can probably be ascribed to the size and strangeness of the animals, and the animals were met with fascination as well as fear and were treated with caution.

Historical strandings of cetaceans have not received a lot of attention in the field of cetological studies, while it holds the potential to answer important questions regarding past

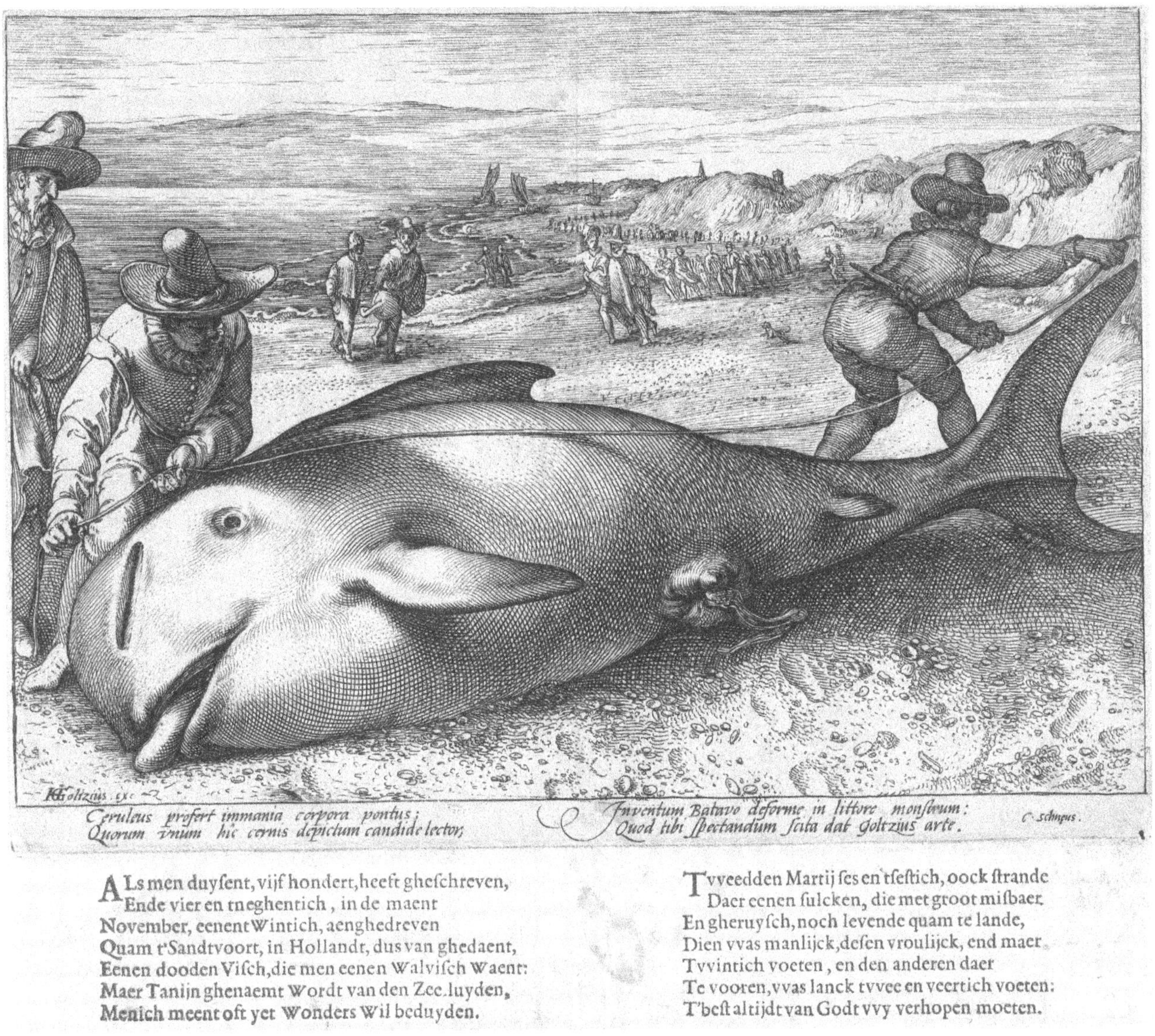

Figure 25. Stranded pilot whale at Zandvoort, the Netherlands in 1594 by Hendrick Goltzius. Object number: RP-P-OB-80.164. Digital image courtesy of the Rijksmuseum Open Data Policy. Available at <http://hdl.handle.net/10934/RM0001.COLLECT.448936>.

species range, stranding occurrences and can be a tool to managers and conservationists trying to evaluate the population status of cetacean species (Sousa and Brito, 2011).

Furthermore, the first knowledge regarding cetaceans was extracted from stranding events and provide a wealth of information regarding the perception of past societies towards cetaceans. Several sources record strandings, however this data is often obstructed by misconceptions and misidentifications. From the historical sources it is often not clear which species are dealt with. One such case is the stranding of a "sperm whale" near Tynemouth, England in 1532, though a description of the whale suggests it to be a large baleen whale instead (Smeenk and Evans, 2018).

Additionally, for many historical strandings, the exact date of a stranding cannot be determined and in more remote areas strandings might have been missed. This leads to an overrepresentation of strandings in more densely populated areas over remote areas. For some cases the length of the stranded whale was also determined, but this was often not done in a professional way, were merely estimates that were often greatly exaggerated, or the measuring unit used (e.g. inches, ells, feet, etc.) is not clearly described.

These aspects make historical strandings an unreliable source for comparison with modern stranding data, however individual cases can still provide valuable information. Smeenk and Evans (2018) attempted to give an overview of all sperm whale stranding in the North Sea area through time by assessing numerous historical sources (figure 26). The oldest identifiable sperm whale stranding in the area dates back to 1254/57 in Stavoren or Enkhuizen, the Netherlands. They were able to determine that a majority of the sperm whale strandings happened in the winter months (November-March), which is still the case for the majority of modern sperm whale strandings in the North Sea area.

Figure 26. Stranded whaled at Beverwijk, 1601 by Jan Saenredam. Objectnumber RP-P-OB-4635. Digital image courtesy of the Rijksmuseum Open Data Policy. Available at: <http://hdl.handle.net/10934/RM0001.COLLECT.337734>.

Moreover, historical stranding data can be used to identify patterns. Pierce *et al.* (2018) tried to correlate the strandings of sperm whales with various factors, including the numbers of sunspots and the sea and land surface temperature signals. According to their research there might be a correlation with the sea and land surface temperature signals, but the underlying mechanism for this is unclear. They suggest that multiple phenomena and factors affect sperm whale strandings.

In other instances, the great value of the resources the carcasses had to offer were deemed to be extremely valued and stranded cetaceans were more than welcomed. Szabo (2005) stated that famine is common topic in many of the Icelandic family sagas. In these sagas a large whale (usually a rorqual) stranded on disputed or common land, after which a dramatic human conflict arose over the division of the animal. An example of this is the already discussed *Grettir's Saga*, though this is purely literary source.

Whale strandings that relieved famines are not topics exclusively seen in Icelandic sagas. A rather extraordinary event was record in the *Life of St. Philibert*. St. Philibert founded the monastery of Noirmoutier in the year 674, located on the island with the same name just off the French coast. At some point in the following decade a group of 237 "marsuppas" (porpoises) stranded on the island. The timing could not have been better as there was a famine in the region at the time. Many of the brothers of the monastery, as well as the people living in the area exploited the stranded cetaceans (Arnold, 2016).

This might be the earliest reference to a mass-stranding event (Arnold, 2016). However, harbour porpoises are not known to mass-strand in such high numbers, suggesting that it might have been one of the members of the Delphinidae family, such as the long-finned pilot whale, striped dolphin short-beaked common dolphin, Atlantic white sided dolphin, or common bottlenose dolphin, which are known to mass-strand (Shirihai and Jarrett, 2006, 21-222). It has been suggested that these "marsuppas" were in fact pilot whale (Guizard, 2018). Guizard (2018) however notes that zooarchaeological findings of pilot whales have not been done in France and even calls it "*Le silence des globicéphales*" and argues that more zooarchaeological research is needed.

Another mass-stranding recorded in the historical records is a stranding of 65 whales in Cornwall in February 1386. The coastal community appears to have quickly been present at the scene and started to butcher the animals and take parts for themselves. However, this was against the law and the king ordered that all "evildoers" were to be arrested (Gardiner, 1997. Modern equivalents of mass-strandings are rarely reported, and the historical record seems to indicate that this was a rare phenomenon in the Medieval Period as well (Arnold, 2016).

Another stranding event connected to a famine can be traced back to 1131 Dublin whose inhabitants were suffering from a great famine that year. That year a large pod of so called "turlehydes" stranded along the mouth of Dodder (Donebrook). These "turlehydes" were described as fish of 30 to 40 feet long and more than 200 were caught and killed by the inhabitants of Dublin (Gifford and Seidmann, 1974, 59). This is probably a description of a large pod of cetaceans that stranded in Dublin, which relieved the inhabitants of Dublin from the on-going famine.

An even older stranding for Ireland can be dated to roughly AD 655, reported by the Irish Augustine in his treatise *De mirabilibus sacrae scripturae* (Smyth, 1996, 84-85). It was described that the animal had hard lumps on its pectoral fins as well as lower jaw and on its snout. Smyth (1996, 84-85) suggested that this description fits the humpback whale. It is interesting that this early record described the animal in such a naturalistic manner that with some certainty the animal can now be identified, as for most other (early) medieval sources this identification cannot be undertaken as the descriptions are not naturalistic and often can even be called fanciful, bizarre, or fantastic.

This is for example the case for a stranding that also occurred in Ireland that was record in the *Annals of Ulster*. This stranding is dated to AD 752 and the animal was reported to have three golden teeth, weighing 50 ounces each. The golden teeth are obviously fanciful, but the stranding of the whale itself might be accurate (Smyth, 1996, 84).

The *Canons of Adomnan*, dating to seventh/eighth century AD Ireland, are a set of dietary laws. The first canon of these dietary laws deals with a rare exception to a ban on eating carrion, more specifically to stranded cetaceans. "Marine animals cast upon the shore, the nature of whose death we do not know, are to be taken for food in good faith, unless they are decomposed." This indicates that the while the consumption of stranded cetaceans was at first frowned upon, this changed early in the Medieval Period already, up until the point that it even became associated with a high-status diet. This complete turnover in perception and associated status of the consumption of cetaceans is extraordinary and should receive more attention (Smyth, 1996, 84).

Sousa and Brito (2011) noted that in the case of Portugal, strandings were not frequently recorded in the Medieval and Early Modern Period. In the few cases that exist, all are concerned with large whales, which could be an indication that only large species would retain the interest and inquisitiveness of people. Sousa and Brito (2011) also noted that on many occasions, the already weak or sick whale was led into shallow waters by coastal communities who subsequently killed it or waited for the animal to die in order to take advantage of the resources. Indicating that the initial fear for the animals was replaced by interest.

In this way, cetaceans existed in a multiplicity of meanings in the medieval and early modern thought (Sousa and Brito, 2011). They were highly prized for the great value of the resources that could be extracted from just one single carcass but encountering one could be catastrophic and could have disastrous repercussions. Strandings regularly occurred in the Medieval Period and appear to have been frequently exploited. However, the exploitation of these stranded individuals and the consumption of cetacean meat for many northern and western European regions, was restricted to a selected few.

2.5. Cetaceans, Status, and Diet

Shortly after AD 1000 historical sources suggest that European society was subdivided based on the philosophical feudalism concept of the three estates of the realm. These three estates were called "*orant*" ("those who pray"), "*pugnant*" ("those who fight"), and "*laborant*" ("those who work"). They are also called clergy, nobility, and peasantry respectively (figure 27). This subdivision was perceived as a divine concept in which each order benefited from the others. The system was based upon inheritance, in which the people of the peasantry and nobility stayed in their respective orders, while the members of the clergy came out of the nobility order (Ervynck 2004).

Almost every aspect of life was different for the three orders, as was their diet. Dietary practices have been studied extensively based on both historical and zooarchaeological sources. An overview of the differences between the three orders will be provided here. Emphasis will be placed on the variation of animal exploitation undertaken by the three orders, and most predominantly the exploitation and consumption of cetaceans.

Of course, the geographic region this study is concerned with is substantial. As a result, dietary practices vary greatly on a geographic as well as temporal scale. However, several trends can still be noticed in the dietary practices in medieval Europe. The spread of Christendom had an especially large effect on the dietary practices of medieval people and will be analysed as well.

2.5.1. Ecclesiastical Diet

Of the three orders, the ecclesiastic diet is the most complex, as hierarchical differences within this order are abundant. Distinctions between the diets of monks, priests, bishops, and nuns are frequently mentioned. Though most members of clergy lived by strict dietary rules, abbot were often exempted from food rules when they were receiving guests. Additionally, several bishops were indeed clerical leaders but lived a similar lifestyle to noble lords with hunting rights and privileges upon land use. Moreover, differences between monastic orders (e.g. Cistercian, Benedictines,

Figure 27. Cleric, Knight, and Workman. Image courtesy of the British Library. Manuscript number: Sloane 2435, f.85. / Wikimedia Commons / Public Domain.

Franciscans, etc.) also existed, making it ever harder to clearly identify the ecclesiastic diet. Cistercians during the early twelfth century for example, followed a more austere diet in comparison to Benedictines.

Zooarchaeology can also be a valuable tool in the reconstruction of the monastic diet. Ervynck (2004) assessed the ecclesiastic diet of several monastic communities in Belgium and argues that remains of large game, as well as pig remains are extremely rare. He furthermore indicates that meat of mammals was only occasionally consumed and when this was done it was almost exclusively mutton or beef. Moreover, fish remains are frequently identified at monastic sites.

The consumption of fish can be associated with fasting periods. Fasting periods were a time of penance, in which the body and the soul remember sin and death. Christians prayed for forgiveness of their sins during these periods. Many members of clergy followed the Rule of Saint Benedict that became adopted by all monasteries in Western Europe from the Carolingian period onwards. These rules meant that all monks that were not weak or ill, were to abstain from eating dairy products, eggs, and meat of four-legged terrestrial mammals. The Christian calendar had multiple fasting days. Every Friday, in remembrance of the death of Jesus on a Friday was a commonly practiced fasting day. Saturdays, in honour of the Virgin Mary, and Wednesdays, the day Judas accepted money to eventually betray Jesus, were also fasting days. For a particular period for each season, fasting days were practiced more seriously. These days were called Ember Days. In Summer these Ember Days were practiced shortly after Pentecost, in autumn right after September's harvest season, and in winter during the Advent in December. The most well-known Ember Days is the period of Lent (Johnston, 2011, 232-233).

Lent is the period of 40 days prior to Easter and always begins on a Wednesday. Any meat, dairy products, or eggs people still had, had to be consumed on the Tuesday prior to the start of the fasting period. This Tuesday became known as Shrove Tuesday, Fat Tuesday, or Carnival. During the full period of Lent, only one meal, in the evening, was supposed to be taken, but through the major part of the Medieval Period this one meal was taken at midday, with another snack right before bedtime (Johnston, 2011, 232-233).

While the diet of most peasants was restricted and most of them did not have access to alternatives for meat, the clergy as well as nobility put considerable effort in finding substitutes. Fish was the main substitute for meat. Recipe books specify that bass, pike, salmon, cod, herring, trout, lamprey, eel, bream, tench, perch, dogfish, flounder, sold, haddock, whiting, and mackerel were frequently consumed, but also cuttlefish, crayfish, shellfish, and molluscs were substitutes for meat during fasting periods.

During the Medieval Period disagreement in regard to the nature of particular animals based on the morphology of their bodies was a common thing (Delaunay, 1997). Animals with scales were by some considered to be fish, while others thought that only body parts that contained "scales" were considered fish. The beaver for example was by some considered to be fish as its tail has "scales", but there was disagreement whether the entire animal or just the tail could be consumed (Wilson, 1973, 38). Other compared beaver to seal, noting that it was half fish as its tail is in the water (Bartosiewicz, Gyetvai and Küchelmann, 2010).

Besides beaver, seals, water rats, hippopotamus, otters, amphibians, and turtles were classed as fish, and therefore allowed to be consumed during fasting periods (Delaunay, 1997, 27). Even more strange is that new-born rabbits, as well as barnacle geese (which were by some thought to be born from barnacles at sea) were by some also consumed during fasting periods (Johnston, 2011, 232-233). Water birds, like cormorants, coots, and grebes were also part of Lenten diet. As these birds spend most of their life in the water, they were classed as fish (Storå, 1968).

Most importantly as part of this study, cetaceans were classed as fish as well. The Chef of the Mainz Elector, Marx Rumpolt published in 1581 what was deemed to be the perfect cookbook "under privilege of the Holy Roman Emperor". This book listed anything that could be cooked and consumed allowed by Catholic dietary rules. Among numerous other animals, he lists the harbour porpoise and states that "what is from the sea, can be prepared on Lenten days" (Herman, 1887, 128; Born, 1989, 111).

While cetacean consumption was often associated with the members of the royal family, members of the clergy were also known to appreciate cetacean meat. This is recorded extraordinarily well for medieval England. For the installation of Archbishop George Neville, that took place in Cawood Castle, Cawood, in AD 1465, an extraordinary large banquet was organized to demonstrate the power and the riches of his family. About 2500 people were invited, including several bishops, knights, abbots, members of his family and numerous others. For this banquet thousands of animals were butchered, including 4000 pigeons, 2000 chickens, 400 herons, 1000 sheep, 2000 pigs, and a large number of other mammals, birds and fish, but also 12 porpoises and seals (Mitchell and Leys, 1950, 250-257). Furthermore, the bishop of Swinfield is known to have consumed porpoise meat (Du Bouëtiez, Clavel and Ravoire, 2013).

2.5.2. Noble Diet

As mentioned in numerous historical sources, cetaceans appear to be part of a high-status diet for major parts of medieval northern and western Europe. This appears to have been especially the case for the eleventh-thirteenth century AD. Indeed, the noble diet has often been associated with the consumption of large game, such as red deer (*Cervus elaphus*), brown bear (*Ursus arctos*), wild boar (*Sus scrofa*), and various wild bird species such as heron (*Ardea cinereal*), crane (*Grus grus*), and swan (*Cygnus* sp.). Cetaceans were part of these "high-status animals".

In regard to the three main domesticates (cattle, sheep/goat, and pig), the nobility can frequently be identified based on a dominant frequency of pig remains. However, this is less well expressed for the coastal regions, for which salty regions are not ideal for pig herding. Overall domesticates still make up the major part of noble diet, while wild animals did not yield a quantitatively significant addition to noble diet (Ervynck 2004).

However, to reaffirm their power, members of the nobility order frequently organised banquets. In these banquets large game such as deer and also marine mammals were served, showing the benefits of their privileges. This serving and eating of large quantities of meat were in line with the original Germanic tradition (Ervynck 2004).

That cetaceans were part of a high-status diet is especially well attested for England. Henry III (King of England from 1216 to 1272), as well as Richard II (King of England from 1377 to 1399) are known to have consumed porpoise meat. At the wedding of Henry V and Catherine of Valois in 1420, roasted porpoise meat was also served and was still regarded as a dish of high taste. At the coronation feast of Henry VII in August 1485, porpoise meat was also still consumed. Some of the recipes that porpoise meat was used in have been preserved and include roasts and pies. Even Elizabeth I (Queen of England from 1558 to 1603) was known to have loved porpoise meat, indicating that porpoise meat was still in great favour until the sixteenth century (Du Bouëtiez, Clavel and Ravoire, 2013).

During the Medieval Period, stranded cetaceans were treated as a "wreck of the sea" and therefore belonged to the King. The King however sometimes granted the "wreck of the sea" rights to local lords. The King frequently however retained the right to the tongue, which was perceived a delicacy. This has been recorded in several records. In the County of Lincoln, the head and tail of royal fish appear to have been reserved to the King, and at Hoggesthorpe, William son of Robert Wyleghby claimed wreck including whales and all royal fish, saving the head and tail for the King (Moore, 1888, 84). The bishop of Durham was challenged in AD 1280-81 for taking wreck at Ellerker, Wakington, Hoveden, and Welleton. The bishop claimed that he was granted these rights by King Egfrid and that these rights were confirmed by the Conqueror and King Edward I. He furthermore claimed that he only claimed whales and served the head and tail to the King (Moore, 1888, 81).

Additionally, the Prior of Spalding claimed wreck of sea on his land within the manors of Pinchbeck, Sutton, Eston, and Spalding by prescription. Though he saved the head and tail of whales for the King, the King's attorney argued that a prescription for exploiting stranded whales is not sufficient warrant and that a special prescription explicitly mentioning whales is needed in order for the prior to claim whales. The jury did however not agree with this and the prior was discharged (Moore, 1888, 82).

This indicates that ecclesiastical institutions were also occasionally granted the "wreck of the sea" rights, though these rights did not always include stranded cetaceans. The Abbot of Tavistock claimed wreck of the sea in the Isles of Scilly, but this claim did not include whales, gold, and cloths of scarlet and ermine (Moore, 1888, 109).

Historical records also attest that the "wreck of the sea" claim was often attested. When a whale came ashore at Wainfleet, as recorded in the *Quo Warranto Rolls* in 1280, Philip de Kyme claimed it because the foreshore belonged to him and it is clear that a whale cannot be taken above the high-water mark. Though royal fish was considered to belong to the King, Philip reasoned that it was his land and therefore his whale (Moore, 1888, 82).

In another record, Roger de Lascelles was challenged for exploiting a whale stranded in Fulestowe. Though he saved the head and tail for the King, the King's attorney did not agree with this and claimed the whole whale belonged to the King. Roger however claimed his ancestors had always claimed whales that stranded on their land, and no decision was reached to whom the whale belonged (Moore, 1888, 82).

Another intriguing case was the claiming a whale stranding on the land of Arundel. At the time Arundel did not have a lord, as the heir was underage. As a result, Arundel was in the custody of the Crown. Walter le Harpur, lord of the manor of Middleton (subinfeudated out of the land Arundel), claimed a porpoise that stranded on his land. Walter le Harpur was fined for taking the porpoise on his land, as though it was his land, the wreck of the sea did not belong to him. This entry makes it clear that there is a difference between the right to the soil of the foreshore and the wreck of the sea. Furthermore, certain rights to land that was subinfeudated could be held by the feudal lord who subinfeudated his land (Moore, 1888, 76-77).

Cetaceans appear to not always be part of the "wreck of the sea". In the Quo Warranto Roll for Somersetshire in 1280, during the reign of Edward I, John de Tregoz is summoned for taking wreck in Burnham and other parts of his land. He claimed that he had the right to do this and the jury agreed with this, with the exception of wreck of wine or royal fish (Moore, 1888, 79). In another entry, Matilda the widow of Walter de Waynham claimed wreck with the exception of wine and whale in Kingston. The jury found that she and her ancestors indeed held these rights, after which she was discharged from the hearing (Moore, 1888, 79).

Yet another entry indicates that Gilbert de Gaunt and Richard Malebise were challenged for claiming a whale, while they did not claim wreck of the sea. However, they saved the head and tail for the King, as had their ancestors. No judgment was entered up (Moore, 1888, 81). Gilbert de Gaunt is challenged in another case for claiming a whale stranded at Hunmanby. Again, he claimed that he did not claim wrecks, but just a whale, and that he and his ancestors always saved the head and the tail for the King. He was discharged by the jury (Moore, 1888, 82).

These discussed cases are just a small portion of all the registered cases that deal with stranded cetaceans, indicating that they were extremely valued. This was also the case on the case on the other side of the English Channel, in France.

Porpoises, dolphins and whales were considered royal fish in medieval France and occasionally stranded on certain estates up the river and just like in England, in France they also became the property of the king or local lord (Ardouin, Hadjouis and Arroyo, 2009). And just like in England, the tongue was perceived as a delicacy, reserved for the nobility and clergy. A record indicates that in 1565 two quintals of whale tongue provided to King Charles IX of France and his mother Catherine de Médicis (Proulx, 1986, 16). During the Norman rule in northern France, a stranded cetacean had to be handed over to the Duke and for an actively caught cetacean, a portion had to be gifted to the Duke (Proulx, 1986, 10).

For many other regions in medieval Europe, similar laws were enforced as those described for England and France, though these are often well less recorded. Even for regions for which historical sources are lacking or understudied, cetacean exploitation has still been ascribed to a high-status diet. This is based on the context of the zooarchaeological remains of cetaceans. This was for example the case for the Iberian Peninsula. As cetacean fragments have been identified at several high-status sites from the Iberian Peninsula, Grau-Sologestoa (2016) has suggested that these fragments prove that cetacean exploitation was associated with the social elite from the Iberian Peninsula. This demonstrates that zooarchaeology can be a valuable source in the reconstruction of past foodways.

Outside of medieval Europe, cetaceans are also frequently associated with high-status. In other cultures, whaling was the prerogative of a select group of people, which conferred upon them considerable prestige. This was the case to the Native Americans on the Northwest Coast of America, in

Oregon and Washington, including the Nuu-chah-nulth, Makah, and Ditidaht. Even the ability to get a carcass to cause to drift ashore was perceived as a prestigious skill. This prestige placed upon cetacean exploitation probably derives from the risks inherent in the operation (Losey and Yang, 2007; McMillan, 2019).

In Japan, a long-standing tradition of whale consumption is apparent. During the Nara period (710-784 AD) several emperors forbade the killing and consumption of four-legged animals. This was forbidden by Buddhist principles. The 40th Emperor Temmu forbade the killing of animals and eating their meat in AD 676 and the 44th Emperor Genshou forbade the killing of animals and falconry activities in AD 721. There are several more emperors that forbade these practices, even as late as the 82nd Emperor Gotoba in AD 1188. Buddhism came to its height in Japan during the Nara period and as a result, fish and cetaceans became an important food source. Whale was not a restricted food source, because it was called "isana" (brave fish or large fish) in the "Man'yoshu" (a collection of 10.000 Leaves which compose thousands of ancient poems edited in AD 770 (Japan Whaling Association, 1987)).

Additionally, also in Japan, in the Tanou ruins at Tanounakanotsubo, Amagasaki City in Hyogo Prefecture, from the older period of the Yayoi Period (3000 BC – AD 300), a whale bone was discovered as a grave gift. It is believed that the grave belonged to a person from a high rank. This suggests that whale meat was probably given to the deceased as a food offer, suggesting it was a valuable food source and associated with people from a high social class (Japan Whaling Association, 1987).

During the Muromachi/Azuchimomoyama Period (1573-1600) whale meat was often perceived as a high-status food source as well, though during later periods whale dishes were consumed by people from various ranks. During the "Shikisankon" ceremony (a ceremony of exchanging cups of sake, an important ceremony for the samurai society) a whale dish was served as part of the seafood dishes. Additionally, during so called "Shirukou" parties, the host family prepared soup, while all the guests brought their own rice. Mushroom, geese, pheasants and whales were all used to make soup, while meat of four-legged animals was never used for this soup.

These examples from Japan can be perceived as an analogy to the situation in Europe, cetacean meat was perceived as a high-status food source. Eventually, with the introduction of new hunting techniques, cetacean meat became more common and available to people from other social strata as well. Japan has a long history of whaling and is still one of the countries that is actively hunting cetaceans for consumption.

For New Zealand, in the Te Ao Māori view, all life is intrinsically interlinked. Whales represent richness, abundances and were regarded as chiefly animals. Early Māori welcomed stranded cetaceans and exploited the carcasses for meat, bones, and oil. In modern times this utilisation has almost ceded and the majority of contemporary Māori now prefer to give the whales a *tangi* (a traditional Māori funeral rite) and bury them (Lowe, 2012). This clearly demonstrates that cultural practices in regard to cetaceans are dynamic and subject to change over time.

2.5.3. Peasant Diet

The peasant diet is less well recorded in the historical record, but Ervynck (2004) argues that the scarce information available indicates that their diet consisted mainly of sheep/goat or cattle meat. Pigs are not frequently associated with peasantry and neither is large game. Hunting rights were for many regions restricted to members of nobility and severe penalties were installed to peasants whom conducted poaching, and in some areas and periods even the consumption of game was perceived a crime (Jarnut, 1985).

A similar situation was valid for cetaceans. For many regions of medieval Europe, waterways were considered "uncultivated lands", solely restricted to feudal elite. As a result, stranded cetaceans (or the monetary value of their sale) belonged to the lord or ecclesiastical institution that had jurisdiction over that piece of coastline (De Groote, 1999). Cetaceans were in this way restricted to the clergy and nobility as well. Though in the case of England and Wales, the Crown occasionally granted stranded whales to the landowner in exchange for the tongue and the head of the animal (Fagan, 2006, 36). In this way whale meat might still have been available to those lucky peasants on whose land the whale stranded. In other areas some sort of finders, keepers were in place. This was for example registered in the "Den Jyske Lov" (Law for Jutland) in Denmark in which the finder of a stranded whale was allowed to take a portion himself (Schnall, 1993; Hybel and Poulsen, 2007, 55).

Though whales and royal fish appear to have often been the legal property of local lords or in other cases the King, peasants are also known to have illegally exploited stranded whales in similar ways that game was also sometimes poached. Several of these instances are recorded in the *Calendar of Patent Rolls*, for medieval England (Boynton, 2016). In these records the peasants are often fined for breaking these rules. In AD 1281, during the reign of Edward I, the lord of Snetsham, Norfolk, recovered damages he suffered after trespassers took a whale that stranded on the shore of his manor in Snetesham (Moore, 1888, 86). Additionally, during the reign of Edward III, in 1342, Alice countess of Lincoln sued several people for illegally taking a whale in the precinct of her manor in Sutton. The defendants were outlawed (Moore, 1888, 151).

Eventually with the rise of urban centres a shift can be noted in regard to peasantry. Urban centres gave rise to a rich class within the peasantry. Eryvnck (2004) called them the *nouveaux riches* and they attempted to imitate the nobility, but they used their money and not their power to achieve this. They simply bought game through their contacts with noble lords in an attempt to copy their lifestyle.

Indeed, cetacean meat could be purchased at several markets in medieval urban centres, such as eleventh and twelfth century Boulogne, Calais, and Damme, as well as sixteenth century Amsterdam (De Smet, 1981; Ypma, 1962, 30). Furthermore, porpoise meat was sold on the market of Newcastle until 1575, when it ceased to be sought (Du Bouëtiez, Clavel and Ravoire, 2013). This rise of the *nouveaux riches* capitalistic entrepreneurs eventually meant the disappearance of the power of nobility and clergy and meant that cetacean meat was no longer restricted to them.

2.6. Medieval Marine Resources Exploitation in Context

Trends in the exploitation of marine resources in the Medieval Period might be useful for the identification of whaling practices. It is likely that fishermen fishing for marine fish first encountered cetaceans of various sizes and eventually adapted their fishing strategies to target cetaceans.

Prior to the High Medieval Period, marine fish sources were of little importance to the economy of north-western Europe, probably only being exploited for subsistence but not major commercial reasons. This was however not the case for Northern Norway and Iceland where by the late ninth century marine fish was exploited and was part of a well-connected and developed chiefly social network which transferred fish from the coast, dried it and transferred it to inland consumers (Barrett, Locker and Roberts, 2004a; van Neer and Ervynck, 2004; Barrett *et al.*, 2011).

At this time a wide range of species were exploited, and the processing practices were diverse. This all suggests that this network was small and was non-economic with the local elite using their power to obtain personal obligations. However, it still generated considerable wealth for the elite in Iceland at least. The Nordic elite were thus already exploiting marine fish in more inland regions, making use of a complex social web and provisioning networks.

While marine fish appear to have been frequently exploited in the most northern part of Europe already by the late ninth century, this was certainly not the case for the rest of Europe. This soon changed and around 950 AD the so called "Fish Event Horizon" (FEH) instigated in north-western Europe (especially in England, the northern part of France, Belgium, the Netherlands and north-Germany) where herring and cod became increasingly more important, even for more inland sites. This reliance on marine fish continued to grow in the eleventh and twelfth century. Cod first became an important component of the catch in the eleventh and twelfth century, following which its proportions declined again and was joined by haddock, ling, saithe and hake as important species as well. It probably first started in York between 975 and 1050, prior to the tenth century in Northampton, between 1050 and 1070 in London and prior to 1030 in Southampton/Hamwic, late tenth to late eleventh century for Norwich, and late eleventh to early twelfth century at Eynsham Abbey (Barrett, Locker and Roberts, 2004a; van Neer and Ervynck, 2004; Barrett *et al.*, 2011).

Zooarchaeological analysis of fish remains from northern Scotland, in the pre-Viking Age (Pictish) period, shows modest numbers of bones from small fish. Only from the ninth and tenth centuries onwards, large cod, ling, and saith are being identified in larger numbers, probably due to the introduction of new food preferences by the new Norse migrants. Even larger numbers of these species are identified for the eleventh and twelfth centuries, consistent with the English and wider European trend of an increase in marine fish exploitation (Barrett, Locker and Roberts, 2004b; Perdikaris *et al.*, 2007, 51-62).

The Fist event Horizon occurred rapidly in the North and Baltic Sea areas around AD 1000, but for example in the case of Flanders, an increase in marine fishing can be tracked over centuries rather than decades. An increase in marine fish in Flanders has been associated with the widespread introduction of floating nets and gill nets. The Flemish fishing activities thrived in the high and Late Medieval Period and waters far north of Flanders were exploited (Ervynck, van Neer and Pieters, 2004). For the largest part of northern and western Europe the fish event horizon took place between AD 850-1050 (Barrett, Locker and Roberts, 2004a, 2417-2421; Barrett, 2016). This increase in marine fish exploitation over Europe and was triggered by various economic, social, and environmental aspects.

The Fish Event Horizon has most predominantly been linked to the Medieval Warm Period around 1000 AD which led to an agricultural expansion. Even though the Medieval Warm Period probably resulted in lower numbers of cod and herring in the Northeast Atlantic, a rapid human population growth might have triggered the fishing for marine species, especially in urban areas which happened during the "urban revolution" around this time as well (Barrett, Locker and Roberts, 2004a, 2417-2421).

Additionally, a decline in freshwater species populations might have triggered an intensification of marine fish exploitation as well. This decline in freshwater species was probably the result of siltation from the intensive agriculture, the proliferation of mill dams, increase in nutrient loads from growing populations of humans and overfishing itself. As a result, freshwater species decreased in importance from the eleventh century onwards (Barrett, Locker and Roberts, 2004a, 2417-2421).

The intensification of marine fish consumption has also been connected to religious food rules issued by the Church (Ervynck, van Neer and Pieters, 2004). Resulting from these dietary rules, marine fish species were more highly sought after than before. The development of preserving marine fish using salting has also been argued to have led to an increase in marine fish exploitation and allowed marine fish to be transported further inland (Barrett, Locker and Roberts, 2004a, 2417-2421, Fagan 2006, 15-57).

Zooarchaeological analysis by Van Neer and Ervynck (2004), has indicated that herring consumption in the Early Medieval Period was restricted to the coastal regions of Picardy in northern France to the Baltic area. Only from 1000 AD did the coastal population supply nearby inland consumers with herring. From the twelfth century onwards herring exploitation increased rapidly and began to be transported further inland. In the Late Medieval Period herring was a relatively cheap fish species and was sold in large quantities. The interaction between supply and demand changed a lot during the Medieval Period. For example, during the plague epidemics around AD 1350 a far lower demand resulted in a price decrease, and even political difficulties had an effect on herring prices.

Van Neer and Ervynck (2004) suggested that while herring might have been a cheap product for some periods and regions, this might not have been the case for others. Indeed, herring remains have been uncovered from a variety of high-status and ecclesiastical sites but based on this it is still hard to determine whether herring was actually part of a high-status diet. It might merely have been a staple food that did not contribute to the luxurious nature of the local menu. Additionally, the remains from high-status and ecclesiastical sites might not actually have been consumed by nobility or clergy, but by the servants and labourers servicing at the high-status or ecclesiastical institutions.

Van Neer and Evynck (2004) suggested that different forms of preservation (e.g. fresh, gutted, or salted herring) might have been deemed luxurious for some areas and periods, but that this is hard to identify based on zooarchaeological data. Furthermore, the distance from the supplier also influenced prices. In fifteenth century Flanders, herring was relatively cheap, while in central Europe herring was expensive. For example, in fifteenth century Konstanz in southern Germany herring was almost as pricey as the most expensive freshwater fish species.

The period between AD 950 and 1050 can now be addressed as a critical period of which the today's fishing crisis is the result (Barrett, Locker and Roberts, 2004a, 2417-2421). The fisheries at this time where undertaken in local waters, especially the southern North Sea, rather than being a supply-driven trade with fisheries being undertaken on a long-range trade aspect. The demand for fish increased and by the thirteenth and fourteenth century, large cities could not be provided with their demand anymore just by local fisheries. Therefore, a "globalisation" of fishery took place, with fish being traded over long distances. These fisheries went farther north, to the Scotland and the northern Isles, Norway and Iceland. After the demographic crash as a result of the Black Death, the fisheries stabilised for some time, but with the continuing growth of the population going on in the fifteenth and sixteenth century fisheries expanded again (Barret *et al.*, 2011, 1516-1524).

Holm *et al.* (2019) have recently proposed the concept of the "Fish Revolution" to demarcate the increase in fisheries in the North Atlantic after the end of Medieval Period (circa AD 1500). This event was instigated by globalisation and climate change, and developed with a transatlantic thrust of cultural, economic, and political significance. Though whaling was of lesser importance during the initial stage of the "Fish Revolution", the subsequent seventeenth century faced a dramatic increase in cetacean exploitation and the depletion of numbers whale stocks on an unprecedented scale.

2.7. Conclusion

Several medieval European cultures have been associated with cetacean exploitation, more predominantly the Basques, Norse, Normans, and the Flemish. Several other cultures also appear to have undertaken whaling as well, but probably on a smaller scale. From the historical sources it appears that the North Atlantic right whale was the main target for most of these groups, and the species was relentlessly hunted over several centuries.

From the historical sources, it appears that whaling was less frequently undertaken from the twelfth or thirteenth century onwards. The Normans appear to have stopped whaling from the thirteenth century onwards, the Flemish performed it less frequently from that time onwards as well, while the Basques went in search for new whaling grounds during this period. This might be a suggestion that North Atlantic right whale populations declined in European waters as a result of whaling practices being undertaken along the European coastlines.

Strandings also happened during the Medieval Period, and several sources suggest that these were frequently exploited by coastal communities. Cetacean meat was in fact a welcome substitute to meat from terrestrial mammals. As a result of the spread of Christianity in Europe, fasting rules were set in place, that limited various food sources. Whale meat was however allowed, and the social elite (both the clergy and the nobility) seized any opportunity they had to get access to cetacean meat. In large parts of Europe, stranded cetaceans were by law the property of the King, Queen, or the local elite. This clearly indicates that cetacean meat was a high-status food sources, not normally within the reach of peasants. However, many historical sources seem to indicate that peasants poached stranded cetaceans to the dismay of the social elite. The elite attempted to tighten their grip on cetaceans even more, setting in place numerous laws ensuring their rights to them.

Furthermore, the increased use of salting during the High Medieval Period, allowed cetacean meat to be transported further in land. This made it possible for the elite located further in land to regularly consume cetacean meat as well.

3

Cetaceans in Zooarchaeology

Cetacean remains are frequently encountered in the archaeological record; however, several factors make research on these remains problematic. This chapter will consider the factors that affect research on and understanding of cetacean material. This is a vital aspect to base any interpretations on and will answer the sub-question: "How can zooarchaeological cetacean remains be studied?".

First, all the taphonomic factors will be considered. Taphonomy is an important aspect of archaeology as whole. In this study, these factors include all the factors and processes in between the procurement of cetacean material by past cultures, until the analysis being performed by archaeologists and the publishing of those results.

The most major problem that arises when studying zooarchaeological cetacean remains is taxonomic identification. Identification to the genus or species level can provide valuable historical biogeographical information on the species (e.g. past species range and responses to climate change), as well as provide insight into the palaeo-environmental conditions of specific regions (Murray, 2008). Additionally, it might help to distinguish seasonality and acquisition strategy, as to whether active whaling or opportunistic scavenging was practiced. Furthermore, zooarchaeological material provides information on the history of cetacean exploitation and the role we had in the extinction of various species and populations.

However, cetacean bone is notoriously hard to identify to the genus or species level and is affected by a large number of taphonomic processes, often resulting in the fragmentation of cetacean bone material. In the past archaeological cetacean material was therefore frequently ignored and little research was carried out on it. Many of the archaeological cetacean remains are merely classified as "cetacean", "whale", "dolphin", or even "marine mammal". As a result, cetaceans might be the least understood mammal group in the field of zooarchaeology.

Furthermore, new advancements in research on cetacean material will be considered. All of these are concerned with optimizing identification of zooarchaeological cetacean material. These include: aDNA research, Zooarchaeology by Mass Spectrometry (ZooMS), Trace Element Analysis (TEA), organic residue analysis, the use of osteological reference collections, and the use of osteological reference manuals. All these techniques have advantages and disadvantages and can lead to different results and therefore create biases in zooarchaeological research. Taking into account the differences between the various methods is vital to our understanding of past cetacean exploitation.

3.1. Taphonomy

As previously mentioned, the field of zooarchaeology is concerned with the reconstruction of past human subsistence patterns and the reconstruction of palaeo-ecological conditions. Both of these goals require taxonomic identification of faunal remains (Lyman, 1987). The taxonomic identification, however, is often rendered problematic by various factors. The factors that affect the archaeological material, post-mortem, pre- and post-burial, are known as taphonomic processes (Lyman, 2008, 264-265). To successfully perform research on zooarchaeological cetacean remains and to fully understand the history of cetacean exploitation, these taphonomic processes should be considered. Quantitative data (e.g. taxonomic abundances, meat weights, and the frequencies of skeletal elements) as well as the distribution of bones (e.g. some bones might disappear from the archaeological record, while other might enter older or newer layers) influence zooarchaeological interpretations and can create unwanted biases (Lyman, 1987). Therefore, knowing which factors have had an effect the zooarchaeological record, all the way from the acquirement of the material (e.g. hunting, killing, or scavenging of the animal) to the discovery of the material by archaeologists and even the eventual publishing of the data, are important aspects that should be considered before interpretation should takes place (Davis, 1987, 22; Lyman, 1987).

Davis (1987) and Lyman (1987; 1994) both produced important taphonomy studies and gave a general overview of the taphonomic processes that influence zooarchaeological material. However, the taphonomic processes for cetaceans are different to those of (large) terrestrial mammals and have not yet received a lot of attention in the field of zooarchaeology. Various taphonomy table overviews have been suggested for zooarchaeological remains, but these are not adequate for cetaceans. Therefore, one has been created as part of this study (table 3). This diagram is based on a diagram created by Davis (1987, 22), but has been adapted to consider cetaceans.

The first stage of taphonomy, is the modifications made by humans and are known as first-order changes. First order changes are those that influence cetacean material, over which archaeologists have no control. The first of these is the actual procurement of a cetacean. The processes that bring along the death and deposition of cetaceans are known as thanatic processes. These processes can be caused by humans, but also by other predators, diseases, old age or natural circumstances (O'Connor, 2000, 20-21).

Unlike most other mammals, cetaceans come from an aquatic environment and therefore their procurement is different to those of terrestrial mammals. There are three separate scenarios in which cetaceans could be acquired by humans. The first of this is active whaling – going actively into the sea and pursue and kill a cetacean individual or group. The second option is exploiting a live stranding. In this situation an individual or a group strands alive along the coast and are killed there by humans or die naturally. The third option is a cetacean dying at sea and subsequently washing up at the shore. The washing up of this carcass can happen within a couple of hours or even weeks. During the time at sea the animal starts to rot and attracts scavengers. As a result of this disintegration, many resources might be lost or sink to the bottom of the sea, resulting in the washing up of only a small portion of the whole carcass at the shore.

Following the procurement, the next step is the exploitation of the actual resources of the cetacean(s). Cetaceans have various resources that can be exploited from their carcasses, depending on the species these include: meat, baleen, bone, teeth, ivory, oil, spermaceti, skin, ambergris, and blubber. Depending on where this exploitation took place, it is likely that people only took those resources that were of use to them and leaving the rest behind at the processing site. If a cetacean is caught at sea, this process might already happen at sea itself and all the waste can be dumped back in the sea. It is unlikely that these remains will be found by archaeologists.

In most (pre-industrial) cases processing was undertaken at the coast. This is the place where our understanding of cetacean exploitation is most severely obscured. The invaluable products left at the processing sites most likely include the osteological remains, while the remains that often do not survive in the archaeological record, e.g. meat, oil and blubber are taken to site and cannot be traced back by archaeologists. This is however dependant on the size of the cetaceans. Small species, such as the harbour porpoise, can be exploited and the entire carcass might be transported to the settlement and butchery of the animal might take place at the settlement itself. This means that their bone remains will be discarded at or close to the settlement. While in the case of a blue whale, which can reach a length of up to 35 meters and can weigh between 50 and 136 tonnes, butchery probably took place at the processing site and the people might have decided to leave the bones at the site, unless value was placed on these bones, e.g. for the production of artefacts or the extraction of oil. This would result in an underrepresentation of larger whales at settlements. This process is known as the "Invisible Whale" (Smith and Kinahan, 1984). The size of cetaceans furthermore means that it is possible to butcher the animal and transport useful products, leaving little trace in the archaeological record (Mulville 2002b).

In the year 1568, seventeen whales stranded near Ipswich, England. The Ipswich corporation is recorded to have arranged the whales being brought up to the quay edge and additionally recorded that the fins and the tails were eventually disposed (Gardiner, 1997). The removal of the meat was not recorded, but the disposal of the fins and tails is valuable information, as this will lead to an underrepresentation of those skeletal elements in the archaeological record.

Processing sites are rarely encountered by archaeologists, as processing is most frequently undertaken at the coast, which is a dynamic zone with lots of erosion and sedimentation. Bone material is often damaged is these contexts and if it survives, archaeological excavations are rarely undertaken in these contexts as sedimentation has covered the remains with meters of soil.

In some optimal conditions processing sites are however encountered. A notable example of this are the two partial North Atlantic right whale carcasses found at the Dengemarsh, England. This site provided a wealth of information and suggest that in Anglo-Saxon England whaling was already practiced. However even in this case the possibility of opportunistic exploitation of two stranded individuals cannot be ruled out. For archaeologists only the bones, teeth and ivory are likely to be recovered from these processing site, which can still have signs of butchery on them, which was also the case for the carcasses at the Dengemarsh site (Gardiner, Stewart and Priestley-Bell, 1998).

Another processing site is the site of Red Bay, Canada. Here numerous cetacean remains have been discovered in the bay, suggesting that the animals were brought there and subsequently butchered (Grenier, Stevens and Bernier, 2007). Other processing sites include some sites in the Arctic and Antarctic region, where the conditions are optimal as well and bones remains remain untouched for hundreds of years (Savelle, 2000; Patton and Savelle, 2006).

Bones were probably most frequently left at these processing sites. However cetacean remains are also frequently found at settlements. The bones that are encountered at these sites often number below five specimens in medieval European contexts and were probably bones that were transported to the settlement as meat was still attached to it or bones that were worked into artefacts or tools. These bones therefore represent butchery waste, artefacts, tools or working waste. Moreover, some bones might be taken to the site for oil extraction.

The process of killing, skinning, butchering can leave marks on bone, though as stated previously, it is also possible to do this without leaving any butchery marks (Mulville, 2002b, 40). Cooking furthermore removes organic constituents and alters the aspects of the isotopic and elemental composition, sometimes resulting in the bone not surviving in the archaeological record. Burning might happen when the bone was (accidentally) exposed to the flames during cooking, the result of trash disposal, or, which is especially the case for cetacean bone, when

Figure 28. Selection of Bronze Age dolphin remains from Ras al Hadd, Oman. Some specimens show signs of burning. Photo courtesy of Dr Elena Maini.

used as a fuel source. Bone, when exposed to flames, shrinks and changes colour, often turning black, as the result of low temperatures which results in organic components becoming carbonized. When exposed to higher temperatures bone becomes extremely brittle and can turn white (figure 28). This was frequently done on cetacean bone remains as their bones contain oil that could be used for illumination or cooking purposes.

After the discarding of the bones, other factors start to influence the preservation of the material. Cetacean bone, in contrast to bone from other large terrestrial mammals, is extremely friable. It is made up of a thin external cortical layer of bone with the internal part of the bone being composed of oil filled cancellous bone (Speller *et al.* 2016). Cetaceans need this bone structure, because if the bone was solid, the animals would be too heavy and lose their buoyancy. However, as a result of this structure, cetacean bone tends to break up more easily than other mammals in the archaeological record. The material can get trampled by humans and animals. Additionally, scavengers, particularly dogs, cats, rats and other rodents, pigs, birds, and insects leave their traces on the bone and damaging it. Furthermore, the larger animals, like dogs can even digest bone, where it is exposed to acids and enzymes, though this they can only do with smaller bones of chunks of larger ones. Trampling, plant roots, burrowing animals, bacteria and fungi are also a factor that should be considered. Due to all these factors, cetacean bone often breaks up in small non-diagnostic fragments. The processes that affect the bones before they are finally incorporated into a forming deposit are known as perthotaxic process (O'Connor, 2000, 20-21).

Furthermore, occasionally cetacean bones appear to be intentionally buried and this can also be viewed as a form of a perthotaxic process. An example is the burial of a harbour porpoise vertebra in a grave at Skateholm I, Sweden dating to the Mesolithic. This vertebra was probably buried with the deceased as a food source intended for the afterlife (Larsson, 1990). Another intriguing find is what appears to be a harbour porpoise grave found at the small tidal islet of Capelle Dom Hui, near Guernsey, dating to the fourteenth or early fifteenth century AD (Pers. Comm. Philip De Jersey 2017). The reason why this particular harbour porpoise was buried there remains unclear. Furthermore, a harbour porpoise mandible was buried as part of the St. Ninian Treasure. It was stained with copper oxides and it remains a mystery why this mandible was buried as part of the treasure (O'Dell *et al.*, 1959).

When the zooarchaeological material is finally incorporated into the forming deposits, it is affected by soil condition. Often optimum soil conditions are not present, and material will decompose in the soil or can be affected by abiotic factors, e.g. soil erosion, displacement by water or wind or by biotic factors such as plant roots or animals. In addition, various materials deriving from cetaceans need different environmental conditions. Permanently wet conditions are good for the preservation of bone, because of the anaerobic conditions. The ideal pH conditions for the bone mineral hydroxyapatite is around 7.9. The least calcified elements, such as infant and juvenile animals are the first to go into solution. Baleen requires environments unfavourable to aerobic fungal activity (e.g. an environment that lacks oxygen, low temperature, limiting pH and very low relative humidity) and can potentially survive in environments where bone does not and *vice versa* (O'Connor *et al.* 2015: 393-417). All these factors that bring about chemical or physical changes in the soil are known as taphic processes (O'Connor, 2000, 20-21).

Occasionally, following deposition, bone material is re-exposed to perthotaxic processes. However, as the bone has been affected by taphic processes, its structure and/or composition has changed and the perthotaxic processes might accelerate, redirect or halt decomposition. This re-exposure to perthotaxic processes is known as anataxic processes (O'Connor, 2000, 20-21).

Certain cetacean products, such as meat, blubber, oil, and spermaceti do not survive in the archaeological record as a result of perthotaxic, taphic, and anataxic factors. However, it may be possible to recover archaeologically visible evidence of these products, by analysing residues left on pottery used to store cetacean remains. Molecular-based approaches can potentially be used to identify the original function of the pottery, for example storage of whale blubber or oil (Heron and Craig, 2015, 707-719).

Archaeological excavations subsequently also play a role. All aspects controlled by archaeologists themselves are known as second-order processes. Choice of excavation area, recovery methods (e.g. sieving) and analytical methods all are important factors and are known as sullegic processes (Lyman, 1987; O'Connor, 2000, 20-21).

When cetacean bone is recovered, in many occasions the material is in a bad condition, often fragmented, weathered, or shows signs of burning. In such instances, archaeologists might be able to identify the remains as being cetacean, but often further identification to the species or genus level is rendered impossible. Even where diagnostic fragments are preserved, morphological variation among several species is minimal, and this may complicate identification, even more so as several dolphin and whale species are comparable in size.

Additionally, the lack of expertise of many zooarchaeologists with cetacean bone, the lack of high level cetacean osteological collections and the lack of aDNA, ZooMS analysis, or other identification methods performed on cetacean remains, all decrease the likelihood of remains being identified to species level. Furthermore, intraspecies variation, sexual dimorphism, inaccessibility of extensive reference collections, and (recent) body size shifts, all problematize identification attempts (for a more detailed overview see the appendix). Evidently, all these aspects limit our understanding of past cetacean exploitation (Sabin, 2002; Speller *et al.*, 2016).

Eventually, even decisions regarding sorting, recording and publishing play a taphonomic role, as are funding sources. Data might get lost at this stage or never published or remains in grey literature, seriously effecting the accessibility of the data. These factors are known as trephic factors (O'Connor, 2000, 20-21).

Collectively, the taphonomic factors outlined above create substantial gaps and biases in regard to zooarchaeological remains. Understanding which factors and processes affected zooarchaeological remains is vital to zooarchaeologists and should proceed any interpretations made. Indeed, one of the most problematic issues concerning the zooarchaeology of cetacean is still taxonomic identification. However, several methods can be applied to identify cetacean remains to the species level and recently several of these methods have resulted in an enrichment of our knowledge of past cetacean exploitation.

3.2. Identification of Cetacean Remains

Recently, research on archaeological cetacean material has increased and they have received more attention by the (zoo)archaeological community. This is partly due to the advancements that have been made in analytical methods that help zooarchaeologists identify zooarchaeological material to the family, genus, or species level.

3.2.1. Ancient DNA (aDNA)

Since identification to species level is not often possible based on morphology alone, other techniques have to be applied to accomplish this. Molecular studies have looked at ancient DNA (aDNA) as a mean to identify zooarchaeological specimens to the species level. However, nuclear DNA (nDNA) from archaeological samples is often severely degraded and is present in low quantities (Linacre and Tobe, 2011). Therefore, studies have targeted the mitochondrial DNA (mtDNA). mtDNA is a double-stranded circular molecule present within the mitochondria organelles. There are multiple copies of the mtDNA per cell, while nDNA only contains two copies per cell. mtDNA additionally tends to be better preserved over longer periods of time in comparison to nDNA, making mtDNA the more suitable of the two for DNA analysis (Linacre and Tobe, 2011).

The cytochrome *b* gene of the mtDNA is the gene most frequently used for species identification and forensic research. This gene codes the proteins and RNA molecules that are involved with cellular respiration and has very little intraspecific variability within a species. The fragments are relatively short and can be compared with a databank of all known species, making it possible to identify the zooarchaeological samples to the species level (Linacre and Tobe, 2011).

Table 3. Taphonomy factors that affect zooarchaeological data, adapted from Davis (1987, 22), and made relevant to cetacean remains. A circle denotes the loss of data.

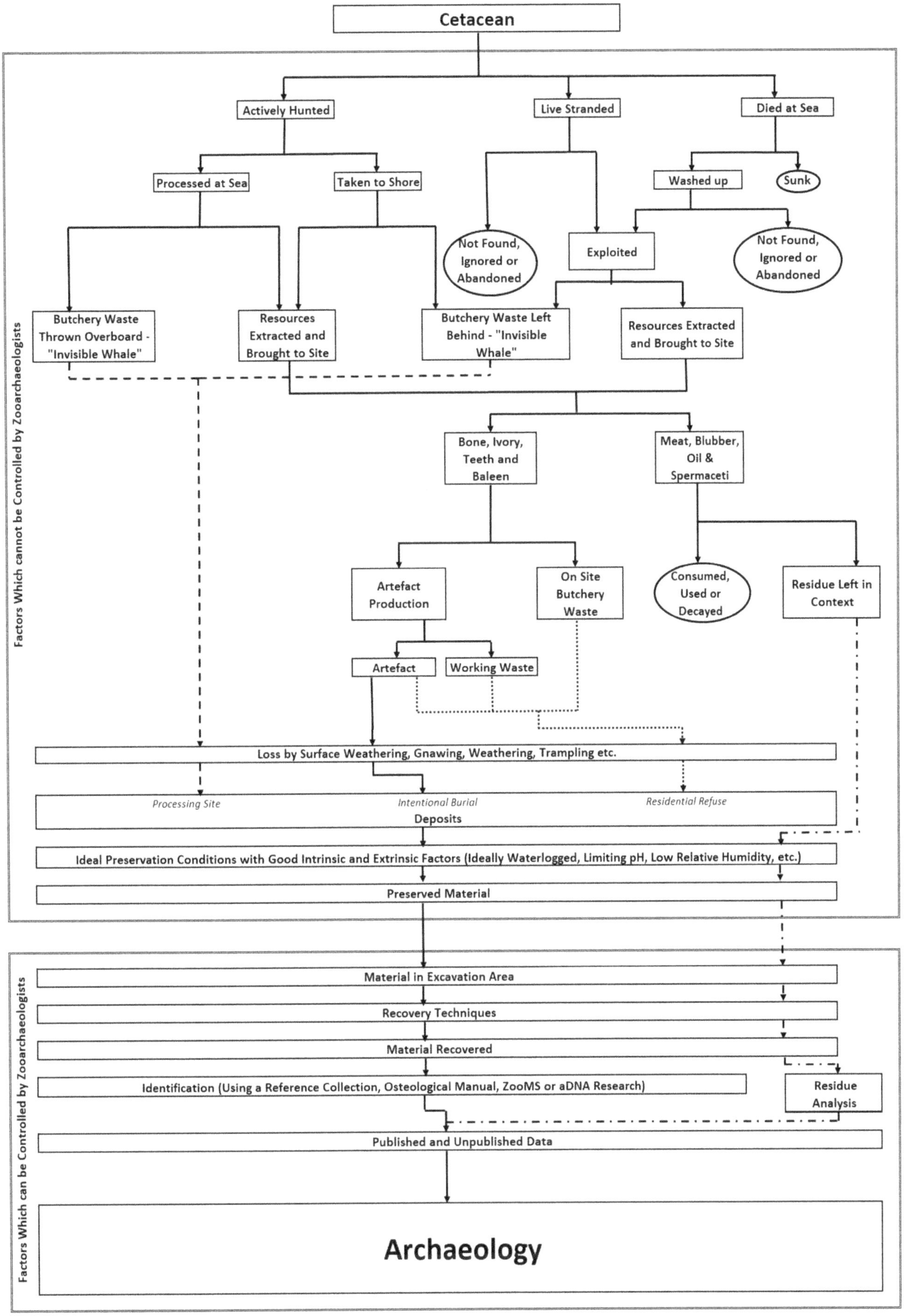

aDNA analysis has been practiced by numerous studies and has enhanced our understanding of past cetacean exploitation. However, most of these studies are restricted to North America. Very little aDNA research has been conducted on archaeological cetacean material from Europe. An example of this is a study by Béland (2012). She performed aDNA analysis on cetacean material from the Nuuchah-nulth First Nations of Vancouver Island and showed that the people exploited both grey whales and humpback whales regularly.

Furthermore, morphological identification of 21 humeri found at Red Bay, Canada, indicated that eight of these were bowhead whale and 13 North Atlantic right whale (Cumbaa, 1986; Cumbaa, Brown and White, 2002). However, DNA research by Rastogi *et al.* (2004) determined that just one of the 21 humeri belonged to a North Atlantic right whale. McLeod *et al.* (2008) also performed research on Basque whaling remains and they determined that during the sixteenth and seventeenth century, Basques whalers mainly targeted the bowhead whale in the Western North Atlantic. It was previously thought that they targeted the North Atlantic right whale, but aDNA research on 218 samples determined that only one sample was from a North Atlantic right whale, while 203 were identified as bowhead whale.

Seersholm *et al.* (2018) conducted an aDNA survey of numerous zooarchaeological specimens from all over New Zealand. Previously, based on morphological identification, pilot whale and dolphins (Delphinidae spp.) were relatively frequently identified in the archaeological record. However, their aDNA survey identified killer whale, true dolphins (Delphininae sp.), Cuvier's beaked whale, fin whale, and southern right whale (*Eubalaena australis*). The diversity of identified cetaceans suggests that the large whales were probably opportunistically exploited through natural strandings, but that the smaller species might have been exploited by the Māori by driving them into shallow waters and spearing or harpooning them. This study shows the potential aDNA analysis can have on our understanding of past cetacean exploitation and additionally highlights the problems of morphological identification.

DNA research has also been conducted on archaeological baleen. Since baleen is a keratinous tissue, it contains DNA which is potentially useful for the identification to the species level. Studies by Gilbert *et al.* (2008) and Sinding *et al.* (2012) have demonstrated that it is possible to extract DNA from archaeological baleen sources (baleen remains as well as baleen artefacts) from settings up to several thousand years in age. It should however be mentioned that the success rate of DNA analysis application appears to be lower for baleen than it is for bone. Additionally, it appears that analysis on baleen that has been worked by humans, undergoing more stress and heat than baleen sources not altered by humans, has a lower success rate as well (Sinding *et al.*, 2012, 3750-3753).

These studies show the great potential aDNA analysis can have on our understanding of past cetacean exploitation. It has even been attempted to completely replace zooarchaeological morphological research by aDNA analysis. While traditionally zooarchaeology requires extensive excavations, excavating large volumes of sediment, Seersholm *et al.* (2016) practiced a less intrusive, metagenomic approach, sampling a range of organic sources, including bone, keratinaceous material, meat and skin from four midden deposits in Greenland. Subsequent DNA analysis identified a wide variety of species, including bowhead whale and narwhal. They concluded that the Saqqaq culture heavily relied on marine mammal exploitation, in particular the bowhead whale. They also found several other species, also already identified by previous zooarchaeological morphological analysis, showing that aDNA research has potential to replace zooarchaeological morphological analysis.

aDNA research has also been practiced on archaeological and sub-fossil samples to understand past species ranges and the effect that climate change has on this. A study by Foote *et al.* (2013) analysed sub-fossil remains of the bowhead whale and analysed its population range during climatic changes of the glacial cycles. Their study indicated that the bowhead whale survived Late Pleistocene climate changes and habitat shifts. They subsequently tried to predict the effect climate change will have on the modern bowhead whale populations and though these changes are hard to predict, Foote *et al.* (2013) suggest that the core suitable habitat of the bowhead will be halved by the end of the twenty-first century.

DNA research has also been practiced estimating the past population size of cetacean species. The International Whaling Commission (IWC) moratorium has halted commercial whaling and allows populations to recover to their pre-whaling size. However, these pre-whaling populations sizes are often calculated using inaccurate historical records and logbooks or are based on DNA substitution and gene flow that can be widely uncertain. Genetic analysis on grey whales in the Pacific Ocean have been undertaken in a study by Alter, Rynes and Palumbi (2007) to estimate population sizes prior to the start of commercial whaling and can help with future cetacean population managements. Their study showed that the current population of around 22.000 individuals, represents only about 22-58% of the pre-whaling era.

While aDNA analysis can certainly enrich zooarchaeological studies, it has several disadvantages as well. Firstly, aDNA analysis is a costly and time-consuming practice. Secondly, aDNA has the disadvantage of degradation accumulating over time. Thirdly, archaeological material that is charred is not suitable to perform aDNA analysis on. Furthermore, aDNA analysis has a highly unpredictable likelihood of success and the success rates are poor in warm environments (Larson *et al.*, 2007). As a result of all these factors morphological analysis will still be necessary and will not likely be replaced by aDNA analysis in the distant future.

3.2.2. Zooarchaeology by Mass Spectrometry (ZooMS)

Another method used for species identification is the relatively recently developed technique of Zooarchaeology by Mass Spectrometry (ZooMS). This is a proteomics-based method that analyses proteins from zooarchaeological remains for the purpose of species identification (figure 29). ZooMS has several advantages over aDNA analysis. First of all, it is cheaper and less time-consuming method. Secondly, proteins are more abundant in zooarchaeological material than DNA. Lastly, proteins survive longer than DNA. The protein ZooMS analyses is collagen. It is the most abundant protein in the vertebrate kingdom and is also used for radiocarbon dating as well as stable isotope analyses. Collagen is composed of a triple helix (COL1a2 chains), of which the alpha 2 chain has high sequence variability. Based on this variability, discrimination of closely related genera can be established, making these collagen peptides a useful source within zooarchaeological research (Buckley *et al.,* 2014).

The divergence time between vertebrate species is an important factor for performing ZooMS. In general, a divergence of 8 million years is needed to make distinctions between vertebrate groups. However, ZooMS could distinguish between blue whale, fin whale and sei whale, while these species diverged 5-6 million years ago.

ZooMS analyses has initially often been practiced on domesticates, though later wild taxa and especially marine mammals were frequently analysed. For the North Atlantic region several species can be identified using ZooMS. The species that can be identified to the species level are common minke whale, sei whale, fin whale, blue whale, humpback whale, grey whale, and sperm whale. Baleen whales are relatively easily identifiable using ZooMS. However, samples of the North Atlantic right whale and the bowhead whale cannot be separated and can only be identified as Balaenidae (Buckley *et al.*, 2014).

For the beaked whales, no distinction can be made between the bottlenose whale and the Sowerby's Beaked whale. The other beaked whale species present in the North Atlantic are not assessed just yet. Furthermore, within the Delphinidae family the species (as well as the harbour porpoise) cannot be separated using ZooMS, instead several groups can be identified. The first group includes the bottlenose dolphin, striped dolphin, common dolphin, and white-beaked dolphin. The second group includes the killer whale and the white-sided dolphin. The third group includes the so called "Blackfish" or the subfamily Globicephalinae, which includes the Risso's dolphin, long-finned pilot whale, short-finned pilot whale, and the false killer whale.

A study by Buckley *et al.* (2014) performed ZooMS on over 50 samples of marine mammals from Iceland and Scotland, identifying a range of different marine mammals, including seals, porpoises, and several of the baleen whales. A considerable number of the ZooMS samples from Scotland were identified as Balaenidae, suggested to represent North Atlantic right whales. These remains were considered to be a valuable source in reconstructing the genetic history of this heavily exploited species. Furthermore, several humpback whale remains were also identified, but Buckley *et al.* (2014) suggested that these probably derived from stranded individuals or potentially from opportunistic whaling. The same was suggested for Iceland, as the remains from that region encompasses a wide range of species, which does not suggest an organized whaling culture.

Another study by Evans *et al.* (2016) analysed cetacean remains from the Lanashuaia locality, Tierra del Fuego. Both ZooMS and aDNA analysis was undertaken and a wide variety of cetacean species were identified, including: blue whale, southern bottlenose whale, southern right whale, humpback whale, and sei whale. Previously, these remains were assumed to represent only one species. Evans *et al.* (2016) suggested that active whaling was not practiced but that cetaceans regularly stranded in the area. Whenever this happened, an aggregation event of people in the area occurred, who subsequently shared the meat and distributed the resources that could be extracted from the carcass. Active whaling was probably not undertaken as a variety of species were present amongst the material and additionally several of these species were very large and fast and only came within the reach of human exploitation with the onset of industrialized whaling.

Speller *et al.* (2016) conducted ZooMS analysis on 17 cetacean remains from the Mediterranean (southern France, western Italy, and Sardinia). These 17 samples, include five that were previously identified as grey whale, using comparative anatomy methods. These five samples are the reason the Mediterranean was assumed to be a nursing and breeding area of the now extinct North Atlantic population of the grey whale. However, ZooMS and aDNA analysis determined that the 17 samples represent North Atlantic right whale, sperm whale, fin whale and Cuvier's beaked whale and not the grey whale. Habitat modelling predicted that grey whale's past range potentially extended as far south as the Mediterranean. The presence of the North Atlantic right whale was an interesting discovery, suggesting that this species was present in the Mediterranean as well.

Rodrigues *et al.* (2018) also focused on zooarchaeological remains of cetaceans in the Mediterranean Sea. While several cetacean species are present in the Mediterranean Sea, little is known about the history of cetacean exploitation in the region. Zooarchaeological remains of cetaceans appear to be rare for the Mediterranean area. Bernal-Casasola *et al.* (2016), were only able to locate 26 archaeological sites within the region for which cetacean remains have been discovered. It is generally assumed that the remains from the region are derived from stranded individuals, which were exploited by various cultures, including the Romans, Gauls, Greeks, Phoenicians, and Mauris.

At least for the Western Mediterranean and the Strait of Gibraltar it has been argued that active whaling was undertaken. Whether this was frequently undertaken or merely opportunistic, remains unclear. Yet the abundance of cetacean remains from southern Spain dating to the Roman period, of which several have been fashioned into chopping blocks, and the presence of iconographic evidence depicting active whaling, suggests that for this region cetacean exploitation was undertaken more frequently. It remains however unclear whether this exploitation encompasses active whaling or opportunistic scavenging. A large-scale organized form of whaling was however not present within the region (Bernal-Casasola *et al.*, 2016).

Rodrigues *et al.* (2018) analysed cetacean remains from the western Mediterranean (Spain and Morocco). This study was able to identify three North Atlantic right whale and three grey whale remains by applying aDNA and ZooMS analysis on elevens zooarchaeological remains. This study suggests that during the Roman period, these two species (which now have disappeared from the region) were abundant in very western part of the Mediterranean and might have utilized the region as a calving or breeding area. Rodrigues *et al.* (2016) suggested that these two species might have been exploited in the western Mediterranean region and as the presence of these two species now has been confirmed by Rodrigues *et al.* (2018), they suggest that a forgotten whaling industry might have been present in this part of the Mediterranean during the Roman period. These active whaling practices might therefore have had something to do with the extinction of both the grey whale and North Atlantic right whale from the region (Rodrigues *et al*, 2018).

ZooMS has also been practiced on material from the twelfth to fourteenth century site of Odense. At the site several bone combs had been recovered, and by the application of ZooMS, one turned out to have been made of sperm whale bone. Another bone was identified as North Atlantic right whale. Both these species are not common visitors to Danish waters, and it has been suggested that the material was shipped to Odense for the creation of various artefacts and tools (Ørsted Brandt *et al.*, 2018).

Solazzo *et al.* (2017) demonstrated that peptide mass fingerprinting can also be applied on baleen. It is often impossible to identify baleen to species, but Solazzo *et al.* (2017) demonstrated that the technique can be successfully applied on both recent as well as archaeological material. They analysed 29 fragments of baleen from archaeological sites in Labrador, Canada and were able to identify these as bowhead whale, indicating a long tradition of bowhead whale hunting in the area.

Yet another example of the application of ZooMS on cetacean material is the study by Biard *et al.* (2017). For this study 30 specimens from Chersonesus in the disputed area of Crimea (both claimed by the Ukraine and Russia) were analysed. On these samples ZooMS, as well as two forms of aDNA analysis (Sanger sequencing and shotgun sequencing) was conducted. Though ZooMS was able to differentiate between the porpoise and the dolphin remains, shotgun sequencing proved to be the mare valuable method in order to identify the remains more precisely to species level (Biard *et al.*, 2017).

Several of the studies described above used a combination of both ZooMS and aDNA analysis. In most of the cases both techniques had similar results, however ZooMS is a little less precise as it is not able to discriminate between several species (e.g. North Atlantic right whale and bowhead whale or the different beaked whale species). These ZooMS studies have however all resulted in interesting and ground-breaking discoveries, showing what ZooMS can enlighten our understanding of past habitat ranges and past cetacean exploitation.

3.2.3. Trace Element Analysis (TEA)

The study of Late Pleistocene and Holocene whale remains from Denmark and adjacent countries by Aaris-Sorensen *et al.* (2010) have shown that Trace Element Analysis (TEA) by Instrumental Neutron Activation Analysis can be used for species identification as well. By plotting the element concentrations of Iron (Fe), zinc (Zn), and chromium (Cr), a distinction can be made between cetacean species that primarily feed on crayfish (e.g. baleen whales) and species that primarily feed on fish and squid (e.g. dolphins, beaked whales, and sperm whale).

Several factors might be the cause for this variation in trace elements between the species. It can be the result of the bone porosity upon the degradation of the bone (and other organic material). Additionally, it can be caused by a difference in the ability of the skeletal organ of the various species to incorporate various trace elements. The most likely is however that the variation is the result of the differences in diet.

Aaris-Sorensen *et al.* (2010) utilised the technique on seventeen samples. The trace element analysis sample from Poulsker, Bornholm, Denmark, indicated that this specimen most likely represents a fish/squid eater. The sample was however morphologically identified as a Balaenoptera sp.. The trace element analysis suggests that this might not be the case and that the specimen is more likely to be from a large fish/squid eater, most likely a sperm whale.

While this study has shown that TEA can be a useful method for the identification of cetacean material, in comparison to aDNA and ZooMS, it is a highly imprecise species identification method. Especially the recent development of ZooMS, made the amplification of TEA obsolete, as ZooMS is cheaper and more precise.

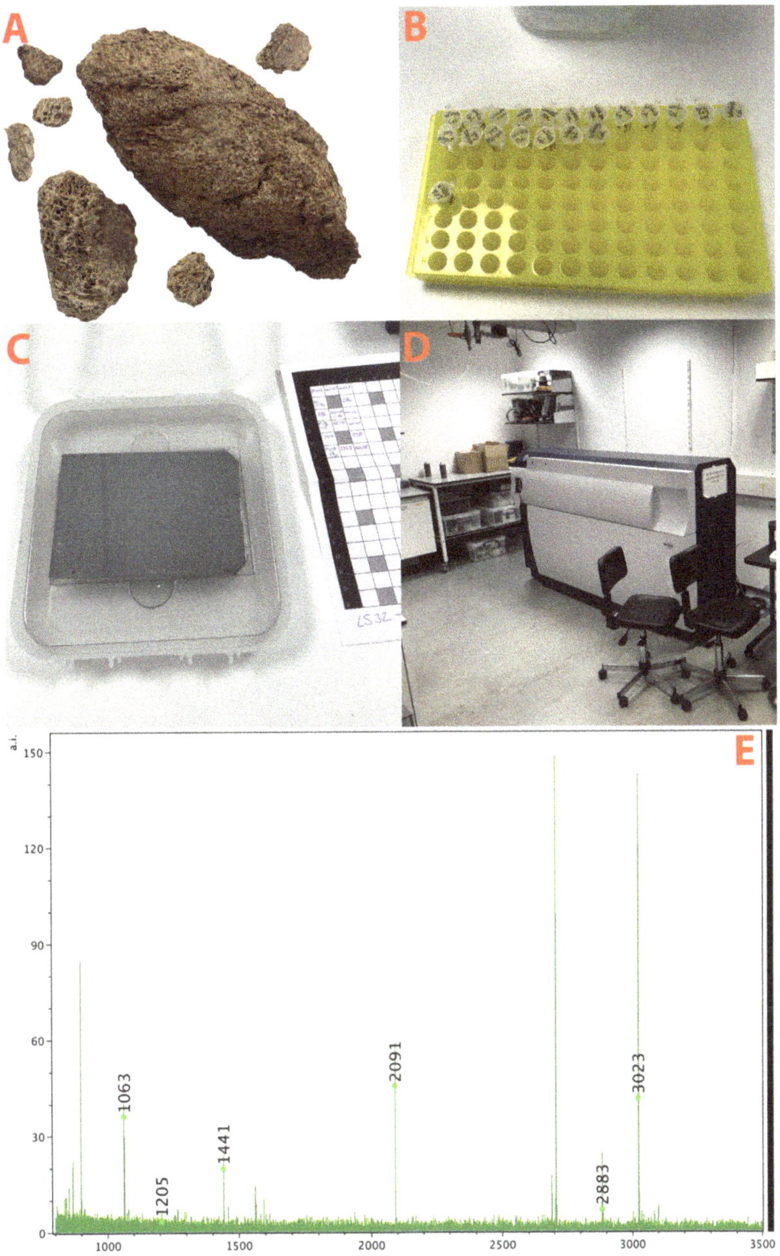

Figure 29. Zooarchaeology by Mass-Spectrometry (ZooMS). A) sample selection. B) Sample preparation in test tubes. C) Sample on the MALDI target plate. D) Samples being analysed using a Bruker ultraflex III MALDI TOF/TOF mass spectrometer. E) Averaged spectra created from the replicates for specimen using mMass software (Strohalm *et al.*, 2018). Photos by author.

3.2.4. Organic Residue Analysis

Unlike bone material, blubber and oil does not survive in the archaeological record. However, molecular-based approaches can be used to identify residue left in pottery and ceramic vessels, and through this technique blubber and oil originated from cetaceans might still be identified (Heron and Craig, 2015, 707-719). This is especially useful when zooarchaeological material is missing. Ancient lipid analysis can potentially provide useful information regarding culinary practices at specific sites, as well as the storage of whale oil (Craig *et al.*, 2011).

Blanco-Zubiaguirre *et al.* (2018) performed lipid profiling and identified the biomolecular markers on residues left in ceramic pottery fragments from the Basque region, dating to the sixteenth to seventeenth century AD. High-Performance Liquid Chromatography – Electrospray Ionization - Quadrupole Time-of-Flight Mass Spectrometry was used to identify triacylglycerol (the main constituent of body fat) and the distribution of it within the samples. Additionally, Gas Chromatography-Mass Spectrometry provided the fatty acid profile and detected the degradation compounds and biomarkers related to marine commodities. This was performed on the archaeological samples as well as on five different cetacean species (fin whale, sei whale, minke whale, humpback whale and harbour porpoise) to serve as reference material. The archaeological material turned out to be most comparable to the rorquals (fin, sei, and minke whale). However, the most likely candidates to which the organic residue belonged (North Atlantic right whale and bowhead whale, and to a lesser extent grey whale and sperm whale) were not analysed, rendering the results inconclusive.

The methods practiced by Blanco-Zubiaguirre *et al.* (2018) proved to be a useful and when reference material from other species is added to the database, has the potential to supply valuable new information in regard to past cetacean exploitation and the storage of exploited material. This is especially useful in regions were osteological remains are rare (as is the case in the Basque region).

3.2.5. Osteological Reference Collections

Osteological reference collections provide zooarchaeologists with the material they need to compare to archaeological material in order to identify these to the species level. Identification of cetacean material is problematic partly due to the fact that many reference collections lack cetacean material. Zooarchaeology is still a discipline that is often only concerned with terrestrial (domesticated) mammals and as a result most zooarchaeology laboratories only comprise a reference collection for those mammals. Some more specialized zooarchaeology laboratories have larger collections that include other animal groups, but cetaceans are not often included in these collections.

Osteological cetacean material is often only present in large natural history museums, like the Natural History Museum in London, UK. Since 1324, "Fishes Royal" or cetaceans, in the territories of England and Wales were a sovereign right, but this was changed in 1913. In this year the National Cetacean Strandings Recording Scheme was set in place, which allowed the museum to have first claims to stranded cetaceans and carcasses to conduct scientific research on them (Sabin, 2002). This allowed the museum to expand their cetacean collection, which now comprises over 10,000 fluid-preserved and osteological specimens from a wide variety of cetacean species.

Sabin (2002) noted that the cetacean reference collection in London is regularly used by taxonomists, zoologists, and conservation biologists. Use by (zoo)archaeologists is increasing as well, as researchers from various European countries travelled to the Natural History Museum with their material to identify them to the species level, including material from Gibraltar and Iceland. The results show the great potential the use of these collections can have on our understanding of past cetacean exploitation.

Other museums with extensive osteological cetacean collections include: Naturalis in Leiden, the Netherlands; the Statens Naturhistoriske Museum in Denmark, Copenhagen (figure 30); the Muséum National d'Histoire Naturelle in Paris, France; and the Natural History Museum, Smithsonian Institution, Washington DC, USA. These collections comprise a wide variety of cetacean species and have great potential for zooarchaeologists. However, with the commercialization of archaeology, zooarchaeologists are often restricted in their time and money and rarely travel to these museums with their material. Additionally, because of the size and the weight of zooarchaeological

Figure 30. Osteological cetacean reference collection part of the Statens Naturhistorisk Museum Copenhagen, Denmark (photo by author).

cetacean material is hard to transport these to the reference collection for comparison. Furthermore, the Convention on International Trade in Endangered Species of Wild Fauna and Flora (CITES, 2019) has enforced strict regulations regarding the transportation of (cetacean) bone material, sometimes limiting possibilities to analyse material abroad. As a result, many of the zooarchaeological remains are merely classified as "cetacean", "whale", "dolphin" or even "marine mammal".

3.2.6. Osteological Reference Manual

Zooarchaeologists often make use of so-called osteological manuals in which the skeletal elements of several species are drawn or depicted. Examples of this are Pales and Lambert (1971), Schmid (1972), Prummel (1987), and Hillson (1992). Manuals like these are widely used by zooarchaeologists, however there are several limitations on the usage of manuals in comparison to osteological reference collections.

Firstly, manuals often show osteological material only from several angles. As a result, several parts of the bone are not clearly visible in a drawing or photograph. 3D scans however make it possible to look at a bone from every viable angle. Secondly, manuals often depict only one bone from each species. Cetaceans have a lot of intraspecies variation, including differences between the two sexes and between populations. If only one specimen is depicted this can be problematic. Thirdly, manuals and 3D scans do not make it possible to touch the bone and feel particular surfaces. This can be helpful for identification purposes.

However, manuals also do have advantages. Firstly, you do not need access to a reference collection, allowing to perform identification wherever, even in the field. Secondly, and this is particularly the case for cetaceans, large and heavy material is hard to work with. Transporting a large cetacean vertebra to a reference collection and comparing it to several specimens is a tough job, while comparison to a piece of paper is a lot easier. Thirdly, as discussed previously, many zooarchaeological reference collections, do not hold an extensive osteological cetacean collection. Manuals can act as replacement for this and fill a gap in our understanding of the history of cetacean exploitation.

Manuals are a useful tool for zooarchaeologists. Unfortunately, the majority of the osteological reference manuals are concerned with terrestrial (domesticated) mammals, birds, fish, and molluscs. Only one substantial manual for cetaceans has been produced. This is the one by van Beneden and Gervais (1868), called "Ostéographie des Cétacés Vivants et Fossils" (figure 31), which was published in several volumes between 1868 and 1880. This means that parts of these manuals are already over 150 years old. The manual comprises 420 pages and 44 plates with drawings. Numerous inaccuracies, especially in the parts considering the baleen whales, are documented in this manual. It however remains an important source to zooarchaeologists working with cetacean material, though the zooarchaeological community will benefit from an updated version (de Smet, 1979).

Besides the guide by van Beneden and Gervais, Benke (1993) also created an osteological guide for cetaceans.

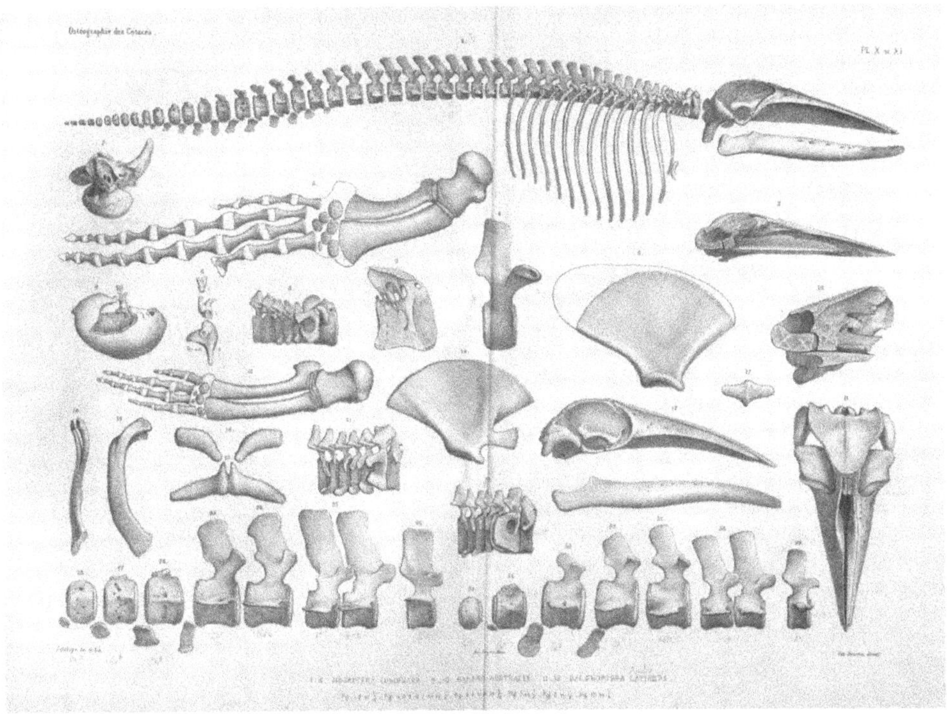

Figure 31. Plate X and XI of the Ostéographie des cétacés vivants et fossiles, comprenant la description et l'iconographie du squelette et du système dentaire de ces animaux, ainsi que des documents relatifs à leur histoire naturelle by van Beneden and Gervais (1868). Image courtesy of the Biodiversity Heritage Library. Contributed by Smithsonian Libraries. Available at: <https://www.biodiversitylibrary.org/item/127737#page/27/mode/1up>.

His work is however only concerned with the pectoral fins and he focuses on both osteology and myology. Regarding the osteology, Benke focused on the scapula, humerus, radius, and ulna and describes the morphology of these elements for each species and additionally he performed several measurements on them as well. The ranges the bones can measure are provided in boxplots and are compared to all other cetacean species, making it an ideal tool for taxonomic identification purposes. However, his sample size is rather small, rarely exceeding five specimens.

The photographs Benke included into this manual only include pictures of the cranium (lateral view), the pectoral fin (lateral view), the scapula (lateral view), and the humerus (ventral, medial, posterior, dorsal and anterior view). No pictures of the radius and ulna (apart from the picture of the entire pectoral fin), the glenoid of the scapula, lateral view of the humerus, or the vertebrae were included into this osteological guide. As a result, many of the diagnostic zones for each skeletal element are not visible. Additionally, the pictures are frequently poorly illuminated and of a poor quality, making it hard to use them for identification purposes.

Indeed, it was not Benke's intention to create an osteological guide that could be aid to zooarchaeologists or palaeontologists. His research focused on functional properties of the pectoral fins and the mobility of it, for which he had to perform research on the osteology and myology. It is still one of the most comprehensive guides there are for cetacean osteology.

Crania are probably the most frequently analysed osteological features of cetaceans. They are frequently the subject of evolutionary biologists trying to reconstruct the evolution of cetaceans (Amaral *et al.*, 2009; Sydney, Machado and Hingst-Zaher, 2012; Berta, Ekdale and Cranford, 2014). A guide to the identification of cetacean crania was created by Jefferson, Leatherwood and Webber (1993). This study denotes all the morphological features that are characteristic for each family group but does not provide morphological information for each species separately. Drawings are made of the dorsal, ventral, and lateral side of the crania for each species and these can be useful for identification purposes. However, for the identification of smaller cranial fragments, the drawings are not detailed enough.

Ekdale, Berta and Deméré (2011) published a guide to the petrotympanic complex (ear region) of several baleen whales. The tympanic bulla and the petrosal bone are pictured from the ventral, dorsal, medial, and lateral side for each species. Additionally, detailed osteological descriptions are provided as well. Though this study on the petrotympanic complexes was undertaken to perform research on the evolution of this body part within the Mysticeti parvorder, the photographs and description are of use to zooarchaeologists and palaeontologists as well.

Another guide was created by Crovetto (1991). This guide, in opposition to the guide by Benke (1993), is only concerned with the vertebral region of the cetacean skeleton, and only with those of several of the large cetaceans (Balaenopteridae family, Balaenidae family, the pygmy right whale (*Caperea marginata*), the grey whale, and the sperm whale). No pictures are included in this study and only a handful of drawings are provided. This study however relies on descriptions of morphology and measurements. It is a useful guide and can be used for the identification of zooarchaeological and palaeontological material, however it is rarely used by zooarchaeologists.

There is a need for a comprehensive osteological guide for cetaceans that discusses both the bones of the pectoral fin as well as the vertebral column and the cranium and should focus on both morphology as well as osteometry. The creation of this guide will most likely result in a better understanding of past cetacean exploitation. Sabin (2002) noted that the Natural History Museum in London, UK was also planning to create an osteological manual to aid identification of zooarchaeological remains. However, Sabin noted that the creation of a manual was still in the planning stages in 2002, and by 2020 an osteological manual has still not been created by the museum.

3.3. Active whaling vs. Opportunistic Scavenging

When studying cetacean remains, it is challenging to distinguish between remains acquired through active hunting and those acquired through opportunistic scavenging. This is even the case for areas where whaling remains a strong living tradition (Losey and Yang, 2007). Signs of harpooning or spearing on cetacean remains are not identified frequently. Cetacean remains often do bear cutmarks, however most of these were probably inflicted post-mortem and do not suggest an active hunting method. It might be that "professional" whalers employed more formal techniques of rendering a whale, leaving a certain signature in terms of cut marks. However, while medieval European cetacean material occasionally shows butchery marks, the sample size is not large enough to differentiate between butchery marks made by "professional" whalers and those by "opportunistic" whalers.

An example of a site with numerous cetacean bones with butchery signs and that showcases the hardship of identifying zooarchaeological cetacean remains that were acquired through active whaling, is the site of Grotta dell'Uzzo in North West Sicily, Italy (Mannino *et al.*, 2015). There a total of 224 cetacean bones were recovered, almost exclusively dating to the Mesolithic-Neolithic transition period. Most of the bones were fragmented, but some could be identified as long-finned pilot whale, Risso's dolphin, and short-beaked common dolphin. Nine skeletal elements with butchery marks were observed, three from large cetacean species and six from small cetacean species. The cut marks are thought to have been produced during the disarticulating process of the carcass and not an indication of hunting. None of the 224 bones appear to bear signs of harpooning, which suggests that the bones derived from stranded individuals. Additionally, there is no solid evidence offshore fishing was undertaken in the Mediterranean during the Mesolithic-Neolithic transition. This study shows that identification of active whaling, merely based on zooarchaeological remains, is almost impossible. This aspect severely clouds our understanding of past cetacean exploitation.

On the Northwest Coast of North America prehistoric whaling was previously only thought to have been practiced by the Makah of the northern Olympic Peninsula and the Nuu-chah-nulth and Ditidaht people of southwestern Vancouver Island. Other groups in the area were assumed to have just exploited stranded cetaceans. However, Losey and Yang (2007) came across a humpback whale phalanx with a bone point made of elk (*Cervus elaphus*) bone, embedded in it at the site of Par-tee in Northern Oregon. This piece indicates that active whaling was also undertaken at Par-tee. However, this whaling was more "low level" whaling, as in opposition to the Makah, Nuu-chah-nulth, and the Ditidaht people who practiced systematic and intensive whaling. The weapons, tools, technologies and practises employed by the people of Par-tee are probably less specialized than for the other groups (Losey and Yang, 2007).

Additionally, ethnographic sources also seem to suggest that whales were opportunistically taken in the region. These sources also seem to suggest that the technology used was not specialized for whaling and that it was only rarely undertaken. However, the ethnographic sources in combination with the humpback whale phalanx indeed seem to suggest that some form of active whaling was undertaken at Par-tee (Losey and Yang, 2007).

The finding of the struck humpback whale phalanx is a special one. Few sites in the Northwest Coast region have been extensively excavated and the zooarchaeological material has often not been fully analysed. Other struck bones might be present at other sites in the region, though these are probably rare. Several have however been identified at the site of Ozette (Losey and Yang, 2007).

Stable isotope analysis has also been used to identify the consumption of cetacean meat. This was done on Dorset remains, Thule-era human burials from northwest Hudson Bay as well as proto-historic burial from Southampton Island, Canada. It was determined that for Modified Thule (those that post-date AD 1350), cetacean meat made up approximately 12% of dietary intake. Though this does not necessarily mean that active whaling or opportunistic scavenging was undertaken (Coltrain, Hayes and O'Rourke, 2004).

Distinction between active whaling and opportunistic scavenging is a contentious issue with many implications for research on the development of whaling (Savelle and Kishigami, 2013). Zooarchaeological information has frequently been used by both opponents and proponents of modern whaling activities, with proponents using zooarchaeological information to argue a long, uninterrupted whaling history of a society. The presence of hunting gear, species representation in the archaeological record, age, and size of the species in the archaeological record, zooarchaeological contextual analysis, and dietary importance based on isotope analysis have all been used to support or deny active whaling activities (Savelle and Kishigami, 2013).

3.3.1. Species Composition

The species represented at an archaeological site may be indicative of whether active whaling or opportunistic scavenging was undertaken. Rorquals, with the exception of the humpback whale are fast swimmers, with the larger species (e.g. the blue whale, fin whale, and sei whale) being able to reach a maximum speed of up to 50 kilometres per hour. Prior to the second world war, catcher boats rarely made more than 25-28 kilometres per hour. The only way to hunt the rorquals was to "stalk" them and hope that they would get within reach at some point. The Norwegians called this method of whaling *luse-jag*. In pursuit medieval whalers were probably not able to hunt the rorquals, however when in coastal or shallow waters, where the speed of these animals was limited, they might potentially come within the reach of whalers.

Post the second world war however, the *Prøyser-jag* method was more frequently practiced. In the twentieth century, vessels were faster, and the whales were forced to swim faster as well and had to come up for air more frequently. When a whale swims fast and comes up for air, it surfaces at an angle, showing a larger part of its body. This was preferred as it generates a larger target area for a harpoon to hit the whale. Many whales were hunted using this method in the second half of the twentieth century (Slijper, 1962, 93-95).

During the Medieval Period the North Atlantic right whale, humpback whale, the grey whale, and the Delphinidae species were the species most frequently targeted. The beaked whales and the Kogiidae are more pelagic species and active whaling on those species is not likely. Even in recent history these species have not been hunted regularly. The bowhead whale, narwhal, and the beluga are Arctic species and can potentially have been exploited in the northern part of Europe, for which especially the bowhead whale would have been an ideal target.

Bone remains of the bowhead whale were regularly discovered at Thule sites in north-eastern Canada and Greenland. The question was raised whether the Thule performed active whaling on the bowhead whale or whether they just opportunistically exploited stranded or ice-entrapped individuals. Savelle and McCartney (1991) compared the difference in age and size of the modern stranded bowhead whales in the region with the material from the archaeological record. Their research indicated that young individuals and yearlings are strongly represented in the archaeological record in comparison to the naturally stranded numbers. This suggests that the Thule did indeed perform active whaling and targeted small and young individuals.

Savelle and McCartney's (1991) research shows the potential the analysis of zooarchaeological material can have. However, in order to perform such an analysis a large archaeological sample size is needed. Furthermore, modern natural strandings have often been caused by anthropogenic factors, such as ship strikes or the ingestion of plastic, making the comparison with archaeological data problematic.

The previously discussed study by Evans *et al.* (2016) on the material of the Lanashuaia locality, Tierra del Fuego, also looked at species composition. While originally the cetacean material was assumed to be derived of actively caught whales, the ZooMS and aDNA analysis indicated that a wide variety of species was present at the site, including some species which were most likely not actively hunted. Evans *et al.* (2016) therefore concluded that unlike previously assumed, active whaling was not undertaken at Tierra del Fuego, but that cetaceans regularly stranded in the area and were subsequently opportunistically exploited.

Wellman *et al.* (2016) also performed aDNA analysis on cetacean material from Par-Tee. They determined that grey whale made up 60.7% and humpback whale 32.1% of the total number of specimens analysed. As previously discussed, the finding of the humpback whale with a bone point embedded in it, suggested that active whaling was undertaken at the site. The species composition suggest that whaling was occasionally opportunistically undertaken. Wellmann *et al.* (2016) base this on the fact that humpbacks rarely strand on the coast of Oregon in recent times. The grey whale was probably more likely exploited when stranded.

These studies prove that species composition can be a useful tool in determining whether active whaling or opportunistic scavenging was undertaken. The disadvantage of this tool is that a lot of material must be analysed and has to be identified to the species level. For medieval Europe sites with large numbers of cetacean remains are rare. The mere quantity of cetacean material itself might be indication of active whaling. In most medieval European contexts however, cetacean material is represented by merely a couple of fragments which is not indicative of an active whaling culture. However, as stated before it is possible to butcher a whale and take the resources, without bringing the bones along (Smith and Kinahan, 1984). Therefore, the low number of cetacean remains at a site does not automatically rule out active whaling.

3.3.2. Ethnographic Sources

There are very few active whaling practices still being practiced in north-western Europe. Only in Norway, Iceland, and the Faroe Islands is whaling still practiced. Those in Iceland and Norway are highly industrialized and show barely any resemblance to ancient whaling practices in those countries. However, in the Faroe Islands the same basic technique is practiced that was practiced there already by the Norse hundreds of years ago (figure 32 and 33). Every year primarily long-finned pilot whale is hunted during the summer months.

The Faroe Islands were an early Norse colony, settled in the mid-ninth century. Zooarchaeological evidence suggests an early reliance on pilot whale, as well as fish, birds, and domesticated animals (Wylie, 1987, 9; Fielding, 2018, 67-97). As the Faroe Islands are not suitable for agriculture, which became even worse with the onset of the Little Ice Age, and the fact that it was incorporated into the Norwegian Kingdom in the thirteenth century and lost its free trading rights, it has been argued that the inhabitants of the Faroe Islands were forced to look for alternative food sources (Fielding, 2018, 85). The inhabitants could have looked at the sea and the pilot whale's abundance in the area and subsequently developed the *Grind* (Szabo, 2008, 99; Fielding, 2018; 85). The first official record regarding the Faroese pilot whaling or the *Grind* dates to 1584, but the practice is thought to date back ever further. The *Grind* appears to have always been a community-based non-commercial enterprise. It targeted the long-finned pilot whale and though it is still unclear when the first *Grind* was practiced it appears to have been a large part of their economy. It provided meat, protein, oil, bone, skin, sinews, stomachs, bone (for both tools and artefacts as well as oil extraction; shoulder blades for shovels and skulls were stacked into fences; Wylie, 1987; Fielding, 2018, 92).

The *Grind*, starts with a person spotting the whales, either from land or on sea (figure 33). This first person to spy the whales, will be rewarded by receiving the largest individual of the kill. Then numerous people join the hunt by boat. Then the boats start to drive the whales to land, into a bay preferably. This is a time-consuming practice and the boats maintain a crescent formation behind the whales and the people on the boats generate a lot of noise to drive the whales forward. The hunters use *hvalvákn* (stabbing lances) to drive the animals forward, *sóknarongul* (hooked iron gaff) which is inserted into the whale's blow-hole to draw the whales closer to a boat or to the beach, and the *grindaknívur* (a long knife) which is used to slice through the spinal cord. The *grindadráp* (harpoon) is only used when a drive is abandoned, and the hunters want to take at least a couple of individuals. However, this also is the end of the communal effort and it is each hunter for himself, resulting in small tax being paid for each whale. Panicked by the noise, the whales strand themselves and are subsequently butchered by the hunters (Wylie, 1987). Other species which are also occasionally taken are the bottlenose dolphin, white-beaked dolphin, Atlantic white-sided dolphin, Risso's dolphin, and bottlenose whales.

Drive-hunting like that practiced during the *Grind* is a hunting method especially useful for hunting dolphin

Figure 32. Whaling in the Faroe Islands as observed in 1854. Image extracted from page 107 of *A Narrative of the Cruise of the Yacht Maria among the Feroe Islands in the Summer of 1854.* Digital image courtesy of the British Library Mechanical Curator Collection / Wikimedia Commons / Public Domain.

Figure 33. Whaling in the Faroe Islands in 2006. Photo by © Erik Christensen / Wikimedia Commons / File licensed under the terms of the GNU Free Documentation License, Version 1.2 or any later version published by the Free Software Foundation; with no Invariant Sections, no Front-Cover Texts, and no Back-Cover Texts.

species. The fact that it seems to have already by practiced by at least AD 1584 onwards seems to suggest that no modern technology is required to practice it and could have already been practiced during the Medieval Period in areas with bays or fjords where the dolphins are driven into in. Though drive hunting looks easy, it still requires skill, strategy and practice (Szabo, 2008, 99).

The Annals of Ulster, dating to AD 828, record "a great slaughter of porpoises on the coast of Ard Cianachta by foreigners". These foreigners were most likely the Norse (Szabo, 2008, 2010-2011). Additionally, in the *King's Mirror* a large group of so called "blubbercutters" (which probably were pilot whales or another species of dolphin, as these live in large pods) were "constantly being caught and driven to land by the hundreds, and where many are caught they provide much food for men" (Szabo, 2008, 110-111). Instances of drive hunting being recorded in historical sources are rare and more frequently large whales are depicted or written about.

3.3.3. Hunting Tools

Using any kind of weapon, catching a large whale is an extraordinary difficult task. large whales have such a thick layer of blubber, only a couple of places were targeted. Whales are typically targeted on the tail, below the pectoral fin, or the back fin, where the layer of blubber is not so thick (Yarborough, 1995, 71). The fact that various sources have noted that multi-purpose weapons were the primary tools used for whaling, makes the identification of whaling equipment problematic.

Lindquist (1995b) has suggested that a wide range of basic, multi-purpose weapons were utilised for whaling by the Norse. The Norse presumably utilised spears to hunt porpoises, dolphins, seals, and potentially even large whales. An early ninth-century grave located in Hundholm in northern Norway contained a large spear point, as well as a hyoid of a large cetacean. The spear point was unbarbed, but Lindquist (1995a, 402) argues that this might be an indication that active whaling was undertaken by this particular person and that the hyoid resembles a trophy. Bone spearheads have also been found in fifth century Norway, but Lindquist (1995b) notes that these were more likely used to hunt seals and maybe the smallest cetaceans, instead of large whales.

Lindquist (1995b) argues that in the case of the Norse, hand-harpoons were only used by the seventeenth century to hunt large cetaceans (figure 34), and during the Medieval Period it appears that lances and spears were the weapons most frequently used. Indeed, iron spear points are frequently found in Norwegian weapon graves and even in the *grind* on the Faroe Islands, multi-purpose weapons seem to be used, with the notable exception of the *grindaknívur,* a special knife used to cut the spinal cord, which is both a cultural artefact and a typical tool used in the Faroese *grind* (Szabo, 2008, 108).

Spears are useful for inshore and bay whaling, like the drive hunting still undertaken in the Faroe Islands, and potentially for old-Norse whaling in Norway's fjord where a whale just has to be struck, an escape route out of the fjord blocked by the hunters, and patience for the whale to drift ashore. For other regions this technique cannot be practiced. By the thirteenth century the *hvaljarn* (whale iron) was mentioned in various sagas, suggesting that, in combination with innovations in boat constructions, larger whale species could be targeted outside of the fjords (Szabo, 2008, 108).

At several Sami sites in northern Scandinavia pickaxes made of reindeer antler have been found. This was probably also a multi-purpose tool, and Odner (1992) associates it with a story written by Knag, the *Matricul*, written in 1694. He wrote that the Sami observed a whale stranding in Mies'kavuodna and the Sami went to the animal by boat with a long piece of wood sharpened to a point, which they stabbed the whale with and killed it (Odner, 1992, 166-167). Whales appear to be generally abundant in the Varanger fjord, with hundreds of whales entering the ford, most of which were probably fin whales, but humpback whales, blue whales, sei whales, bowhead whales, and North Atlantic Right whales were probably abundant too, as well as the beluga and the harbour porpoise. The Sami probably took advantage of strandings of which the pieces were divided along a well-developed tax-system (Odner, 1992, 45-46).

However, in other regions, harpoons seem to be used earlier than the seventeenth century proposed by Lindquist (1995b) for the Norse. Adriaen Coenen in his *Walvisboek* (1585) describes that whalers in sixteenth century Netherlands used a harpoon to hunt whales. They used several small ships to approach the whale and used music and noise to draw the animal closer to the boats. Then a rusty harpoon was thrusted into the whale after which the line of the harpoon was attached to several floating barrels that prevented the animal from diving and fleeing. The whalers then just had to wait patiently for the animal to die.

A similar technique appears to have been used by the Basques. Historical sources dating to the Late Medieval and Early Modern period indicate that the Basques used a barbed iron harpoon attached to a shaft of oak wood. These harpoons were driven into the whale and attached to line were several "drogues" or drags. This prevented the animal from fleeing and made it tired. The tired animal was eventually hit by more harpoons and lances in order to kill it (Szabo, 2012). Then it was towed to shore, where it was immediately cut into several pieces with an axe. The fat was melted in cauldrons on large fires lit on the shore. This was supposedly a smelly business, as the city of San Sebastián forbade the melting of whale fat within walls of the city (Goyheneche Farnie, 1984).

Albertus Magnus speaks a similar harpooning weapon being utilized in the thirteenth century AD, presumably in the northern Netherlands and Germany. He describes a harpoon of which the shaft is fashioned from pine wood and the point of the harpoon is shaped like a sharp barbed arrow to allow easy penetration of the whale's skin. A long rope is knotted to the end of the harpoon (Albert the Great, 1987, 338-342).

To the author's knowledge no clear whaling equipment has been retrieved from a medieval archaeological site. As a large portion of the weapons used served as multipurpose tools this obscures our understanding of medieval whaling and limits our ability to identify whaling centres based on material culture. Furthermore, in the debate regarding active whaling versus opportunistic scavenging it will play a limited role. Only when a spear- or harpoon-head is recovered embedded into a whale bone, it can be argued that active whaling was practiced. Unfortunately, specimens like this are rarely retrieved from the archaeological record. A notable exception is the previously discussed phalanx of a humpback whale with a bone point embedded in it from the Par-Tee site in Northern Oregon (Losey and Yang, 2007).

3.3.4. Boats and Ships

The boats used in the grind are traditional skiffs and fishing boats, that are used for fishing and drive hunting whales. There is not traditional *grind*-boat, again making it hard to identify shipwrecks that were once used in the pursuit of cetaceans. Small vessels were however preferable as they possessed the manoeuvrability needed in the hunt.

It has been suggested that small Norse clinker-built ships, up to six meters, like those accompanied the Gokstad ship burial, were perfect for whale driving. Furthermore, the Skuldelev 6 vessel (a *ferja*), would also have been an appropriate vessel. This vessel measures eleven meters and would have been manned by about twelve to fourteen people. These kinds of vessels were perfect as they do not actually take in the catch but merely drive the animals to shore where they are subsequently butchered (Szabo, 2008, 108-109). However as mentioned before, drive hunting would only have been an option on a coastline with bays or fjords. For other regions, other strategies had to be applied.

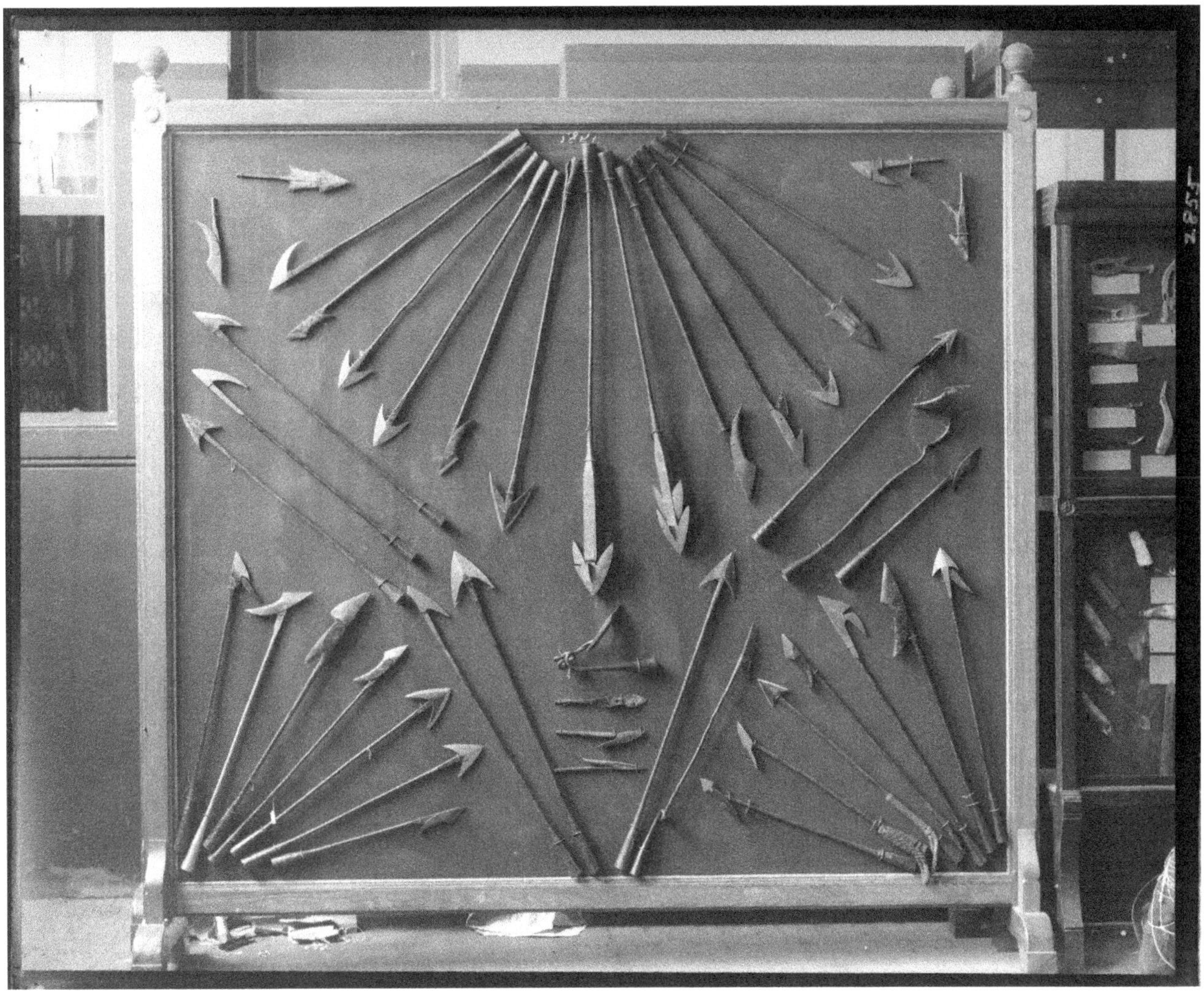

Figure 34. Harpoons mounted for exhibit. Photo created by the United States National Museum Photographic Laboratory. Smithsonian Institution Archives, Acc. 11-066, Box 005, image no. MAH-2855. Image courtesy of the Smithsonian Institution. CC0 License. Available at: <https://www.si.edu/object/harpoons-mounted-exhibit:siris_arc_387364>.

A shipwreck has been discovered in the Mediterranean near Cavalaire, France. The shipwreck has been dated to AD 1480 and had a length of 15.7 meters and was 5.75 meters broad. The fashion in which the ship had been created, including the type of nailing used, suggest it was a Basque ship (Delhaye, 1998). The Basques are known to have had trading enterprises in the Mediterranean (Heers, 1955). This vessel might have been used for that, however on board a large quantity of crossbow bolts and spear tips had been discovered. These weapons might have been used for whaling practices. On board several whale bones were discovered as well, including a vertebra of a young individual. The Basques are known to have frequently targeted young cetaceans. This vessel might have been a whaling ship searching for whales in the Mediterranean. However, the whale bones also show signs of sawing and chopping activities. This might also suggest that the bones were merely tools on board of the ships (Delhaye, 1998). Furthermore, in comparison with the whaling vessels used in Labrador, Canada, the Cavalaire ships seems rather large.

In Red Bay, Labrador, Canada several shipwrecks have been discovered by maritime archaeologist as well. By the sixteenth century the Basques travelled here and set up a port called *Butus* (now called Red Bay). This port was located on the Strait of Belle Isle, which is where the bowhead whales migrated through to get to their calving grounds in Gulf of St. Lawrence and therefore an ideal place to hunt these animals. Several whaling vessels, including at least four Basque galleons and four small whaling crafts were discovered at the underwater archaeological site. The small crafts, known as *chalupas*, formed a major component of the Basque whaling in the region, used for the pursued, killing and eventual towing of the carcass. The *chalupa* had a length of about 8 meters and would have been manned by seven people (including a harpooner, oarsmen, and steersman; Parks Canada, 2017). Its size made it highly manoeuvrable and was perfectly adapted for the hunt on the slow-moving bowhead whales and North Atlantic right whales.

Indeed, the *chalupas* were used for centuries and were first used in the Basque region itself and were launched from the harbours or beaches in the region and could tow the carcasses back to the coast. This technique would leave whale bone material at the coast that could potentially enter the archaeological record (Laist, 2017, 110-113).

Furthermore, these *chalupas* could also be launched from large galleons, creating the possibility to perform pelagic whaling. It has been suggested that this increased the range of the Basque whalers by 1600 kilometres, making it possible to reach Ireland, the English Channel, Great Britain, and potentially even Iceland. The caught whales however had to be butchered on sea, leaving no trace in the archaeological record. Alternatively, the whales had to be butchered at a whaling station, like the one at Red Bay. The galleons primary task was merely carrying the cargo of whale oil exploited in the region and shipping it back, indicating that only the smaller boats were actively used during the hunt (Laist, 2017, 110-113). One of the galleons found at Red Bay, is thought to be the *San Juan*, a 27-meter-long three-master, that was lost in a storm along with a large portion of its cargo in 1565 (Parks Canada, 2017). It is not known when the Basques started to perform this form of whaling, but probably predates 1500 AD.

3.4. Zooarchaeology, Cetaceans, and the interpretation of socioeconomic status

Archaeology for a long time was dominated by discussions that viewed food merely as "diet" or "subsistence" and just biologically necessary. This was especially the case during the rise of New Archaeology and processualism in the 1970s and 1980s. However, modern zooarchaeological studies influenced by postprocessual or interpretive archaeology, are increasingly interested in exploring how politics, ideologies, gender, status, religion, and economies have influence on "foodways" and "cuisine" (Twiss, 2012).

Social diversity is an increasingly important aspects of zooarchaeology and "food", a culturally defined term, makes up an important aspect of it. In many instances, optimization and maximization of energy and nutrition are not the sole drivers of foodways (Hamilakis, 1999). Many studies are focussed on small scale assemblages in order to identify this social diversity (such as the analysis of individual actions, or household specific patterns; Twiss, 2012).

While food is still a biologically necessary aspect of daily life, it is affected by cultural phenomena such as status roles and religious beliefs. Because it such a vital aspect of daily life, it is uniquely well suited for expressing cultural beliefs and ideologies. Food production, food processing, food consumption, and even the discarding of food, are all aspects in which people interact with food, and these aspects are subject to cultural and ideological practices.

Beyond this reconstruction of subsistence, zooarchaeological material is also a valuable tool in the reconstruction of economic differentiation, social strata, ethnicity, racial grouping, gender, ideologies, and religion. In this section special attention will be given to zooarchaeology and the identification of social stratification. This encompasses economic advantages, prestige, as well as political leverage.

Zooarchaeological remains (figure 35) allow to reveal behavioural patterns that were not documented within historical sources. As stated before, food is a biologically necessary aspect of daily life, but it is also a cognitively prominent material culture which allows for the identification of social distinctions (Twiss, 2012). In human society, social inequality is ubiquitous, and this can be noticed in diet as well. The definition of social status is however hard to grasp. It can depend on gender, occupation, political standing, ethnicity, religion and economic standing. The various forms of social status can be inherited or achieved (Ashby, 2002). Diachronic studies of status associated with food has been conducted frequently. In order to assess social status, emphasis is placed on rare, exotic, or labour-intensive products and species (Twiss, 2012). Even butchery marks have the potential to reveal information regarding the social status of the consumers (Foster, 2016, 2).

For a large part of the Medieval Period, society, which was for a major part defined of feudalism, was divided into the three estates of the realm: nobility, clergy, and peasantry. People of the first two realms had a high social status than the people of the peasantry. This can also be noticed in their diet. For medieval Europe the first two realms had more access to wild mammals, while especially the clergy relied more on fish (Ashby, 2012).

This can be assessed by analysing zooarchaeological remains from an archaeological context. In order to access the social status of a site, the field of archaeology relies on a form of spatial analysis known as compositional patterning. Utilizing the analysis, zooarchaeologists assume that zooarchaeological assemblages from a spatially discrete location or context have been used by an identifiable groups or individuals. These groups or individuals are of a particular social status. By analysing the zooarchaeological assemblage, zooarchaeologists might be able to identify the dietary fashion of a particular social class or ethnic affiliation (Driver, 2004).

The utilization of compositional patterning, however, has several shortcomings. Numerous taphonomic factors (like those outlined previously) contribute to the composition of zooarchaeological assemblages. As a result, differences between zooarchaeological assemblages are not necessarily due to social or ethnic differences between the sites but can be the result of taphonomic factors (Driver, 2004; Lyman, 1994). These aspects sometimes hinder the comparison between sites, but often stochastic variation or aberrant results from some sites, is assumed to be compensated for each other as long as the dataset is large enough (Ervynck, 2004).

Therefore, analysis of zooarchaeological remains can be a valuable technique in order to assess the social states of a particular site. The diversity of species in the zooarchaeological record, the relative abundance of the domestic species, as well as the relative abundance of wild taxa are aspects on which social status can be defined. Skeletal element representation and butchery marks can also reflect on the social status (Ashby, 2002).

Especially the ratio of domestic to wild taxa is frequently used as an indication of status in many parts of medieval Europe. Wild animals were acquired through hunting and for many parts of Europe, hunting was restricted to the upper classes. Prestige was particularly placed on wild animals that were rare, mobile, exotic, or even dangerous (Ashby, 2002). The symbolic and social implications of hunting in farming societies is often deemed to outweigh its value as a provider of proteins and minerals (Kent, 1989, 132). This can be ascribed to the high social value placed on meat as a food source (Hamilakis, 2003, 239).

Prior to the Norman Conquest, the hunting of wild boar, bear, and deer was also associated with high status (Wilson, 1973). Indeed, deer bones have been found at high status sites such as Okehampton and Launceston Castle, however Grant (1992) and Crabtree (1990) noticed that high numbers of deer bones were also found at sites with a low status signature. This can be the result of poaching. Zooarchaeological remains might therefore give a false high-status signature.

Deer bones are often found in large quantities at high status medieval sites like Okehampton (Grant 1992) and Launceston Castle (Albarella and Davis 1996). They clearly indicate high status, given the restrictions on hunting noble game discussed above. However, antler is a poor indicator of status, as it may have been collected following shedding, and even butchered antler may have been traded through noble estates to lower class craftsmen (MacGregor 1989). Deer bone, particularly with butchery evidence, is a much more reliable indicator of social position. Interestingly, significant amounts of deer bones are found at sites that are assigned low status based on other evidence (Grant 1992; Crabtree 1990), and so the presence of prestigious animal bone may indicate poaching rather than a high socioeconomic standing.

For medieval England, birds that were particularly valued were partridge, woodcock, swan, plover, and peacocks, while domesticated birds were more associated with peasantry (Albarella and Davis, 1996; Hammond 1998). Thomas (2007) argues that across the era of the Black Death in England nobles began to rely on wild birds as status differentiator, as the lower classes increasingly got more access to buying meat. Furthermore, Sykes' (2004) research indicated that during the Medieval Period the social status associated with swan consumption declined, as the emergent middle classes gained access to them more frequently due to their increased purchasing power.

Analysis of a noble and urban household at Namur conducted by Pigière *et al.* (2004) indicated clear differences between the two. Here the nobility had access to more marine fish, as well as a large variety of wild terrestrial mammals and birds. Species found for example are red deer, brown bear, wild boar, heron, and peacock. Written sources indicate that hunting indeed was a noble privilege and that when these laws were broken by peasants these instances were treated as severe crimes. Furthermore, the nobility had access to larger freshwater fish individuals in comparison to the urban household, which also had access to freshwater species as well as herring.

Interesting to note is that while the consumption of pig is often associated with a high-status diet, this is not the

Figure 35. Selection of zooarchaeological cetacean specimens deriving from a variety of sites in the Netherlands (Photo by author).

case for the wooded environment around Namur, which is very favourable for herding pigs (Pigière *et al.* 2004). This again clearly demonstrates that interpretation of the status of a site based on zooarchaeological data should always be undertaken with caution and should be treated on a case by case basis.

Research on the diet of monastic orders has also been conducted in which the zooarchaeological remains are often compared to dietary rules (De Grossi Mazzorin and Minniti, 1999). In some cases, stable isotope analysis on individuals from different mortuary contexts has been conducted (Le Huray and Schutkowski, 2005). This was for example undertaken on human remains from burials in Greenland dating to the Norse period. The analysis indicated that marine proteins did not make up a considerable portion of the diet of people of a known high status. The remains of a bishop found at Igaluku, Greenland indicated a higher terrestrial protein content. This suggests that the bishop had access to terrestrial food sources, something that is not in line with the Christian dietary practices of that period (Arneborg *et al.,* 1999).

In other zooarchaeological studies, the Christian fasting rules have left their traces. This is for example well portraited in the zooarchaeological assemblage from the former Carthusian monastery Mauerbach in Lower Austria dating to the first half of the seventeenth century AD. Bones of major domesticates were almost completely absent from the assemblage, while fish remains and aquatic animals such as beavers, waterfowl, terrapins, and molluscs are well represented. This food pattern is clearly related to the food rules of the monastic community (Galik and Kunst, 2004, 224).

The comparison between sites might be useful for identification of particular patterns, but people of different social status might have lived together in one settlement and have both produced zooarchaeological refuse that ended up in one context. Van Neer and Ervynck (2004) concluded that interdisciplinary studies, e.g. analysis of ceramics, metal, and glass from an archaeological context in comparison to the zooarchaeological remains will always be vital in understanding the social status of a particular site and stressed that attributing the remains of a particular species (in their case the herring) to a particular social class should be undertaken with caution.

3.5. Conclusion

The identification of cetaceans is hampered by a variety of taphonomic factors, including the often-fragmented state of the material, the large size of the material, and the fact that many of especially the large cetaceans have a similar osteological morphology. Moreover, the lack of extensive osteological cetacean reference collections further limits identification attempts. These factors all hamper our understanding of cetacean exploitation. This makes it clear that there is no uniform answer to the sub-question this chapter was dealing with: "how can zooarchaeological cetacean remains be studied?", and that each cetacean bone should be treated on a case by case basis.

As an attempt to partially solve the identification problems, the Osteological Reference for Cetaceans in Archaeology-Manual (ORCA-Manual) is being created (van den Hurk, forthcoming). This manual will be a valuable tool for the identification of the 35 most commonly found cetacean species in the North Atlantic

Molecular based methods are increasing in interest in the field of zooarchaeology, and aDNA, TEA, and organic residue analysis are all methods that can potentially be useful for the identification of cetacean remains to the species level. Another, recently developed methods, is ZooMS. This method is an on collagen-based identification methods, and has proven to be a cheap, fast, and reliable method. ZooMS analysis was undertaken on several samples discussed in the case studies.

Identification of remains to species level is also valuable in order to understand socioeconomic status of a people at a site. In medieval societies, whales were often deemed to be a high-status food source. Identification of cetacean remains at ecclesiastical and high-status sites, such as monasteries, abbeys, and castles, are useful for the reconstruction of the associated status of cetaceans with this high-status diet.

By far the hardest part of cetaceans in zooarchaeology, even harder than the identification of their remains to the species level, is determining which remains derived from actively caught individuals and which from opportunistically exploited individuals. The species composition of a zooarchaeological assemblage, ethnographic accounts, the presence of suitable hunting tools at a site, and the access to suitable boats and ships, are all potential aspects that might prove that active whaling was undertaken. However, the most reliable indication of active whaling is still the presence of a spear or harpoon imbedded in a bone but known cases of this are extremely rare.

4

Zooarchaeological Research on Cetaceans from Medieval Northern and Western Europe

Excavations at numerous medieval sites located in Northern and Western Europe have yielded cetacean remains. As part of this chapter all data regarding the zooarchaeological cetacean remains dating to the Medieval Period that could be accumulated by the author, were assessed in order the answer the sub-question: "At which medieval sites were cetacean remains found?". Zooarchaeologists have attempted to identify those remains to the species level and identify the skeletal elements, but in only a small number of cases have they been successful.

The collected data was further analysed, and the numbers of species, skeletal elements, and artefacts made of whale bone were assessed to identify patterns in the exploitation of cetaceans and what their raw resources were used for. Furthermore, the contexts the material derived from was evaluated in order to see whether the social elite indeed did try to monopolize the exploitation of cetaceans from the High Medieval Period onwards, as was suggested by Gardiner (1997), or whether cetacean meat was widely available all over Northern and Western Europe.

The acquired data furthermore holds the key in identifying which societies exploited cetaceans. Comparison with previously discussed whaling cultures was undertaken, in order to assess whether the historical and zooarchaeological confirm each other. Furthermore, the zooarchaeological and historical data were compared in order to suggest whether societies relied on active whaling or on opportunistic scavenging of carcasses to get access to cetacean products.

4.1. Methods

As part of this study, zooarchaeological reports concerned with remains dating to the Medieval Period (AD 400-1600) were analysed. These reports were checked on whether the zooarchaeological assemblages contained cetacean remains. All cetacean remains, even small fragments, were considered and included in this study. This data collection was undertaken for the following countries and regions: Iceland, Ireland, the United Kingdom (including England, Wales, Scotland and Northern Ireland, as well the Isle of Man and the Channel Islands (the Bailiwick of Guernsey and the Bailiwick of Jersey), but excluding its overseas territories), Norway (excluding Spitsbergen/Svalbard and Jan Mayen), Sweden, Finland (including the Åland Islands), Estonia, Latvia, Lithuania, Poland, Denmark (including the Faroe Islands, but excluding Greenland), Germany, the Netherlands (excluding its overseas territories), Belgium, Luxemburg, France (excluding its overseas territories), Spain (excluding the Canary islands), Portugal (excluding the Azores and Madeira) and the North-western Federal District of Russia (which includes the Archangelsk Oblast, Vologda Oblast, Kalingrad Oblast, Leningrad Oblast, Murmansk Oblast, Novgorod Oblast, Pskov Oblast, Republic of Karelia, Komi Republic, the Nenets Autonomous Okrug and the federal city of Saint Petersburg). Greenland, though under considerable Norse influence during a portion of the Medieval Period, was excluded from this study as the extent of this study was restricted to northern and western Europe. Greenland is physiographically part of North America.

Unfortunately, many archaeological publications were not published in English, problematizing data collection. This was especially the case for Russia, for which only one site was considered as part of this study. Languages dealt with as part of this study were English, Dutch, French, Spanish, Portuguese, German, Danish, Norwegian, and Estonian. In order to get access to these and other publications, numerous foreign zooarchaeologists were contacted and provided valuable data. Furthermore, archaeological reports dealing with medieval sites in all parts of northern and western Europe present in the collections of the university libraries of University College London, the University of Nottingham, and the University of Groningen were assessed, looking for sites with cetacean remains. Additionally, the words "whale", "dolphin", "porpoise", "medieval", and "archaeology" were translated for the different languages and then an extensive online search was undertaken based on those translations. Data collection has been extensive but certainly not complete or fully comprehensive and many sites were probably unfortunately not identified.

Most frequently the zooarchaeological reports provide data in a NISP-table (Number of Identified SPecimens), giving an overview of all the species identified. However, in some cases cetaceans are excluded from this. This is primarily the case for the larger species, for which the bone material has been crafted into tools or artefacts. As a result, they are occasionally excluded from zooarchaeological sections of archaeological reports, and instead included in a chapter concerned with tools and artefacts (which often discusses the tools per raw material used for the creation of them, e.g. bone, iron, wood, etc.). This hampered the search for cetacean material.

In general, all the cetacean remains were hand-collected during the archaeological excavations at the site, and only a very minimal portion of the cetacean remains were retrieved by sieving. This can be ascribed to the fact that cetacean remains are large, and furthermore as small fragments of cetacean bones are not frequently identified as such. The only two exceptions identified as part of this study are the sites of Clarendon Centre, Oxford, UK, and Koksijde, Belgium, for which respectively 3 and 6 cetacean specimens were recovered through sieving (Douglas *et al.*, 2015; Zeiler, 2018).

The acquired zooarchaeological data was incorporated into a database. This provides a clear overview of the species exploited, as well as the identified skeletal elements. Furthermore, the locality, region and country the material originated from, the date, site type (rural (small settlements, farmsteads, camping sites, etc.), urban (medium to large sized settlements), high-status (castles, royal strongholds, etc.), ecclesiastical (abbeys, cathedrals, monasteries, etc.), grave contexts (inhumation graves, boat burials, etc.), or other) were incorporated into the database. The structure of the database allows for analysis of the meta-data on both a regional as well as a temporal scale.

All additional information regarding the cetacean specimens (e.g. signs of burning, butchery, working, gnawing, root etching, or any other kind of modification by anthropogenic and naturogenic factors) were recorded as well. The state of epiphyseal fusion (unfused, fusing, or fused) was recorded as well as this can potentially provide information regarding the age of an individual.

4.2. Results

As part of the zooarchaeological study, sites with cetacean remains dating to AD 400-1600 have been identified for all countries considered, with the exception of Latvia and Lithuania. A total of 406 sites with cetacean remains have been identified. An overview of the number of sites per country is provided in appendix I. Especially high number of cetacean remains derived from northern Europe, especially Iceland, Norway, England, and Scotland, while few remains derived from the eastern Baltic area (with the exception of Estonia, although all the material for that country originated from just two boat graves). The geographical distribution of the sites is provided in figure 37 and table 4.

For 56 sites the number of identified specimens could not be assessed. This leaves 350 sites for which this data could be extracted from the archaeological reports. Analysing this data reveals that for a large number of sites merely a handful of cetacean specimens were identified and for 202 just 1 specimen was identified (figure 36).

Additionally, 5456 identified specimens have been recorded for the 406 sites. Although, for 72 sites it could merely be stated that cetacean remains, or a particular species was present within the zooarchaeological assemblage without an actual number of these remains being stated in the zooarchaeological report. As a result, the number of specimens for these 72, could be just one or in the thousands. As part of this study the minimal number of one will be assumed, meaning that the total number of cetacean specimens will be put on 5528, but this number could be much higher. These specimens derived from at least 18 different species. Indicating that just over half of the 35 species present in the North Atlantic Ocean were exploited in medieval Europe, although this number may potentially be higher.

4.2.1. Results: Species

Looking at species representation, it can be noted that 4627 of the 5528 could only be identified as "unknown cetacean" (figure 38 and 39). It is likely that the majority of these 4627 specimens belong to one of the large baleen whales as the material is often described as "pieces of whale bone" or "fragments of whale bone". Without any size estimation those fragments are considered to be "unknown cetacean" as part of this study. This category makes up 83.7% of the total number of cetacean material. "Unknown cetaceans" were identified at 257 of the 406 sites (63.3%). This clearly demonstrates that identification of cetacean material is extremely hard to be undertaken and is hampered by the various taphonomic factors outlined in table 3. This greatly delimits our understanding of cetacean exploitation.

Table 4. Overview of number of identified species (NISP) of cetacean bones and number of sites for the countries considered as part of this study

Country	NISP	Number of Sites	Average NISP per site
Belgium	24	6	4,0
Denmark	34	16	2,1
England	622	65	9,6
Estonia	382	2	191,0
Faroe Islands	5	3	1,7
Finland	1	1	1,0
France	34	19	1,8
Germany	52	13	4,0
Guernsey	1	1	1,0
Iceland	2103	29	72,5
Ireland	49	28	1,8
Netherlands	96	44	2,2
Northern Ireland	7	6	1,2
Norway	1472	105	14,0
Poland	24	2	12,0
Portugal	86	5	17,2
Russia	1	1	1,0
Scotland	412	36	11,4
Spain	11	8	1,4
Sweden	109	14	7,8
Wales	3	2	1,5
TOTAL	**5528**	**406**	**13,6**

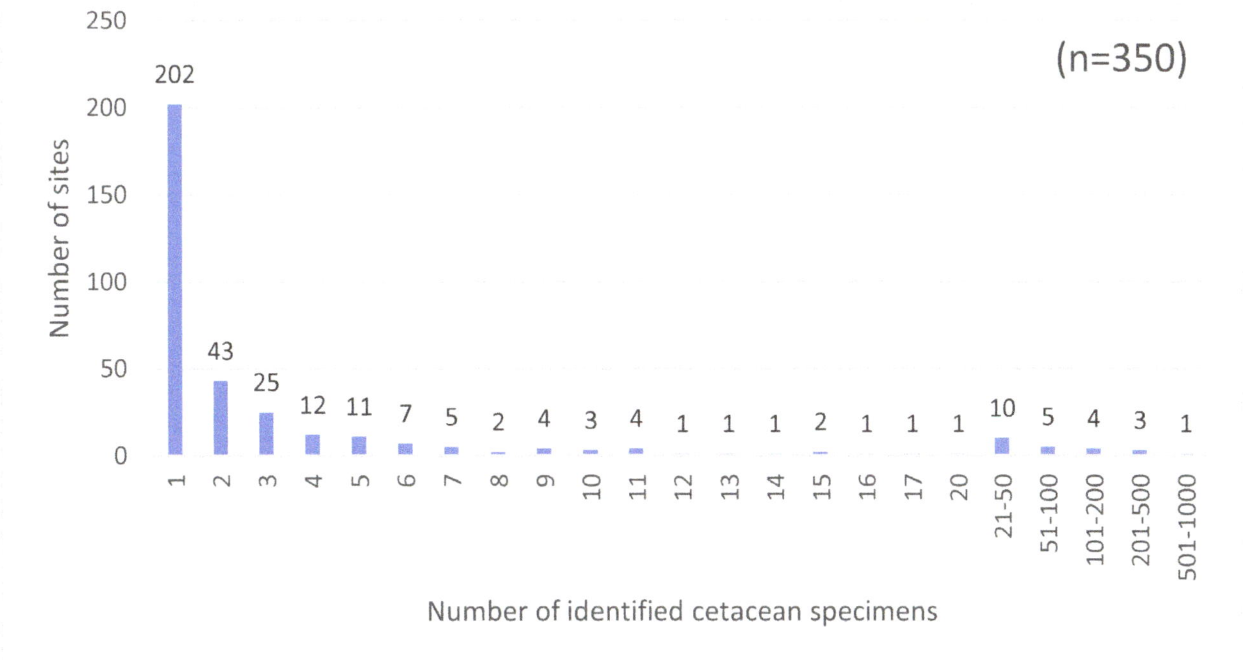

Figure 36. Number of sites with corresponding number of identified cetacean specimens (n=350). This clearly indicates that for a large portion of the sites only a handful of cetacean specimens were identified.

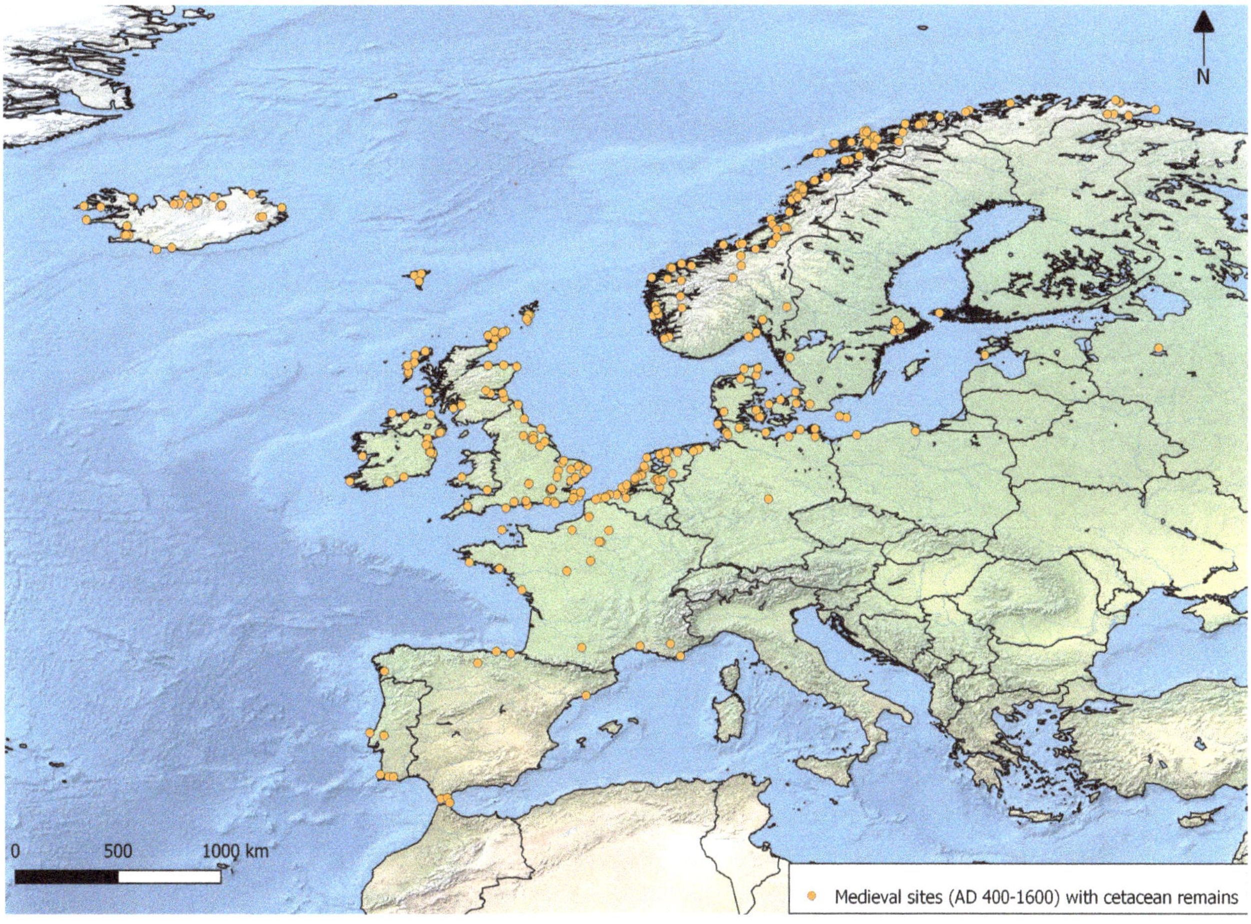

Figure 37. Geographical distribution of medieval sites (AD 400-1600) with cetacean remains deriving from medieval contexts in Northern and Western Europe. Created by author.

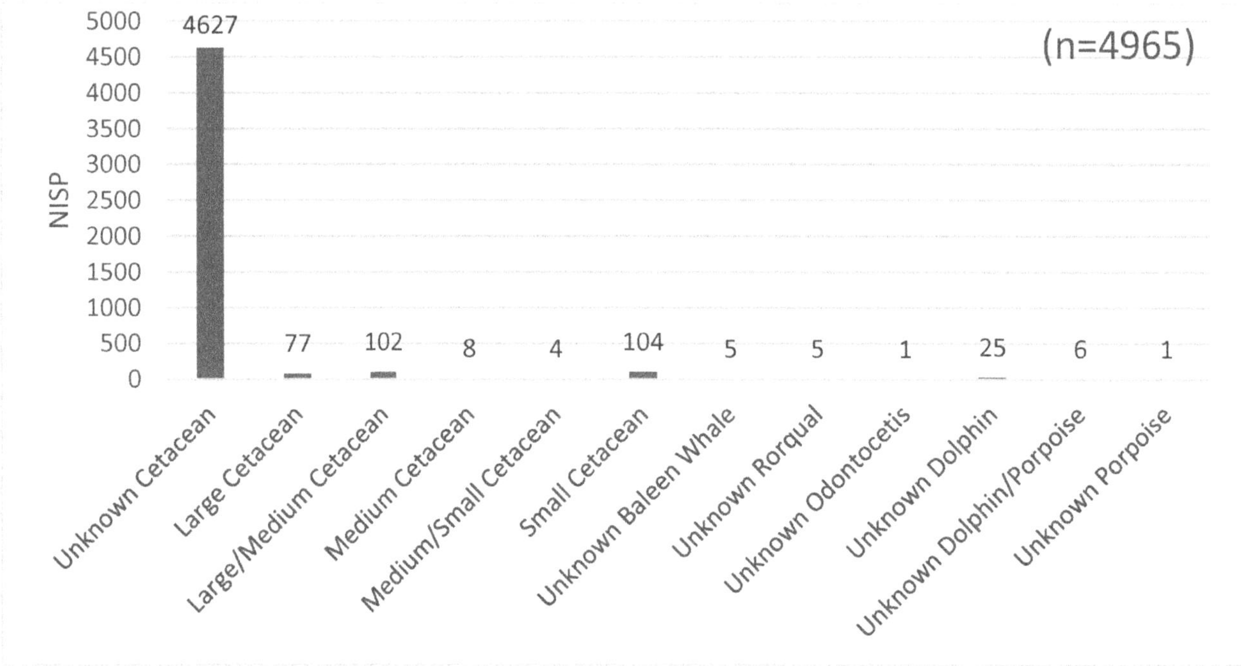

Figure 38. NISP of remains for which species identification was not or only partially possible.

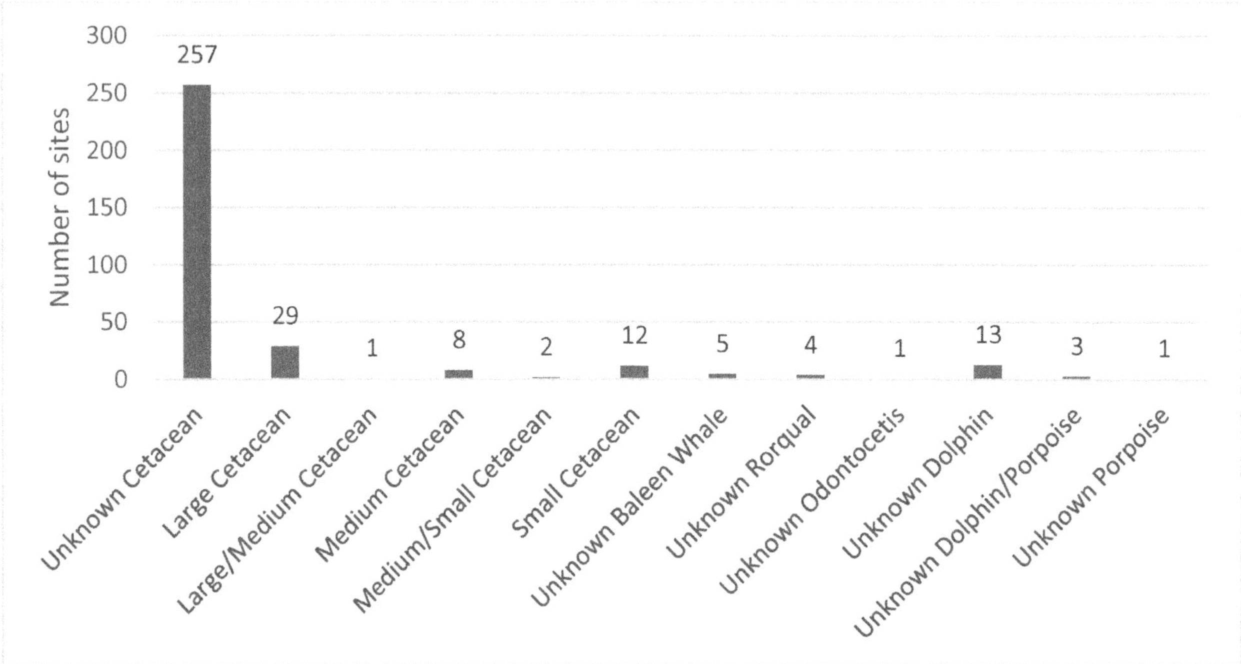

Figure 39. Number of sites at which remains could not be identified to species level.

Furthermore, for several fragments a size indication was provided, resulting in these fragments being included in the "large cetacean", "large/medium cetacean", "medium cetacean", or "small cetacean" categories. Moreover, several fragments could be identified as "unknown baleen whale", "unknown rorqual", "unknown Odontoceti", "unknown dolphin", "unknown dolphin/porpoise", or "unknown porpoise". This last category is an odd one. This specimen derived from Calvert's Building, 15-23 Southwark Street, London, UK (Gardiner, 1997). The harbour porpoise is the only porpoise species living in European waters, suggesting that this "unknown porpoise" must have been a harbour porpoise. Although, the word "porpoise" is sometimes used to included dolphins as well, and it is therefore assumed that this "unknown porpoise" is either a harbour porpoise or one of the smaller dolphin species.

Analysing the remains for the Odontoceti cetaceans (toothed cetaceans; figure 40 and 41), it is clear that the harbour porpoise is the most commonly found species. This species is not only the most common cetacean species in European waters, but also considerably smaller than the other species, making it relatively easy to identify their remains. These factors probably led to the strong representation of the harbour porpoise in the zooarchaeological record.

Two other species that are probably overrepresented as a result of their differing osteological morphology are the sperm whale and the killer whale. These two species are respectively the third and fourth best represented species in the archaeological record based on NISP, and the second and third best represented based on the number of sites. The sperm whale is the only toothed whale that is considered to be a "large cetacean", while the killer whale is by far the largest member of the Delphinidae family. The majority of the identified specimens for both these species are tooth fragments. Their teeth are relatively easy to identify and have often been used for the creation of artefacts, such as the killer whale toothed used as a gaming piece or dice found at Tussøy (Skomsvoll, 2012), or two sperm whale teeth used for the creation of two pawns part of the Lewis Chessmen (Stratford, 1997).

Furthermore, looking at the NISP, the common bottlenose dolphin is also strongly represented, though it should be noted that the majority of these specimens derive from the site of Flixborough (Dobney *et al.*, 2007). The species was only identified at six sites.

The long-finned pilot whale is also strongly represented, though twelve of the nineteen specimens derive from the site of Bermondsey Abbey (van den Hurk, Rielly and Buckley, forthcoming). Interestingly, their remains have also been identified at the site of Kvívík, Faroe Islands, dating to the Norse period (Dahl, 1971). This might be an indication that long-finned pilot whales were already hunted in the Faroe Islands by this time, as it is still being undertaken as part of the *Grind.* Although, the species is known to be abundant in the surroundings of the islands, indicating that stranded species were merely exploited by the Norse settlers on the islands.

Other dolphin species have also been encountered, including the white beaked dolphin, short-beaked common dolphin, Atlantic white sided dolphin, and the Risso's dolphin. The osteological morphology of these smaller Delphinidae species (including the common bottlenose dolphin) is comparable, obstructing the identification of these species.

The number of identified specimens for the beaked whales are low. The northern bottlenose whale is the best represented species, and also remains of the Cuvier's beaked whale and the Sowerby's beaked whale have been

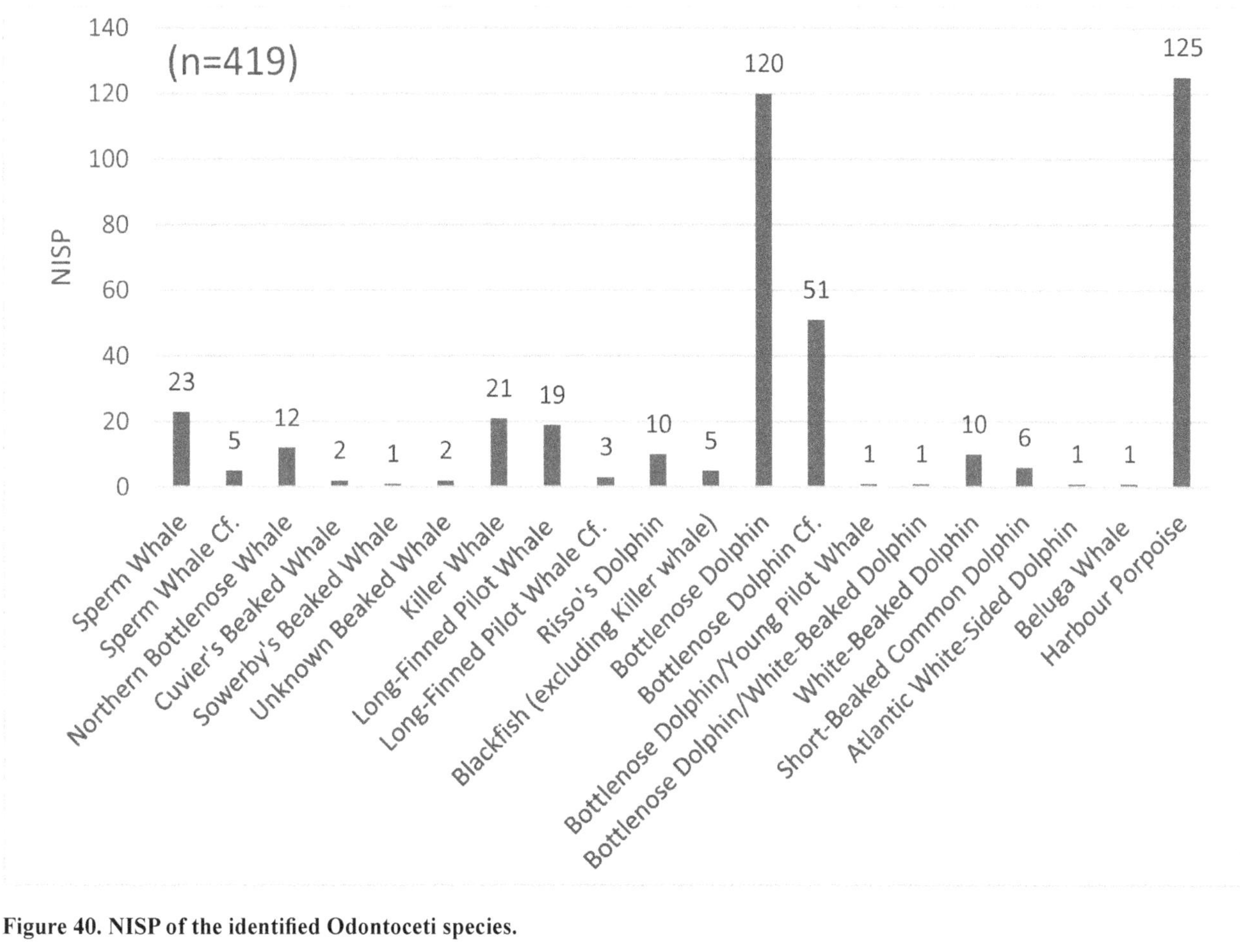

Figure 40. NISP of the identified Odontoceti species.

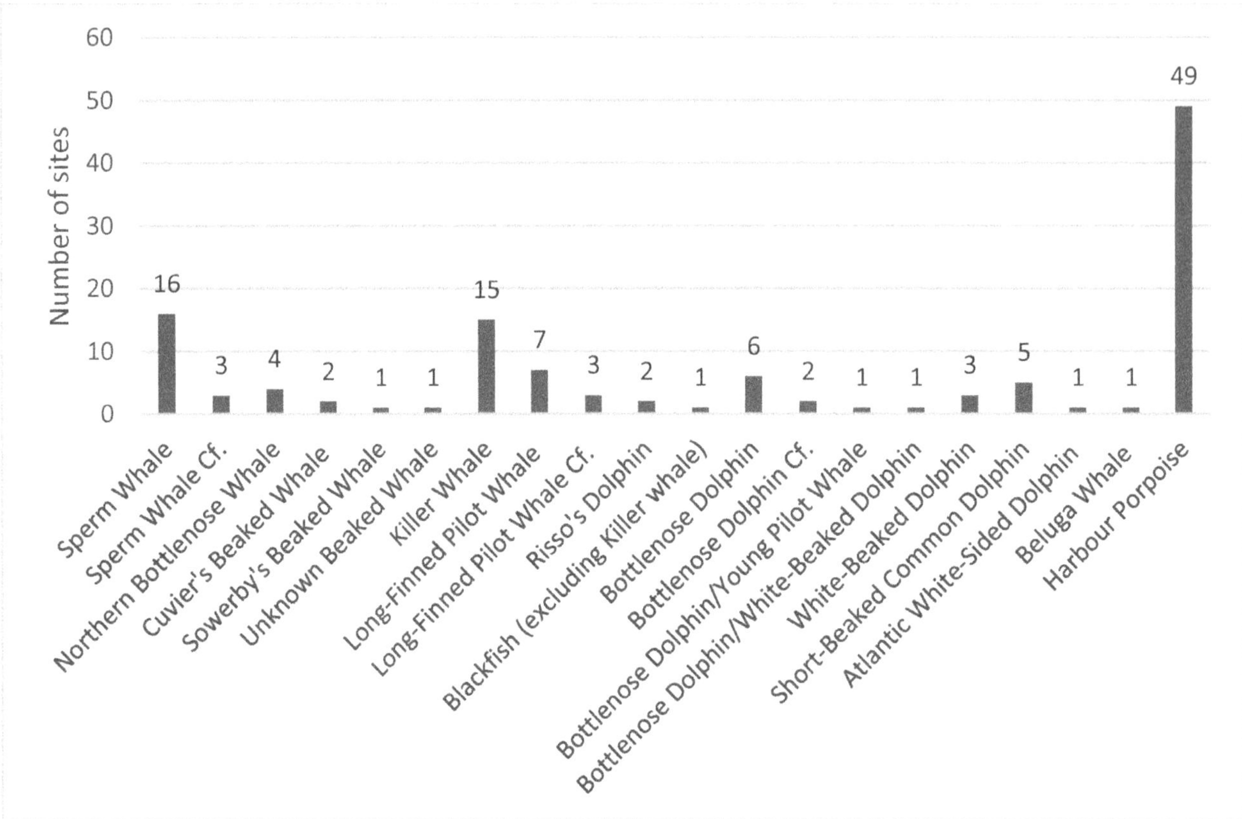

Figure 41. Number of sites the Odontoceti species were identified at

identified. These species are pelagic and are even nowadays rarely encountered and little understood. The specimens identified probably derived from stranded individuals.

Furthermore, one beluga specimen was recovered from Naesholm, Denmark (Mohl, 1961), but narwhal remains have been not been identified even though the tusks are known to have been considered precious artefacts (Pluskowski, 2004). It might however be that the tusks were indeed so precious that they were never discarded and never ended up in the archaeological record.

In regard of the Mysticeti cetaceans (baleen whales; figure 42 and 43), one species predominates: the North Atlantic right whale. Of the 103 identified species, 80 derived from the site of Peniche, Portugal (Teixeira, Venâncio and Brito, 2014). These 80 bones might have belonged to several individuals. This species is known to have been the primary target of the Basque whalers (Aguilar, 1986), and it is likely that other whalers might also have targeted this species. This species tends to float after it has been killed, is a coastal migrating species, tends not to be aggressive, and contains a lot of blubber.

The fin whale is also strongly represented, and remains of this species have been identified at eleven sites. This species is still relatively abundant in the North Atlantic, though industrial whaling has severely depleted the population. It is unlikely this species was frequently exploited by whalers, as the rorquals are known to be fast swimming species. Although, trapping them in a bay or fjord might have been a method in which this species might have been within the reach of whalers. This might have been the case for the common minke whale, and even the sei whale and blue whale as well.

The bowhead whale was probably rarely seen and exploited in European waters as the species is more commonly found in Arctic waters. The humpback whale is also rarely encountered, suggesting this species was not frequently exploited even though it is a coastal species.

Another interesting finding is the presence of the grey whale in the medieval archaeological record. It is assumed that the grey whale went extinct during the Medieval Period on the European side of the North Atlantic and during the seventeenth/eighteenth century on the American side of the North Atlantic. Sub-fossil remains of this species have been relatively frequently encountered, but in recent years ZooMS studies have identified several remains from archaeological contexts as well. A case study will focus on this species.

4.2.2. Results: Skeletal elements

Analysis of skeletal element representation in the archaeological record is a valuable technique to undertake, that allows to investigate whether meat-bearing elements are abundant or rare. This allows to see whether the bones were brought to site for the consumption of the meat still

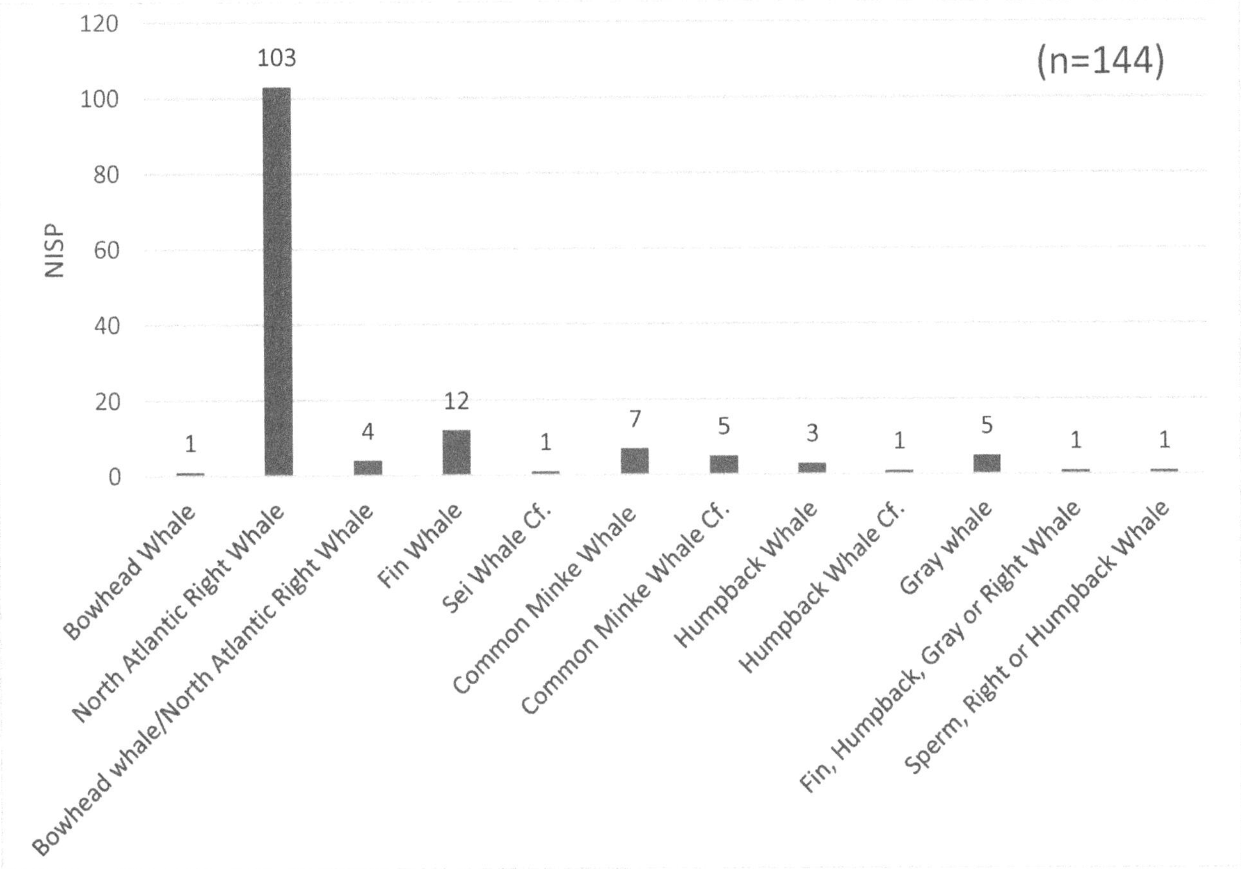

Figure 42. NISP of the identified Mysticeti species.

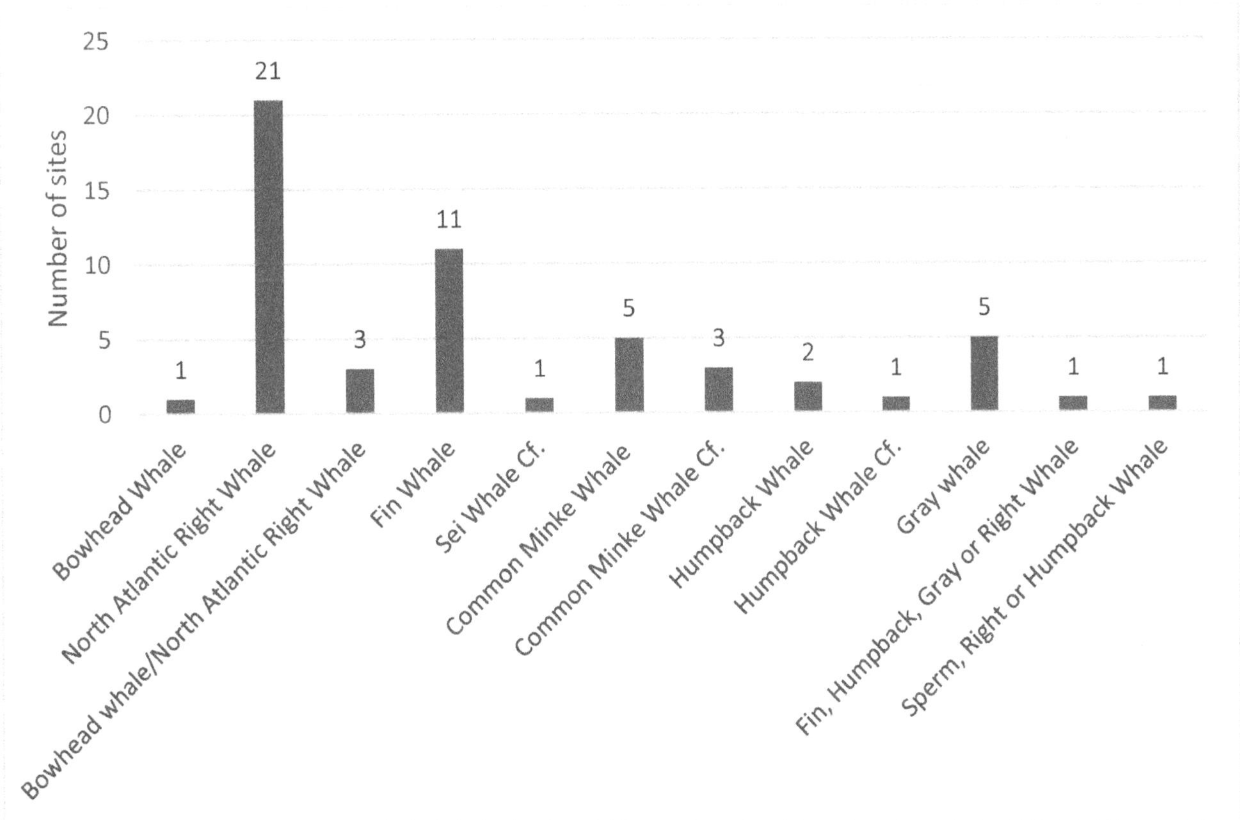

Figure 43. Number of sites the Mysticeti species were identified at

attached to the bone, or whether bones were brought to site for other purposes, such as architectural or artefact/tool creation purposes.

The meat utility index (for harbour porpoise) created by Savelle and Friesen (1996) and for the North Pacific right whale by Omura *et al.* (1969), and the architectural utility index created by Savelle (1997) have proven to be valuable resources in the investigation of past cetacean exploitation. Although these utility indexes are designed to analysis large assemblages, primarily found in the Arctic region. The medieval assemblages assessed as part of this study have generally only contributed very few zooarchaeological cetacean remains, not allowing comparison to the either the meat utility indexes or the architectural utility index.

Moreover, merely 442 specimens (8.0%) of the 5528 specimens assessed as part of this study, were identified to skeletal element (figure 44). This number is potentially higher, but in many of the zooarchaeological reports skeletal element representation was not specified. However, still for a large number the skeletal elements could not be identified, and the specimens were probably large chunks of whale bone not allowing identification of either the skeletal elements or the species. Many taphonomic factors affect whale bone, as outlined in the taphonomy section of this study. Cetacean bone is frequently used as a raw material for the creation of a variety of artefacts and tools, including gaming pieces, weaving swords, combs, and cleavers. As a result of this bone working, identification of the original skeletal element is hampered. Additionally, for many zooarchaeological reports assessed as part of this study, the skeletal elements identified were not provided, even though this information might have been known.

As all cetaceans lack a hindlimb, an overrepresentation of all the other skeletal elements is expected in comparison to most other mammalian species. Based on the number of identified specimens provided in figure 44, it is clear that vertebrae are the most frequently identified skeletal elements. Although, in many of the cases these vertebrae could not be identified to one of the four vertebral sections, with only a couple of vertebrae being identified as either cervical, thoracic, lumbar, or caudal vertebrae. Caudal vertebrae are the most frequently identified. This was expected as these are the most abundant vertebrae in the vertebral columns of all cetaceans. Cervical vertebrae are the least frequently identified, but all cetaceans have only seven of these and for some species several or all of these can be fused together. Even though the number of cervical vertebrae is the same for all cetacean species, there is a large variety in the number of vertebrae between the species, with the white beaked dolphin having up to 94 vertebrae (Reeves *et al.*, 1999a), while the Sowerby's beaked whale only has up to 47 (Martín *et al.*, 2011).

Ribs are also frequently identified, however, just as is the case with ribs of terrestrial mammals, these are hard to identify to species. Furthermore, cranial pieces are also frequently identified, and in some studies, these have been identified as one of the cranial bones, including rostrum, maxilla, and occipital. Although, most cranial elements are fragmented, not allowing the identification of what part of cranium or to what species the bone belonged to.

Pectoral fin bones are rarely identified in the archaeological record. This can potentially be ascribed to the fact that the pectoral fins contain little meat. Of the pectoral fin bones, the scapula is the most frequently encountered element, while metacarpals, carpals, or phalanges have not been

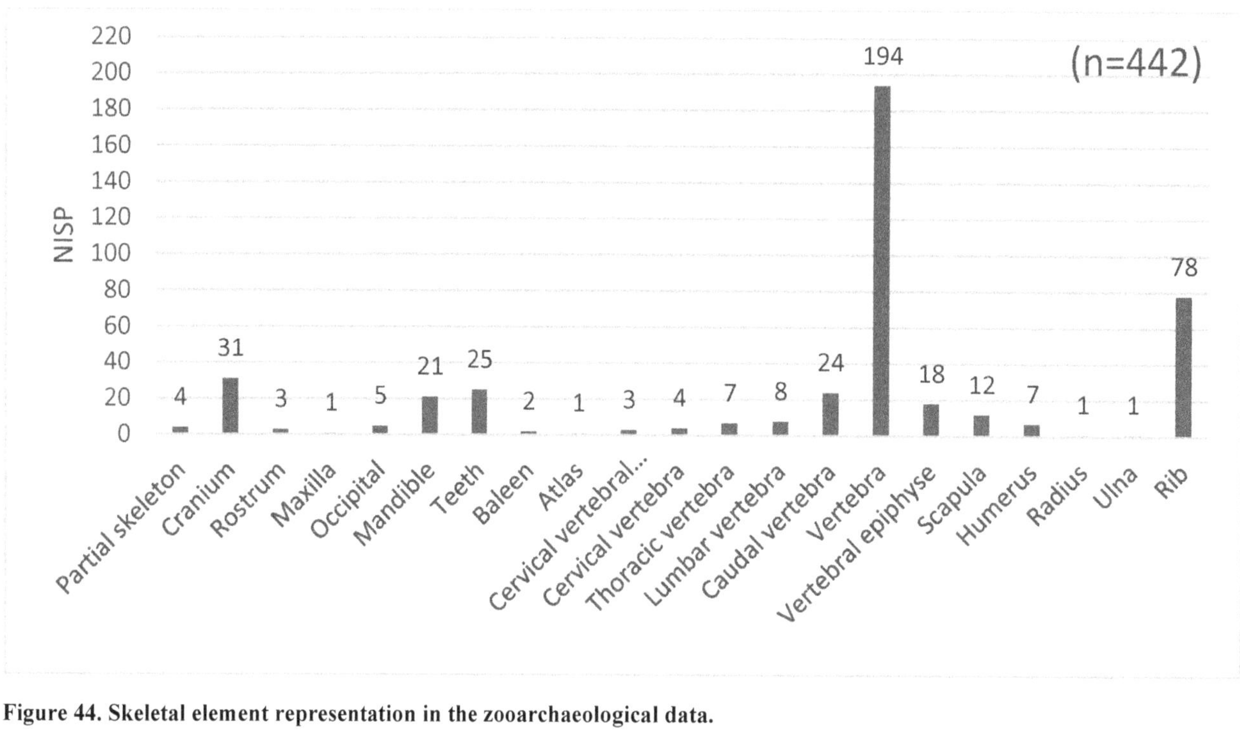

Figure 44. Skeletal element representation in the zooarchaeological data.

identified at any of the sites assessed as part of this study. At Witchampton, Dorset, England several chessmen carved from "flipper bones" were identified, but unfortunately it was not specified which pectoral fin bones these were (Dalton, 1928).

Several teeth were also identified. Most of these belong to sperm whale, killer whale, or Risso's dolphin, though the teeth of several other species have also been identified. The recovered teeth of the sperm whale and the killer whale were in most instances worked and made into artefacts such as four warders and two pawns made of sperm whale teeth, part of the Lewis Chessmen (Brown, 2015, 7).

Baleen, though occasionally mentioned in medieval sources to have been used for the creation of various products, is only rarely recovered from the medieval archaeological record. As part of this study it has only been identified at two sites. Baleen was identified at the Oseberg ship in Norway dating to the Norse period. Baleen was used to fasten the planking of the ship (Shetelig, 1917, 294). Additionally, baleen has also been found at the castle at Perth, Scotland, dating to the AD 1300-1499 (Moffat, Spriggs and O'Connor, 2008). This low number of sites where baleen remains have been found can be ascribed to taphonomic factors that rarely support the preservation of baleen material.

4.2.3. Results: Artefacts and Butchery

Cetacean bone material is frequently used for the production of artefacts and tools. Research on cetacean material has even been conducted in the mountain range of the Pyrenees. From the Upper Palaeolithic period in southern France, several worked cetacean bones, all from Magdalenian contexts have been retrieved. One carved sperm whale tooth was found at the site of Le Mas d'Azil, Ariège commune (Poplin, 1983, 81-94) and 57 worked cetacean bone fragments (including projectile points, foreshafts, wedges and half-round rods) from Isturitz, Isturitz commune (Pétillon, 2008, 720-726). Both these sites are inland sites, showing that the cetacean bone travelled a considerable distance from the seashore where they were likely to be collected from stranded individuals. Further research by Pétillon (2013) indicated that at eleven other sites, cetacean material was also present, suggesting that cetacean material was acquired at the Atlantic coast of France and was transported along the Pyrenees for at least 350 kilometres, indicating it was a valuable raw material.

Artefacts from medieval Northern and Western Europe are often thought to have been made of antler or bone material from large terrestrial mammals. However, recent studies have revealed that artefacts are frequently made of whale bone (Hennius *et al.*, 2018). Of the 406 sites assessed as

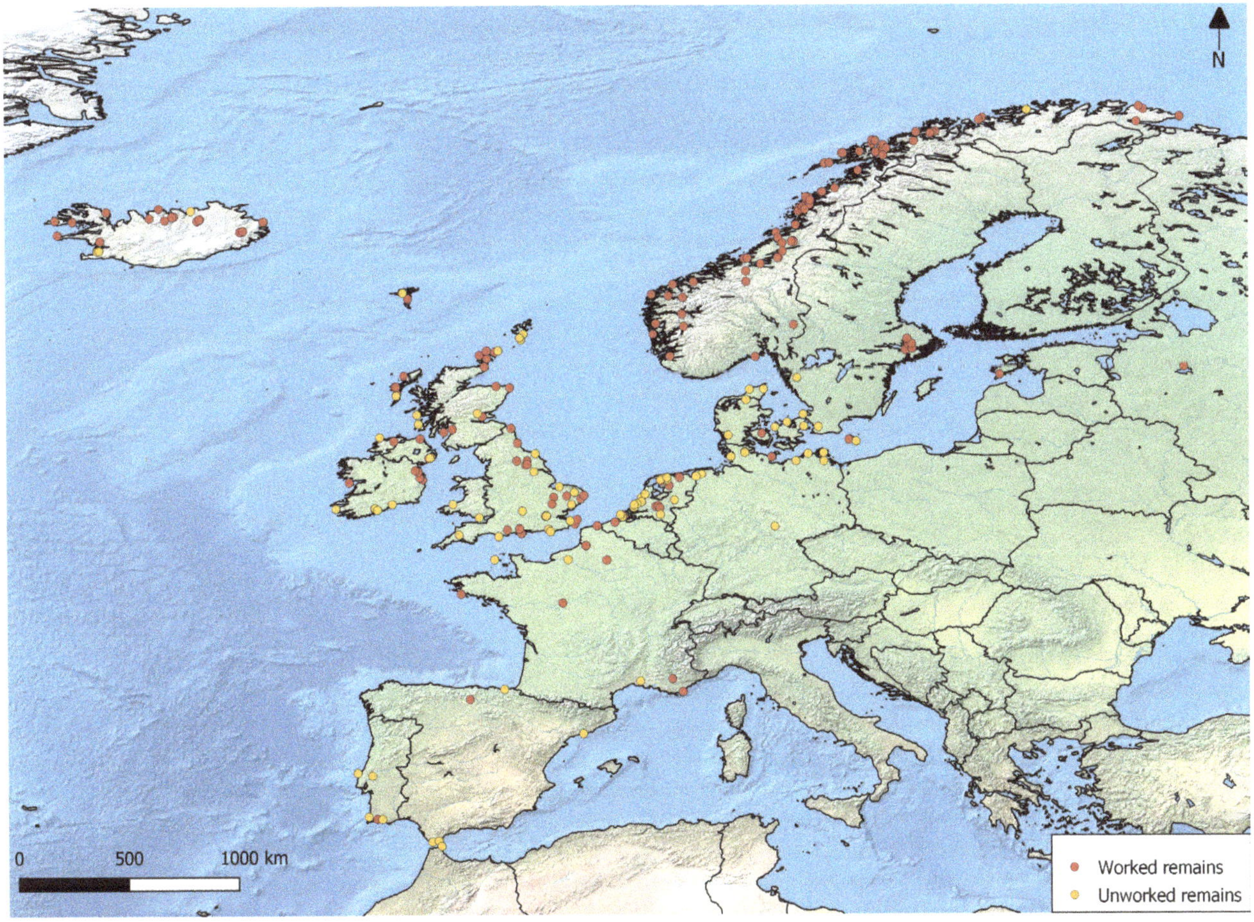

Figure 45. Sites with worked and unworked cetacean remains deriving from medieval contexts in Northern and Western Europe. Created by author.

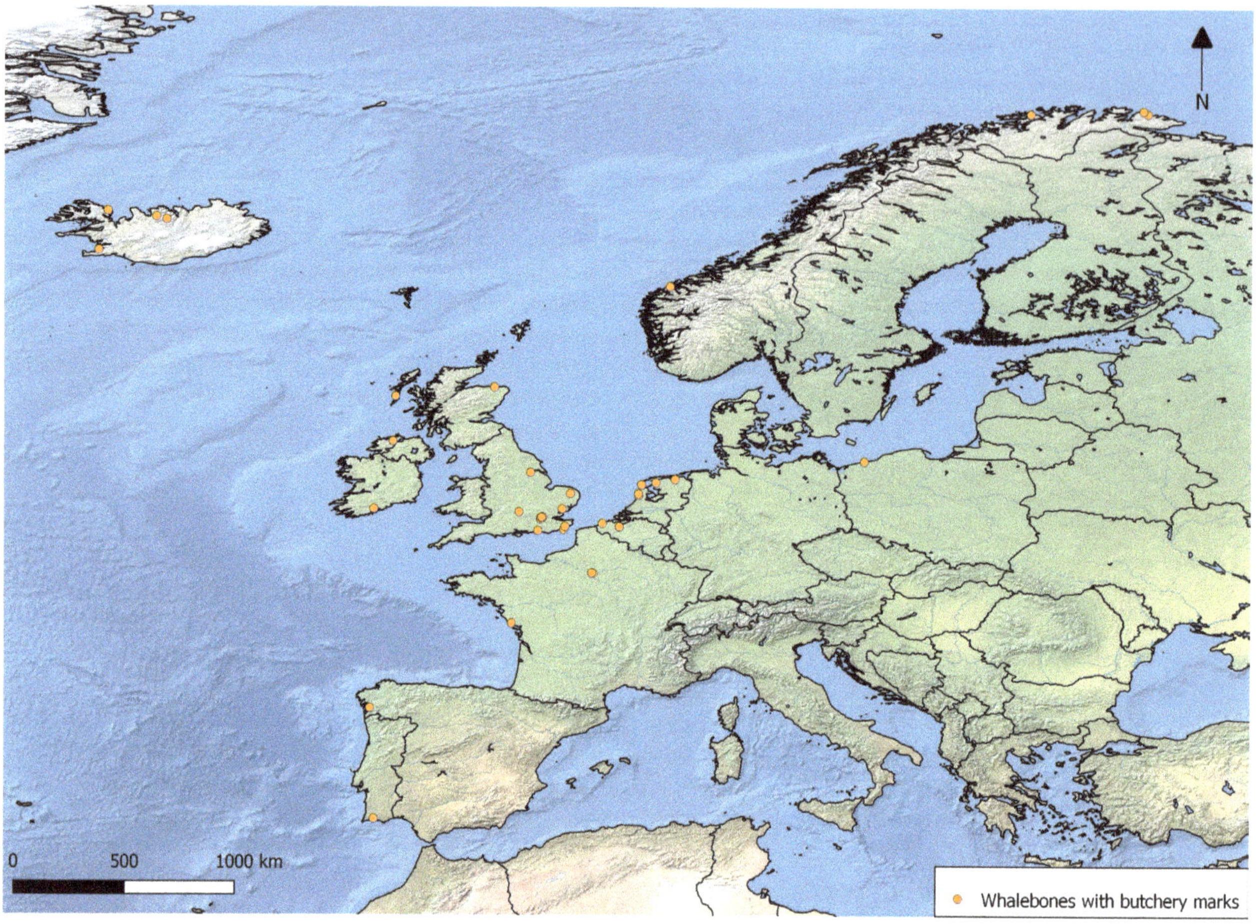

Figure 46. Cetacean bones with signs of butchery deriving from medieval contexts in Northern and Western Europe. Created by author.

part of this study, whale bone used for creation of tools or artefacts have been identified at 189 sites (46.6%), indicating that whale bone was frequently used as a raw material. Artefacts made of whale bone are especially predominant in the former Norse region, with many worked remains coming from Norway, the Shetland Islands, the Orkney Islands, northern mainland Scotland, and parts of Ireland (figure 45). For Norway the majority of these can be dated to the Late Iron Age and the subsequent Norse Period. Whale bone artefacts are rarely found in Early Iron Age contexts (Sjövold, 1974, 249). "Weaving swords", "whale bone plaques", "cleavers", and "forks" are among the most abundant artefacts recovered from Late Iron Age and Norse Period contexts from Norway (Petersen, 1951).

With the creation of tools made of whale bone, whale bone working debris is also created. This debris is also occasionally identified in the archaeological record. However, the signs left on the bones might also be caused by butchery practices. The distinction between signs caused by butchery and those caused by the creation of tools and artefacts is hard. Butchery signs have been identified at material from 45 sites (11.1% of the total number of sites; figure 46). These signs have been identified at a variety of specimens and species, though most of these have either been identified at large pieces of whale bone (which might be misidentified as signs of butchery and are merely signs of bone working), while also several have been identified at dolphin and porpoise species. Bones of these smaller species are not frequently used for the creation of tools and artefacts, making identification of butchery signs easier.

Nevertheless, signs of butchery were only identified at a very small portion of whale bones assessed as part of this study. Mulville (2002b, 40) noted that it is possible to butcher a whale, without removing or coming in touch with any of the bones. This can potentially explain the lack of butchery marks identified.

As part of this study a variety of artefact types made of cetacean bone material has been identified. The creation of several artefact types, such as whale bone plaques and combs, would have required detailed craftsmanship and a lot of time. Other types of artefacts such as chopping blocks, architectural features, wedges, and "notched implements" (identified at Foshigarry and Bac Mhic Connain (Hallen, 1994)), would have required barely any skill and might have been created *ad hoc*. Based on this a clear distinction can be made on artefact types that can be identified as being "high-status" artefacts and those "non-high-status". However, for several regions, especially in the Norse region, both "high-status" and "non-high-status" whale bone artefacts have been recovered, indicating that whale bone was not a raw material used exclusively for the creation of "high-status" artefacts or tools, but was exploited and used for both.

4.2.3.1. High Status Artefacts

A large variety of high-status artefact types can be identified from the archaeological contexts. Some of these were elaborately decorated or would have required detailed craftmanship such as whale bone plaques, combs, and caskets, while for others the context denoted that these were likely to be high-status artefacts, such as the recovering of several weaving swords, cleavers, and again whale bone plaques from high-status graves in the Norse region. The artefacts described in this section are identified as high-status artefacts.

4.2.3.1.1. Gaming Pieces

Gaming pieces appear in Scandinavia during the Roman Iron Age (first to the fourth century AD). The gaming pieces have been associated with the game of Hnefatafl. Hnefatafl is a board game in which a centrally located king is attacked from all sides by the gaming pieces. These gaming pieces are from the Roman Iron Age period onwards strongly linked to high-status weapon graves and are made of ivory, horse, teeth, glass, amber, but most commonly bone or antler. Most of the Roman Iron Age and the subsequent Migration Period pieces are made of elk antler or thick bones of horse or cattle. However, during the transition to the Vendel Period around AD 550, the gaming pieces are made larger and taller, and during the subsequent Viking Period (after AD 790) the pieces are made increasingly spherical and it has been suggested from these periods, whale bone was the most common raw material used (Gustavsson, Hennius and Ljungkvist, 2015).

A recent study by Hennius *et al.* (2018) analysed hnefatafl gaming pieces from more than 200 sites from Sweden and the Åland Islands dating to the Vendel period (550-750 AD) and the Viking Age (750-1050 AD). Most of the gaming pieces originating from Sweden are from the central-northern part, especially around Lake Mälaren. Hennius *et al.* (2018) noted that whale bone gaming pieces are often larger (around 30mm) in comparison to gaming pieces from terrestrial mammalian bone or antler (around 20mm). However, since whale bone is relatively porous, they are less elaborately designed. It appears that whale bone gaming pieces were introduced on a large scale shortly after AD 550 but were replaced as raw material of choice by walrus ivory around AD 1000.

Hennius *et al.* (2018) focussed on whale bone gaming pieces. Most of the gaming pieces discussed (150 of the 200) came from burials spread all over Sweden, the Åland Islands, as well as fifty finds from the medieval town of Sigtuna dating to 300-1200 AD.

The study argued that during the Roman Iron Age or Migration Period (200-550 AD), gaming pieces were associated with elite graves, suggesting it was a prestigious commodity. Whittaker (2006) has suggested that gaming pieces can be associated with high-status people as it can be explained in terms of "conspicuous leisure", in which the social elite distinguishes themselves by performing non-productive activities, such as game playing. This presumably changed during the subsequent Vendel and Viking age periods, during which "common people" were also buried with gaming pieces (Hennius *et al.* 2018).

However, a recent discovery in Estonia, suggests that also during the Vendel and Viking period, gaming pieces were still associated with the social elite. In Salme, Estonia two boat graves were unearthed dating to the AD 550-799. From this boat grave four dice made of dolphin teeth were recovered, as well as a staggering number of 328 gaming pieces made of whale bone (personal communication, Jüri Peets, 2015). Cetaceans (especially large whales) are rare in the Baltic Sea, so access to whale bone was limited. The fact that 328 whale bone gaming pieces were buried as part of this mass grave suggests that it was a high-status grave.

Gaming pieces have been discovered at another boat grave. During excavations at a high-status boat grave at Scar, Orkney, Scotland, UK, 22 whale bone gaming pieces were discovered as well as the well-known whale bone plaque. These pieces have been dated to AD 895-1030 (Owen and Dalland, 1999). Furthermore, during excavations at Washingtonstreet, Cork, Ireland an Anglo-Norman style gaming piece was recovered from a late twelfth century context (Kelleher, 2003). Additionally, at Gufuskálar in Western Iceland, three gaming pieces were found, of which two were made of haddock bone and one of whale bone. These have been dated to AD 1300-1499 (Pálsdóttir and Sveinbjarnarson, 2011).

Moreover, two gaming piece draughtsmen was recovered from Freswick Links, Caithness. It is not clear for what game these pieces were used, but the pieces are circular in shape and decorated with circles. Batey (1987, 342) noted that it was made of whale bone and suggests that similar pieces were found at Bryggen, Norway and Southampton, UK, but it remains unclear whether these were also made of whale bone. The pieces can also not be precisely dated and are dated to AD 0-1299.

In regard to the gaming pieces of Sweden, Hennius *et al.* (2018) have suggested that the whale bone came to all parts of Sweden all the way from northern Norway as part of a trading network. In northern Norway at several sites, slab-lined pits used for blubber processing have been found from archaeological contexts. Gaming pieces have also been recovered from northern Norway. Skomsvoll (2012, 132) has reported the finding of a killer whale tooth from Tussøy, Tromsø kommune, Troms which might also have been used as a gaming piece or dice.

It has been suggested that already from the sixth century onwards the Sami in northern Norway performed active whaling on the North Atlantic right whale and extracted the meat and blubber and transported the whale bone to more southern regions in Scandinavia where it was used for the creation of gaming pieces, but also weaving swords and bone plaques. It has been argued that there is a correlation between the appearance of whale bone gaming pieces and the appearance of other artefacts made of whale bone (such as plaques and weaving swords). This suggests that around the time these artefacts appear, whaling might have been practiced facilitating the need for whale bone material.

Indeed, during the sixth century, an increase in the exploitation of non-agrarian products can be noted for northern Norway, for example fishing or the hunting for terrestrial mammals. The ninth century *Voyage of Ohthere* also seems to suggest active walrus hunting practises being undertaken in northern Norway. In the *Voyage of* walrus tusks were part of a tax system, indicating that value was placed on it (King Alfred the Great, 1855, 9-11).

While Pre-Viking Period gaming pieces have also been found in Hamwic, Prittlewell, and York, UK, these are not made of whale bone. Gaming pieces made of whale bone seem to be primarily restricted to Scandinavia, though the gaming pieces of Prittlewell have been suggested to be made of whale bone and also the pieces from Hamwic and York need to be analysed better in order to assess whether these are made of whale bone (Webster, 2011; Hennius *et al.*, 2018). The large quantities of whale bone from northern Norway, suggest that whale bone was acquired there and transported to more southern regions of Scandinavia for the production of artefacts (Hennius *et al.*, 2018). It has been suggested that whale bone artefacts were already constructed in Northern Norway, as no working debris have been found in Sweden, but whale bone working debris is not found anywhere else in Scandinavia either, making it impossible to determine where whale bone artefacts were actually constructed.

Whale bone gaming pieces of another kind have also been found at other sites in Europe, including France. Chazottes (2017) analysed the use of animal remains (including antler, ivory, horn, but also coral, pearls, tortoiseshells, and whale bone) in the Provence region, France for the Medieval Period up to the modern period. Her study showed that whale bone was used for the creation of gaming pieces as well. These gaming pieces are different to the Scandinavian gaming pieces, as they are flat discs and not hemispherical pieces.

At the castral site of La Moutte, Allemagne-en-Provence two elaborately decorated gaming pieces were discovered. The gaming pieces date to the end of the tenth century or the beginning of the eleventh century and were probably used for the game tric-trac, a form of backgammon. The first one depicts two large birds of prey and the other a four-legged creature (Chazottes, 2017). The context indicates that these gaming pieces were probably used by the social elite of the castral and were highly valued.

Other tric-trac gaming pieces from France have been identified at Rue Saint Wulfran and Quai de la Point, both located in Abdeville, Somme, as well as the site of Tours, Indre-et-Loire and the site of Compiègne, Picardy (Gaborit-Chopin and Bardoz, 2005, 407; Motteau, 1991, 138; Riddler, 2014). This shows that gaming pieces made of whale bone were present at various regions in tenth, eleventh, and twelfth century France. These gaming pieces were of a later period than most of the gaming pieces from Sweden and have a very different appearance suggesting that they did not originate from that region. It remains however unclear where the raw material to create these pieces of derived from.

Other tric-trac, or tabula or backgammon gaming pieces have been identified at Arnemuiden Hazenburg II, Middelburg, the Netherlands dating to AD 1100-1199 (van Dijk, 2018, 87-91). A gaming piece from Townwall street, Dover, UK was dated to AD 1150-1250 (Sabin, Bendrey and Riddler, 1999). Furthermore, two sites in Germany have yielded gaming pieces. The first site is the high-status site of Oldenburg, Schleswig-Holstein (AD 793-1066). At this site gaming pieces of both whale bone as well as walrus ivory have been identified (Gabriel, 1988). The second site is the site of Plessenstrasse, Schleswig, Schleswig-Holstein (AD 1100-1199; Ulricht, 1984, 58). Another gaming piece resembling a tric-trac piece was recovered from Helgøygården, Karlsøy, Troms, Norway. This piece dates to AD 1100-1600 (Skomsvoll, 2012).

Another interesting find is a killer whale tooth presumably used as a gaming piece or a dice coming from Tussøy, Troms, Norway. This piece could not be clearly dated but was probably created somewhere between 800 BC and AD 1600 (Skomsvoll, 2012).

The most well-known gaming pieces from medieval Europe are the Lewis Chessmen dating to the twelfth century. The Lewis Chessmen are a collection of 78 chess pieces discovered in 1931 on Lewis in the Outer Hebrides, Scotland (figure 47). The majority is made of walrus ivory, but four warders and two pawns are made of sperm whale teeth. The production of pieces has often been ascribed to the Trondheim region in northern Norway, as a similar, presumable unfinished, chess piece of a similar style was unearthed at the Saint Olav's church in Trondheim in the 1880s. Furthermore, in Lund, Sweden the front feet of a knight's horse similar to the Lewis chessmen knight's horse was found at Lund, Sweden, in 1817 a chess queen was found in a bog in Ireland, and in 1952 another chess queen was recovered from an Inuit camp in Greenland (Brown, 2015, 7).

Another site from which chess pieces have been unearthed is the site of Witchampton, Dorset, England. At the site multiple chessmen carved from the pectoral fin bones of whales have been found. These pieces are dated to the tenth-twelfth century, but it remains unclear where these have been crafted (Dalton, 1928).

Figure 47. Three of the Lewis Chessmen (three bishops (H.NS 24, H.NS 25, and H.NS 26), of walrus ivory. Image courtesy of the National Museums Scotland. / Wikimedia Commons / Public Domain / CC-BY-SA 4.0.

It is interesting to note that many of the discussed gaming pieces find their origin in northern Norway (figure 48). It is possible that this region was a centre of gaming piece production during a large part of the Norse Period and the High Medieval Period. For many of the pieces, whale bone was used, meaning that it was a prized raw material. Pieces produced in northern Norway might have been transported to other Norse regions including Ireland, Sweden, Scotland, and Iceland. However, this theory needs more research to be validated.

4.2.3.1.2. Plaques

Whale bone plaques are flat pieces of whale bone trimmed to a square or rectangular shape between 20 to 30 cm long each way. Many of the whale bone plaques have on one side a pair of torturing stylized animal heads. These heads resemble the heads of horses. Furthermore, many theories have been proposed regarding the function of the whale bone plaques, but the most likely theory is that the whale bone plaques were used as a kind of "ironing-boards", potentially for linen cloth (Sjövold, 1974, 254-256).

Whale bone plaques have been identified at various sites in especially Scandinavia. Isaksen (2012) lists 40 whale bone plaques originating from Norway, 6 from Sweden, 5 from Scotland, 2 from Denmark, 1 from Ireland, and 1 from England (figure 49). These plaques are probably

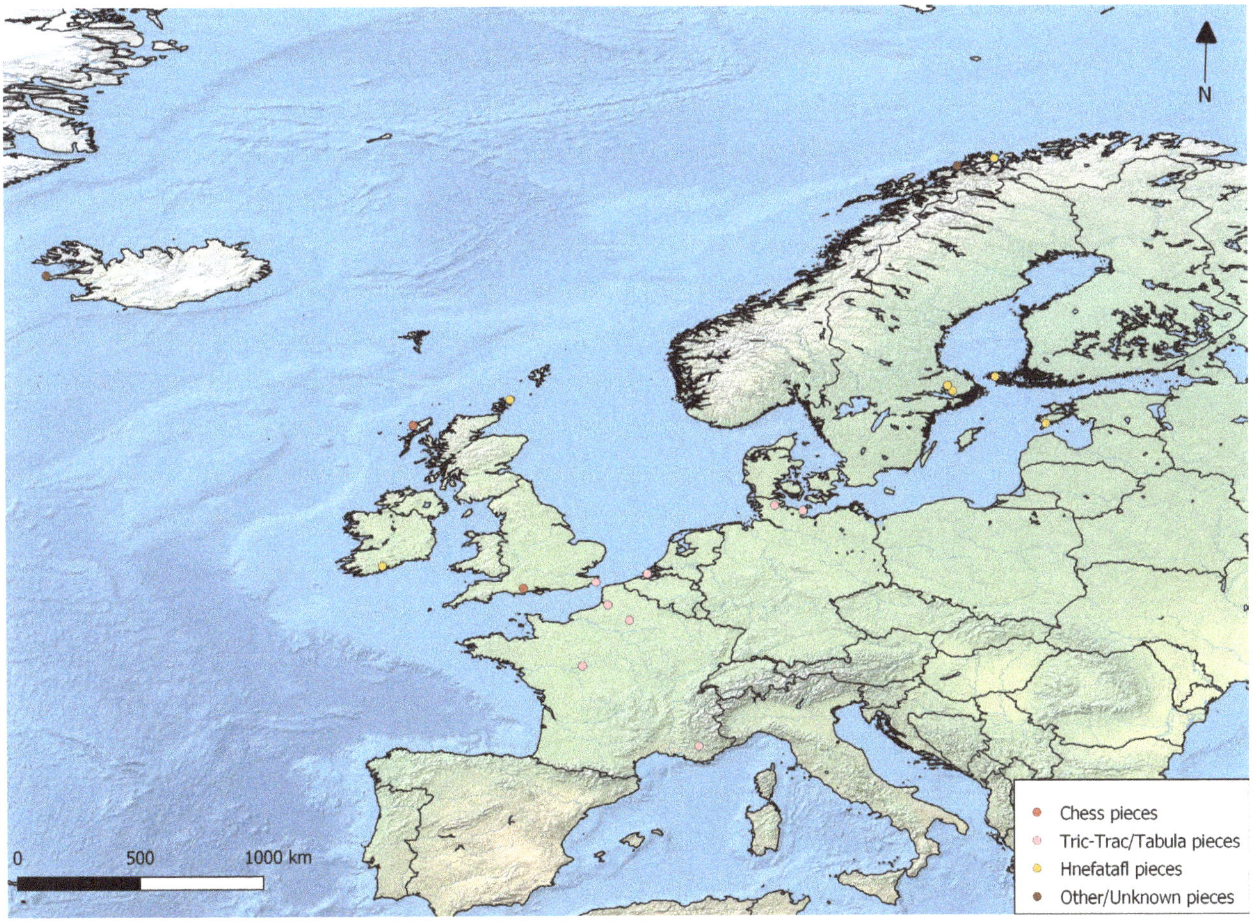

Figure 48. Location of whale bone gaming pieces.

constructed out of the mandibles of large baleen whales. The provenance of the plaque discovered in England is obscure, as it might have been brought to Ely, England by a collector from another region. The remaining 39 are all from Norse areas. The fast majority of the finds from Norway are from Central and Northern Norway. The fact that several whale bone plaques are found outside of Norway, suggests that the plaques were either traded or the tradition was brought to other Norse regions and new plaques were created there

Most of the whale bone plaques from Norway have been recovered from grave contexts, but several are stray finds. Of the plaques deriving from grave contexts, the majority originations from female graves. These graves are often richly furnished, suggesting that they are high-status graves belonging to the economic and social upper strata of society. Linen cloth was probably an expensive product, therefore the deposition of whale bone plaques in high-status graves is in good accordance with the theory that whale bone plaques were also perceived as a high-status artefact (Sjövold, 1974, 254-256).

Sjövold (1974) suggested that the plaques were first developed in the Merovingian Period and their final shape developed in the Late Merovingian Period. This shape was retained until the end of the Norse Period. Unfortunately, most of the raw data regarding the whale bone plaques is not provided by Petersen (1951) or Sjövold (1974). Isaksen (2012) provides a lot of data, but the dates for most of the specimens are not provided. As a result, all the whale bone plaques are incorporated into the zooarchaeological database created as part of this study, and wherever possible the dates were extracted from either Petersen (1951), Sjövold (1974), or Isaksen (2012), but for several the dates are not provided and for an even larger group the context is not known, not allowing a chronological overview of whale bone plaque deposition in graves.

Many of the plaques coming from Norway are fragmented not allowing typological analysis. Sjövold (1974) mentions that of the plaques that could be dated from north-Norwegian gravesites, six could be dated to the ninth century and three to the tenth century. Furthermore, from other contexts one was recovered from a Merovingian Period context (AD 550-800) and several from the Tröndelag County were dated to the eighth century.

Since many of the whale bone plaques are retrieved from rich graves, it has been suggested that the plaques were used as an expression of identity in an area with mixed Norse and Sami populations. Indeed, many of the graves also contained other artefacts associated with the Sami, suggesting an active trade between the Norse and the Sami (Isaksen, 2012). In regard of the gaming pieces, Hennius *et al.* (2018) suggested that the Sami performed

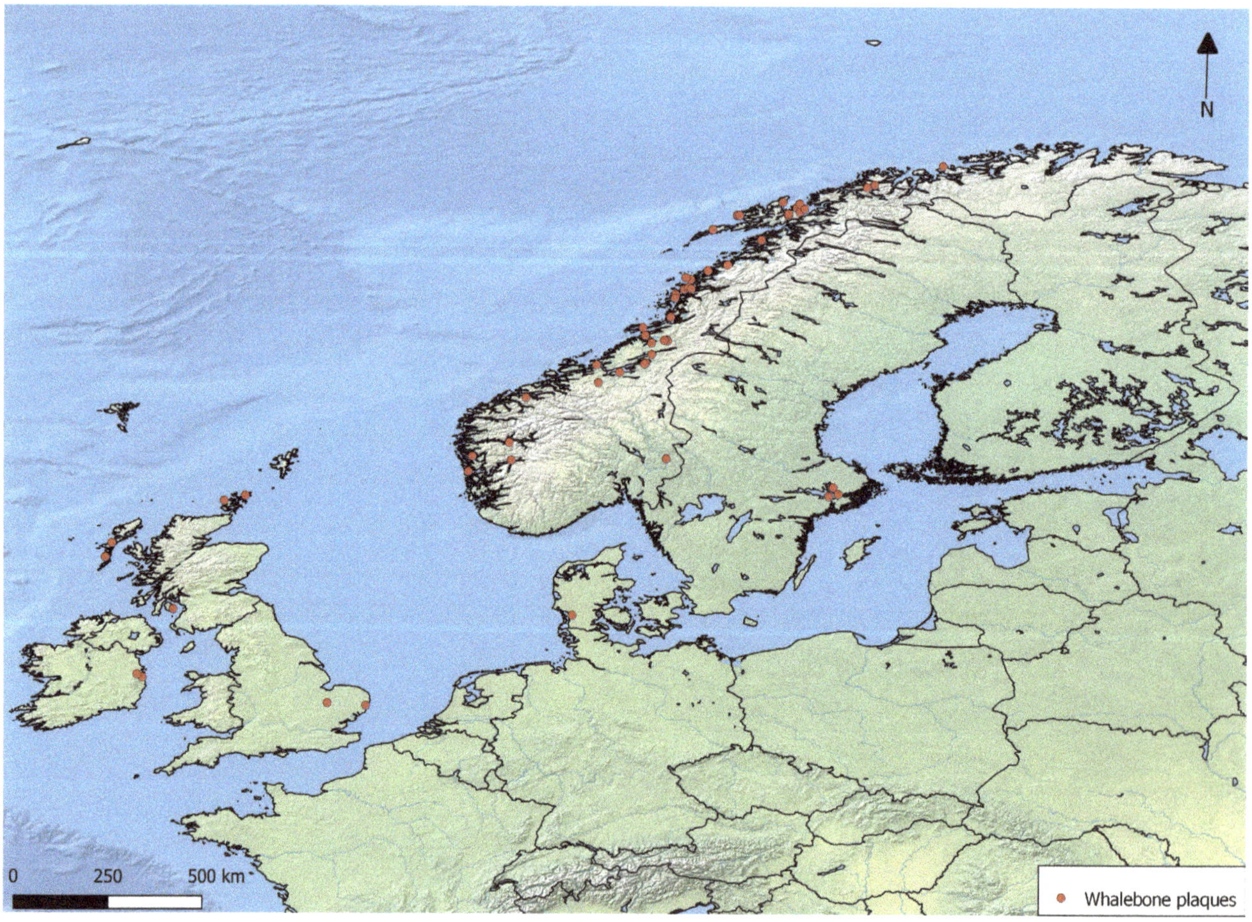

Figure 49. Location of whale bone plaques.

whaling, extracted the whale bone from the carcasses and transported those to other regions. It might be possible that the whale bone exported to the Norse territories in northern and central Norway was used for the production of whale bone plaques. However, it is also possible that the Norse performed whaling themselves in central and northern Norway, or just had waited patiently for a stranding, and extracted the bone from the carcasses themselves.

4.2.3.1.3. Weaving swords

Weaving swords (also known as "weaving battens" or "sword beater") are another artefact type frequently constructed out of whale bone. These swords were used for beating the weft into the warp of the warp-weighted loom (Petersen, 1951). The swords are most commonly found in Viking Age contexts in Norway. They have also been found in other areas of the Norse sphere of influence. Examples are known from Greenland, and one sword was recovered from Kornsá (Mehler, 2007), and yet another one from Quoygrew on the Orkney Islands dating to eleventh-twelfth century AD (Harland, 2012; figure 51).

Two other finds are from the northern part of the Netherlands. These two specimens originate from the terp-sites of Rottum and Leens (figure 50) and are associated with the Frisians. It remains unclear whether these two specimens were brought from Norway to Frisia, or whether the Frisians constructed these themselves. Cetacean remains are occasionally found at terp-sites and have been used for the production of artefacts, however most of these are interpreted to have been used as simple net-weights and do not indicate sophisticated bone working necessary for the production of these weaving swords.

Petersen (1951) identifies 283 iron weaving swords and 73 whale bone weaving swords for Norway. The bone weaving swords are generally longer, broader, and have a longer handle. Most specimens are 60-80 cm long, 4-6 cm in breadth, and have handles 20 to 28 cm long (Petersen, 1951, 290-291). Looking at the dating of the weaving swords (both the iron and the whale bone swords), there are 156 that can be dated. Of these just 13 belong to the Merovingian period (AD 550-800), 89 to AD 800, and 54 to AD 900. However, of the weaving swords made of bone, 4 are dated to the Merovingian period, 8 to AD 800, and 6 to AD 900.

Figure 50. Fragment of weaving sword found at Leens, the Netherlands. Photo by author.

However, Sjövold (1974) provides different data and mentioned just 72 weaving swords. Of these he dates 7 to the Merovingian period and 22 to the Norse Period (of which 8 to the ninth century, 5 to the tenth century, and the remaining 9 could not be dated more precisely), while the other 41 cannot be dated.

Sjövold (1974) argues that whale bone weaving swords are more commonly recovered from the northern part of Norway (especially from Nordland and Troms counties), while in the south weaving swords are more frequently made of iron. Of the 72 whale bone weaving swords Sjövold (1974) lists, 38 were recovered from grave contexts, while the other 34 were stray finds without a clear context. Weaving swords (both whale bone and iron) are most commonly found in female graves (163 known cases) but are also occasionally found in male graves (24 cases). However, of these 24 male graves none of the weaving swords were made of whale bone (Petersen, 1951).

Many of the graves the weaving swords were found in are rich graves with various other artefacts also present. It has been suggested that, just in the case of the whale bone plaques, the weaving swords were an expression of identity in an area with a mixed Norse and Sami population (Petersen, 1951). Moreover, as the production of both the whale bone and iron weaving swords required considerable effort, and some are decorated, and the fact that many of the graves they were found in are high-status graves, the weaving swords might have been a status symbol as well.

Petersen (1951) notes that there appears to be no typological development through time, suggesting that weaving swords were first developed in the Merovingian period (AD 550-800) and remained in use to the close of the Norse Period. Whale bone weaving swords of recent date are also known from the Faroe Islands, and also from western Norway whale bone weaving swords are known to have been preserved until at least 1879, though these were no longer in use (Petersen, 1951).

Unfortunately, most of the raw data regarding these 72/73 whale bone weaving swords is not provided by either Petersen (1951) or Sjövold (1974) and remains deeply buried in (Norwegian) grey literature. Furthermore, both publications are several decades old and more weaving swords might have been recovered recently. As a result, only a small number of weaving swords were incorporated into the database created as part of this study. For a more detailed study regarding the weaving swords, see Petersen (1951) and Sjövold (1974).

4.2.3.1.4. Cleavers

From central and northern Norway several so called "cleavers" have also been found (figure 53). It has been suggested that these were used for chopping meat, but whale bone is not an ideal material for this practice. Petersen (1951) suggests that it was a tool used for tanning. Petersen (1951)

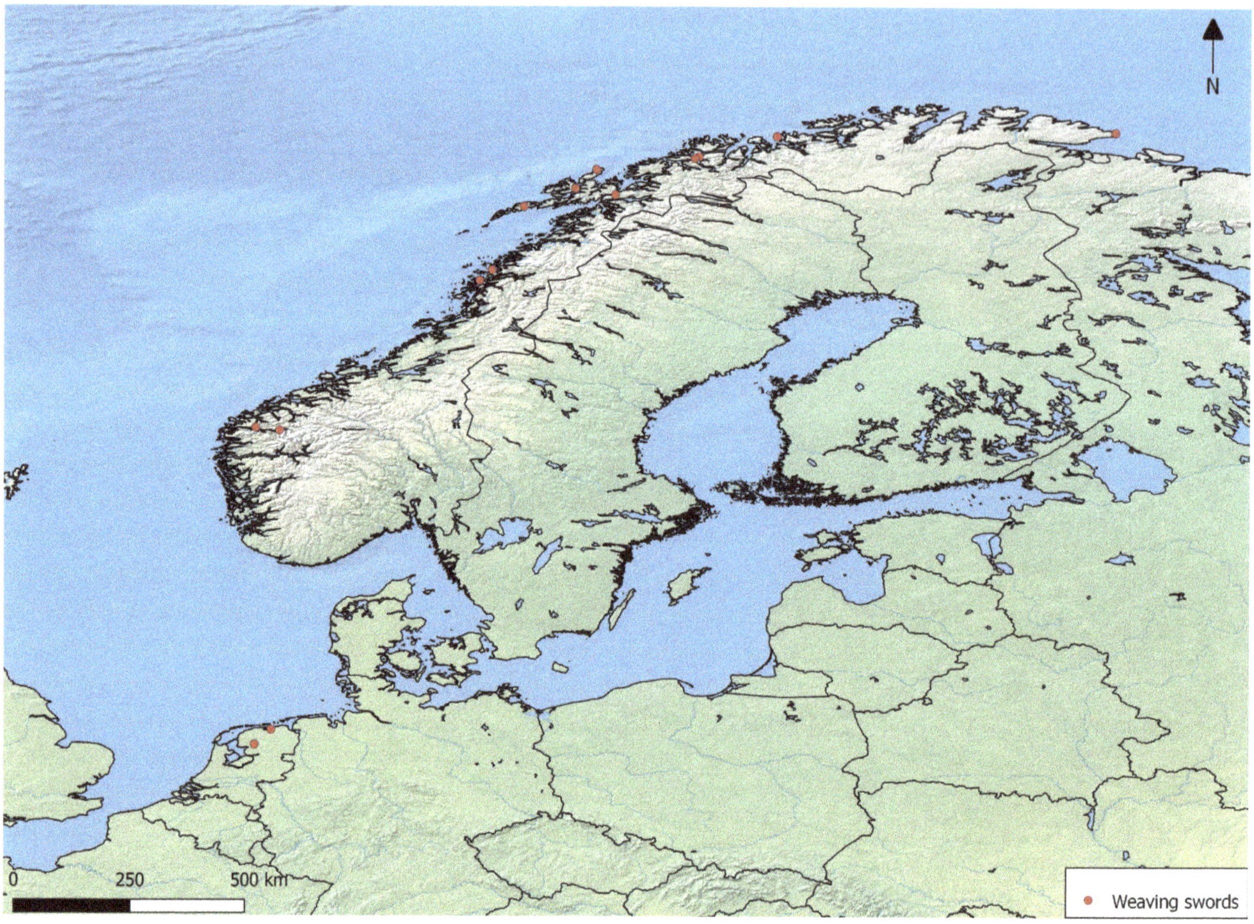

Figure 51. Location of whale bone weaving swords.

and Sjövold (1974) identified several of these, but the data is again obscured by a lack of raw data. The cleavers have been dated from the seventh to the ninth century and just like the weaving swords and the whale bone plaques, these cleavers are frequently found in high-status female graves. Sjövold (1974) furthermore, states that at least three of the specimens are made of reindeer antler, but still the majority of the cleavers are made of whale bone.

Cleavers have also been identified at two other sites outside of Norway. The first one was identified at Drimore, South Uist, Outer Hebrides, Scotland and has been dated to AD 800-1100 (MacLaren, 1974). The second cleaver was decorated with an incised ring-and-dot motif and was found at the Hight Street excavation at Dublin, Ireland and was dated to AD 850-1350 (Ó Ríordáin, 1973, 135-140; figure 52). These two finds therefore also date to the Norse Period, which suggests that cleavers were treated in a similar manner as gaming pieces, whale bone plaques, and weaving swords and were produced in northern Norway and transported to other regions of the Norse sphere of influence.

4.2.3.1.5. Caskets

Several caskets made of whale bone have also been identified in parts of Europe. These have however not frequently been recovered from archaeological contexts and might have been too precious to have been simply discarded. The Franks Casket is an example of one of these whale bone caskets, and was discovered in a private home in Auzon, France. This piece is now on display at the British Museum and has been dated to the Anglo-Saxon period in England (more precisely the early eighth century AD) (Wood, 1990). The texts on the panels of the casket are in Old English and Latin and the scenes display Roman, Germanic, Christian, and Jewish tradition.

Another casket, the Gandersheim Casket, also dates to the eighth century AD, and is now on display in the Herzog Anton Ulrich Museum in Braunschweig, Germany. Like the Franks Casket, the Gandersheim Casket was probably produced in Anglo-Saxon England and transported abroad on a later date (Webster and Backhouse, 1991, 177-178). Yet another interesting find is a whale bone panel fragment, believed to have been a book-cover dating to the late eighth century AD, found near Larling church in the UK (Webster and Blackhouse, 1991, 179).

This clearly indicates that the Anglo-Saxons mastered the working of whale bone and were able to create detailed scenes. The number of these finds are however low, suggesting that whale bone carving might have been a specialized profession practiced by only a few.

Two other caskets have been recovered from Eglinton Castle and from Fife, Scotland. These are however of a

Figure 52. Whale bone cleaver from the High Street excavation at Dublin, Ireland (left) and another whale bone plate from the same site (right) on display at the National Museum of Ireland (photo by author).

later date. The caskets were probably made in the fifteenth or early sixteenth century Western Highlands region. Similar caskets can be observed on grave slabs at Mull and Iona (NMS, 2019a, 2019b). These caskets are however different in shape than the Anglo-Saxon caskets. It might be that the creation of these later caskets was influenced by the Anglo-Saxon caskets, but as few remain it is impossible to test this hypothesis.

4.2.3.1.6. Swords

Sword pommels, grips, or hilts made of whale bone have also been recovered from the archaeological record. Several of these have been dated to the British Iron Age. From South Cave, North Humberside, UK, 3 swords have been recovered. For these swords, at least parts are made of whale bone or sperm whale teeth (O'Connor,

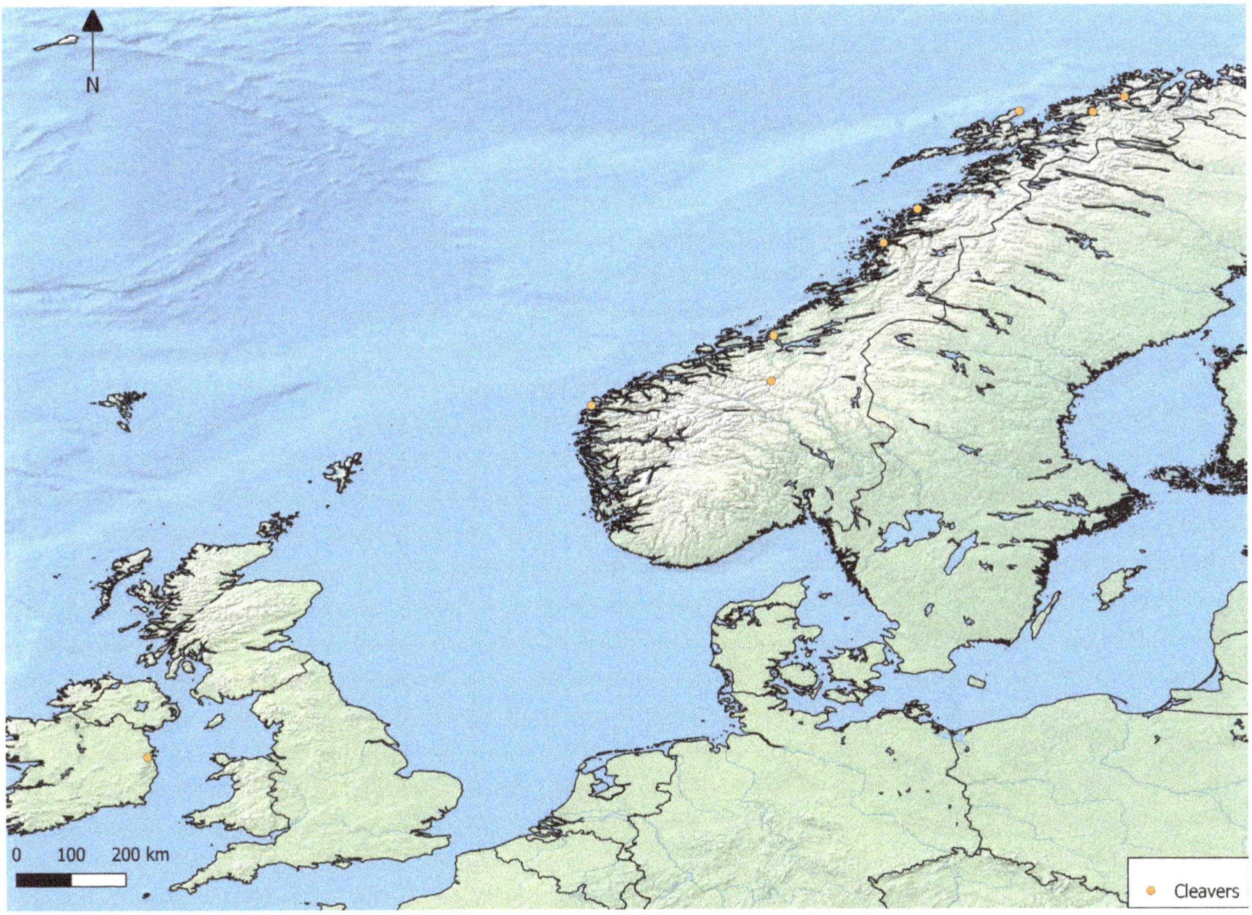

Figure 53. Location of whale bone cleavers.

2013). Furthermore, from the Broch of Gurness, Orkney, Scotland, a sword guard carved from a sperm whale tooth was recovered dating to the Iron Age (Hedges, 1987).

Whale bone sword elements have also been recovered from the Medieval Period. These artefacts are often recovered from grave contexts and are an indication of a high-status burial. One example of this is the finding of whale bone sword hilt from Collierstown, Leinster, Ireland dating to AD 427-606 (O'Hara, 2010, 14). Another sword pommel was identified during excavations at 16-22 Coppergate, York, UK, which was dated to AD 875-975 (Bond and O'Connor, 1999).

Whale bone is only rarely used for the creation of swords and was probably only used for ceremonial swords and where not used for fighting. The creation of swords made of whale bone must have required detailed craftmanship, resulting in these swords being categorized here as "high-status artefacts".

4.2.3.1.7. Combs

Combs made of bone or antler are frequently recovered from medieval contexts all over Europe. In some instances, however, whale bone was used for the creation of these combs. The identification of this material is hard and only a couple of whale bone combs are known. It has been proposed that the site of Hamwic, UK, was a whale bone working centre, where combs, but also other whale bone tools and artefacts were produced (Riddler and Trzaska-Nartowski, 2014).

Whale bone combs have also been recovered from Foshigarry and Bac Mhic Connain (Hallen, 1994), 15-16 Bedfordstreet, London, UK (Riddler and Trzaska-Nartowski, 2013), 16-22 Coppergate, York, UK (O'Connor, 1989), Christchurch Place, Dublin, Ireland (Ó Ríordáin, 1975), and Odense, Denmark (Ørsted, Haase and Collins, 2018).

This suggests that whale bone was widely used as a raw material for the production of combs and it is probable that whale bone was used in more instances for combs, but these might have been misidentified as either antler or bone of other large mammals. Again, the production of this combs would have required detailed craftsmanship and must have been a commodity available to just the social elite.

4.2.3.1.8. "Forks"

Another peculiar type of tools made of whale bone are "forks". Three of these specimens have been identified from Norway. The forks have just two tines. The specimen from Brönnöy was extracted from a male boat grave dating to the ninth century, the specimen from Sömna was a stray find, but probably originated from a grave as well, while the third specimen comes from Rogaland and also originates from a male boat grave dating to AD 900 (Sjövold, 1974). Just like the other whale bone tools discovered in various graves in Norway, these "forks" were probably perceived as high-status artefacts.

4.2.3.2. Non-High-Status Artefacts

Besides being used for the creation of high-status artefacts, in some regions whale bone is also used for other tools and artefacts, not associated with the social elite *per se*. Examples of this are chopping blocks, which appear to have been used not just widespread all over Europe, but all over the world. Moreover, in the Norse region, particularly in regions without a lot of wood, whale bone was also used for the production of a variety of other tools.

4.2.3.2.1. Chopping Blocks

Another aspect that especially the vertebrae of large cetaceans appear to have often been used for are chopping blocks (figure 54 and 55). Chopping blocks have especially frequently been retrieved from areas where wood is a scarce resource. However, chopping blocks also appear in areas where this was not the case. The vertebrae that have been used for these purposes often display chopping and cutting marks on both the cranial and caudal side of the vertebral body. The transverse processes and the spinous process have frequently been chopped off, leaving just the vertebral body.

Besides vertebrae, some cranial elements also appear to have been used for chopping purposes. This is not an

Figure 54. Vertebral chopping block from Dokkershaven, Zeeland, the Netherlands, displaying many chop- and cutmarks. Photo by author.

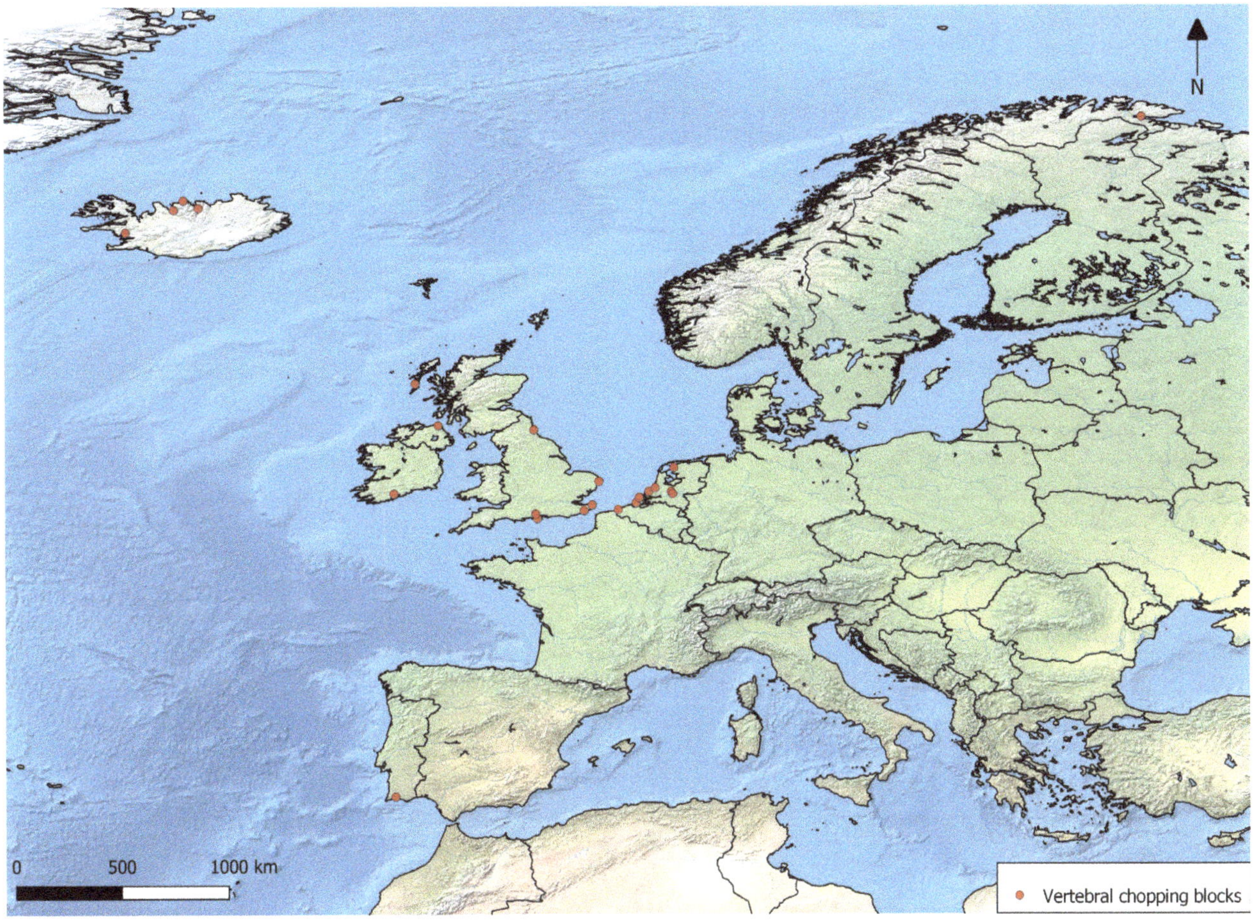

Figure 55. Location of vertebral chopping blocks.

activity restricted to medieval Europe. Archaeological research conducted by Monks (2001) on Toquaht sites on Vancouver Island in British Columbia, Canada indicated that even ulnae, scapulae, mandibles, and humeri were generally used as chopping blocks or cutting boards. At the native American site of Par-Tee in Oregon, USA two whale ulnae also appear to have been used for filleting purposes (Wellman *et al.*, 2016).

Even from more recent sites have skeletal elements of whales with signs of chopping and cutting been recovered. An example is the excavations at Great Island in Cape Cod, USA where a tavern was unearthed dating to the late seventeenth to early eighteenth-century AD. At this tavern a whale vertebra was uncovered displaying multiple chop- and cutmarks (Lombardo, 2010). Furthermore, at the site of Three Saints Harbor on Kodiak Island in Alaska, USA, dating to the eighteenth/nineteenth century AD, at the time under Russian influence, a whale vertebra with cutmarks was identified (Crowell, 1997, 199).

This indicates that the using of vertebrae as chopping or cutting surface is a widespread practice, not exclusive to a geographical region or period. This seems to also have been the case for medieval Europe, though several have been identified in the Netherlands and south-eastern England.

4.2.3.2.2. Architectural Features

An architectural utility index was created by Savelle (1997). This was created based for Thule sites in the Canadian Arctic. Savelle has suggest that crania, mandibles, maxillae/premaxillae and cervical vertebrae were used for the construction of various architectural features. Skeletal elements that have a high meat utility are underrepresented in the Thule archaeological record, suggesting that the skeletal elements that are brought to the site, were used for architectural purposes.

This architectural utility index was created for the Thule region in the Canadian Arctic and was based on osteological features of the bowhead whale. Savelle (1997) has proposed that the an "architectural approach" to zooarchaeological remains from other archaeological contexts might prove useful. However, as zooarchaeological cetacean remains from the fast majority of medieval European sites have minimal osteological specimens of cetaceans, this approach is not suitable for this region. The only exception is the northern part of the research area, including Scotland, Norway, and Iceland where indeed higher numbers of cetacean remains are found.

Olaus Magnus in his *A Description of the Northern People* also describes the practices of people in Northern Europe to utilize osteological remains of cetacean as architectural features (Olaus Magnus, 2010, 1105). He describes houses, walls, doors, roofs, benches, and even tables constructed out of whale bone. He furthermore states that doors were made of whale skin, though this seems unlikely as whale skin is thin. More probable is the usage of seal or walrus skin for this purpose. Olaus moreover states that it is not clear whether the whale bones have been derived from actively caught animals or from stranded individuals that were opportunistically scavenged for their bones. Whale vertebrae were used for the construction of several medieval buildings in Bayeux (Musset, 1964).

However, many of the sites incorporated as part of this study have only yielded a handful of cetacean remains, suggesting that cetacean bone was only rarely used for architectural features. As part of this study, cetacean bone has only been identified to have functioned as some sort architectural feature in a handful of localities. At Akurvík, Iceland, a vertebra of a North Atlantic right whale was used as corner support for a building (Krivogorskaya, Perdikaris and McGovern, 2005) and at Vatnsfjörður, also in Iceland, whale bone was recovered from a foundation context of a building (Edvardssen and McGovern, 2005). Furthermore, at Keel West, Ireland, a cervical vertebra was built into the wall a building. This bone is identified as a humpback whale (Barton, 1943).

Just like as is the case with artefacts, cetacean bone used for architectural features, are mostly predominant in the Norse region. Especially in Iceland, were few trees grow, whale bone probably was a welcome addition as raw material used for construction purposes.

4.2.3.3. Other Artefacts

In numerous parts of the Norse region whale bone artefacts and tools are frequently found. This is especially the case for the Orkney Islands, Shetland Islands, and the Outer Hebrides, all part of Scotland, as well as from Ireland. An example is the finding of several pins, a notched peg, and several more unidentified whale bone artefacts and tools found at the Brough of Birsay, Orkney (Curle, 1982). Furthermore, at Freswick Links, northern Scotland, two whorls, seven snecks, two gaming pieces, and four other worked whale bones were identified (Batey, 1987). Even more worked whale bone was recovered from the excavations at Foshigarry and Bac Mhic Connain, including three unidentified whale bone plaques, fifteen vertebral containers with lids, 35 notched implements, one wedge, one pinhead, one comb, one mirror handle, and several more whale bone artefacts and tools (Hallen, 1994). This material is dated from the Scottish Iron Age to the Norse period and Medieval Period, though few artefacts are dated to the period following the Norse Period.

A variety of these tools and artefact might be interpreted as high-status, while others appear to be *ad hoc* tools not associated with the social elite. This indicates that the Norse used whale bone for the creation of a variety of artefacts and tools, and that the raw material was not restricted to the social elite exclusively. It appears that artefacts that would have required great craftmanship as well as time, were associated with the social elite, while whale bone was also used for simple tools.

The historical records suggest that active whaling was occasionally undertaken in almost all regions of the Norse sphere of influence and the zooarchaeological assemblages suggest that cetacean bone was frequently used for the creation of several tools and artefacts. Although it still remains unclear to what extent the Norse relied on active whaling to obtain whale bone to create these tools and artefacts. There is still a possibility that opportunistic exploitation of stranded cetaceans still supplied the Norse with the greater part of the whale bones.

Furthermore, besides being used for artefact or tool production, a recent study by Hambrecht and Gibbons (2018) on over 3000 individual cetacean specimens from Gröf, southern Iceland, dating to the seventeenth and eighteenth centuries AD, indicated that whalebone was also used for the oil inside as a fuel source. All the identified cetacean bones showed signs of burning, while for other sites, such as Akurvík (a site with also a large number of cetacean bones) only 32% of the specimens showed signs of burning. This clearly demonstrates that whale bone, besides being used as a raw source for the creation of tools or artefacts, was also used as a fuel source by the Norse (Hambrecht and Gibbons, 2018).

In other regions than the Norse region, whale bone was also used for the creation of artefacts and tools, but in much lower numbers. An interesting piece is a decorated belt buckle found at Santa María de Hito, Spain, dating to AD 900-1000 (Gimeno, 1978). Additionally, at Novgorod, Russia whale bone was used for the production of knife handles (personal communication Dr Liubov Holden, May 2016). Another, rather impressive artefact made of whale bone is the "Adoration of the Magi" currently stored in the Victoria and Albert Museum, London, UK (Museum number 142-1866). This piece has been dated to AD 1120-1150 and was probably made in Northern Spain (Victoria and Albert Museum, 2017). It has been proposed that a radius of a rorqual was used for the creation, however, looking at the back side of panel seems to show a curve in the bone, suggesting that it was made out of a mandible instead of a radius.

Furthermore, at Hamwic, UK, numerous whale bone fragments have been identified and it has been proposed that this was whale bone working centre (Riddler and Trzaska-Nartowski, 2014). The case study concerned with medieval England will focus more on Hamwic.

4.2.4. Results: Contexts, Status, and Trends

High concentrations of sites with cetacean remains are located in the Netherlands, western Belgium, eastern England, the Orkney Islands, the Shetland Islands, the

Outer Hebrides, northern and eastern Iceland, northern and western Norway, eastern Denmark, and the Dublin area in Ireland (figure 56). It is surprising that very little zooarchaeological material has been recovered from the Basque region (both the Spanish and French parts), as well as from Normandy. The Basques and the Normans are two of the medieval cultures most frequently associated with whaling. The lack of zooarchaeological cetacean remains deriving from those regions might potentially be explained by fewer archaeological excavations being undertaken in the region, or the Basques and Normans might have had little interest in using cetacean bone remains and might have left those at the shore or potentially thrown them overboard where there is small chance they can be recovered by archaeologists. This has been recorded for example for late nineteenth century Salvador, Brazil. Whalers were forced to throw the bones of the whales they had caught in the sea, away from land, or risk a penalty of eight days in prison (Posturas, 1873; Garcia, 2020, 83-96). Additionally, the site of Strákatangi in Northwest Fjords, Iceland, is historically identified as a seventeenth century whale processing site, possibly of Basque origin. Excavations have revealed no whale bones at the site, however underwater, off the site, whale bone has been identified, revealing that the bone was discarded by the whalers (Hambrecht and Gibbons, 2018). These sources are of a post-medieval date, but the practice of discarding whale bones in the water might potentially have been undertaken during the Medieval Period as well.

Another potential reason why cetacean bone is less abundant in some regions, is because the author was unable to cross a language-barrier. As stated, numerous archaeological reports and studies are still only published in native languages, preventing more detailed intercountry analysis from being conducted.

Recently however, more cetacean remains have been recovered from the Normandy region, dating to the Roman and Medieval Periods. These remains are currently being analysed using ZooMS and will potentially reveal more details regarding the history of Norman whaling practices (personal communications Dr Tarek Oueslati of the Université de Lille, April 2019).

In regard to the Basque region, Grau-Sologestoa and García-García (2018) have noted that in recent years there is an increase in medieval zooarchaeological studies being conducted on sites from the Basque region and Spain as a whole. Grau-Sologestoa, Albarella and Castillo (2016) and Grau-Sologestoa (2018) are examples of zooarchaeological studies conducted on remains from the Basque region, although cetacean remains are not mentioned in these studies. Remains of cetaceans appear

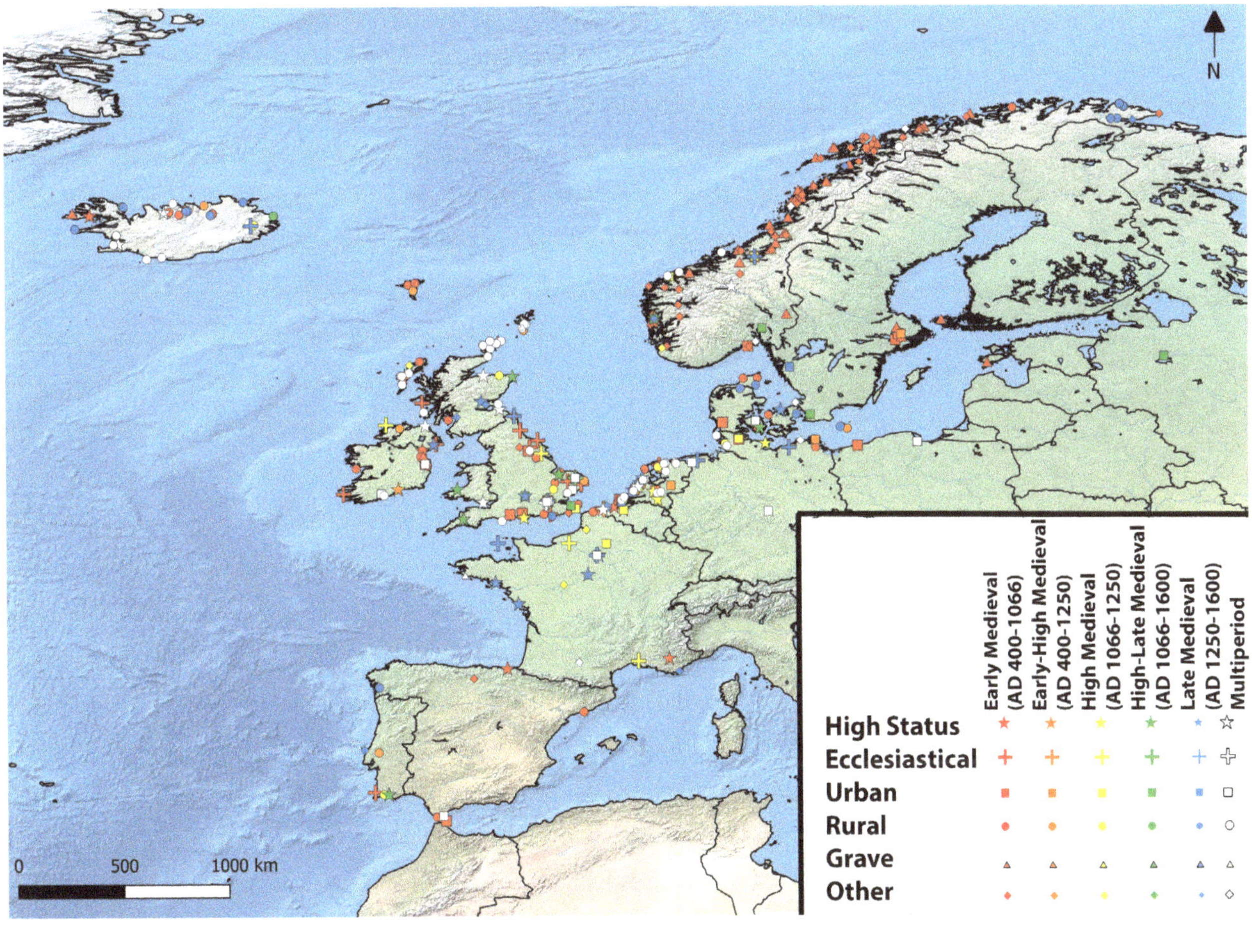

Figure 56. Geographical distribution of medieval sites with cetacean remains.

to be rare in the region (personal communication Grau-sologestoa, 2016). Historical sources seem to indicate that whaling was frequently undertaken and that whaling enterprises were an intergral part of Basque culture, but the lack of zooarchaeological cetacean remains is intruiging.

Though potentially not as abundant in every region, cetacean bones from archaeological contexts and zooarchaeology hold the potential to identify periods in which cetacean exploitation was more important than in others. In order to assess this, the number of archaeological sites with cetacean remains were analysed. Most sites have a dating that spans several centuries, making it hard to identify any temporal patterns. In order to overcome this problem, the dating of the sites was divided in 25-year time periods. Each 25-year period a site covers is subsequently counted as one entry in the temporal data, displayed in figures 57 and 59. For example, a site that covers the period AD 850-924, is counted as one for the periods: AD 850-874, AD 875-899, and AD 900-924. Two graphs were created based on the temporal data, the first one (figure 57) is based on the site types, and the second one is based on the countries the sites are located in (figure 59). Furthermore, figure 61, displays the temporal data of the site types proportionally.

Furthermore, all the sites with cetacean remains have been chronologically plotted by assessing the number of identified specimens and dividing those by the length of the date range of that site to give an estimate of frequency density across that range. This will provide estimated frequency distribution in a similar manner as also performed by Orton *et al.* (2014) whom analysed cod remains deriving from London. In opposition to this method, the sites were divided into site type categories and countries the material derived from. Twenty-five-year intervals from AD 400-1600 were used to produce an overall distribution (fig 4). For example, a site with two identified cetacean remains dating to the period AD 850-924, is counted as 0.33 for the periods: AD 850-874, AD 875-899, and AD 900-924. In order to counteract overrepresentation an entry for a period for a particular site cannot exceed 3. Therefore, a site with 200 specimens again dating to the AD 850-924, would only record 3 specimens for each of the three 25-year periods. This was again done based on site types (figure 58), as well as for the separate countries the sites are located in (figure 60). Figure 62 displays the temporal data of the sites types proportionally.

The overall temporal overview indicates that there is a gradual increase in sites with cetacean remains from AD 400 to AD 950, with an especially rapid increase in site numbers for the period of AD 750-800. Based on the frequency density however there is a decline initiating around AD 750, reaching its lowest point around AD 850, and then rapidly increasing again, reaching a peak around AD 975. It should be noted that AD 950 is the period that is generally perceived as the period the Medieval Warm Period initiated. Mannino *et al.* (2015) have indicated that during periods of climate variability, cetaceans are subject to (mass-)strandings. The peak at AD 950/975 might be explained by an increase in strandings as a result of the climatic variability caused by the Medieval Warm Period. Coastal communities might have exploited the stranded cetaceans.

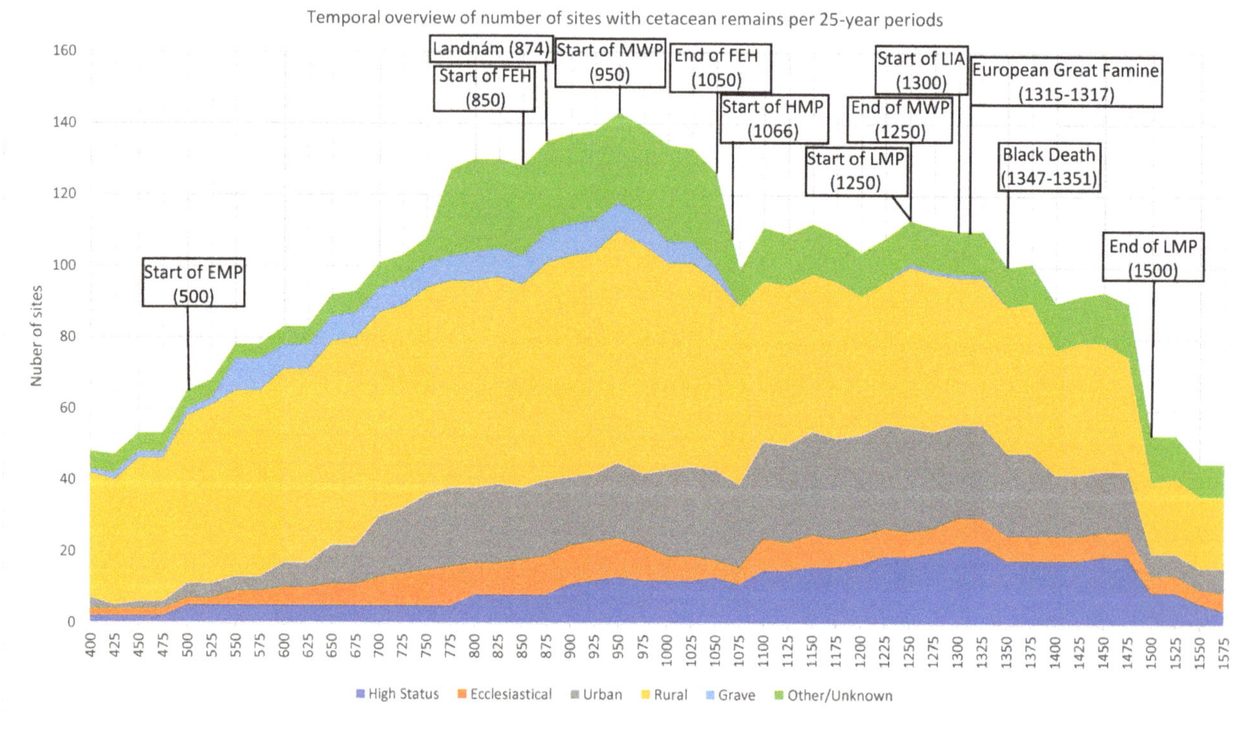

Figure 57. Temporal overview of number of sites with cetacean remains 25-year periods, distribution based on site types. EMP: Early Medieval Period. FEH: Fish Event Horizon. MWP: Medieval Warm period. HMP: High Medieval Period. LMP: Late Medieval Period. LIA: Little Ice Age.

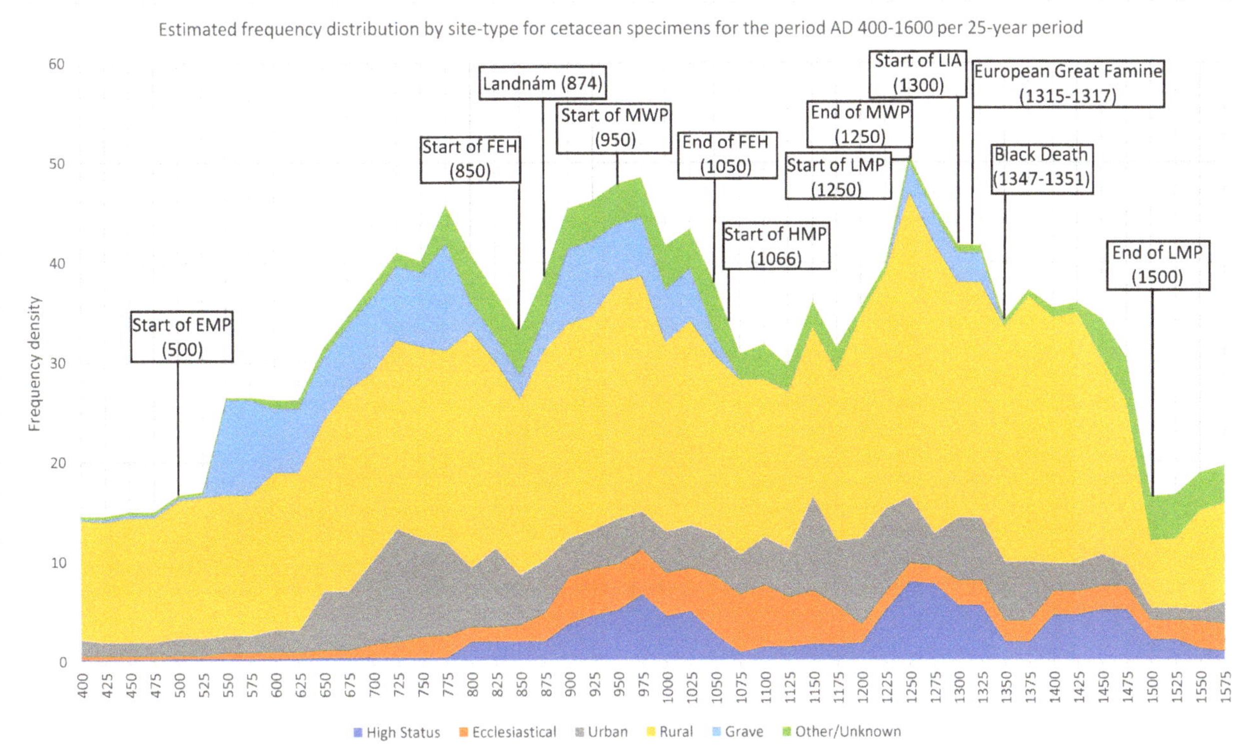

Figure 58. Frequency density overview of cetacean specimens per 25-year periods, distribution based on site types. EMP: Early Medieval Period. FEH: Fish Event Horizon. MWP: Medieval Warm period. HMP: High Medieval Period. LMP: Late Medieval Period. LIA: Little Ice Age.

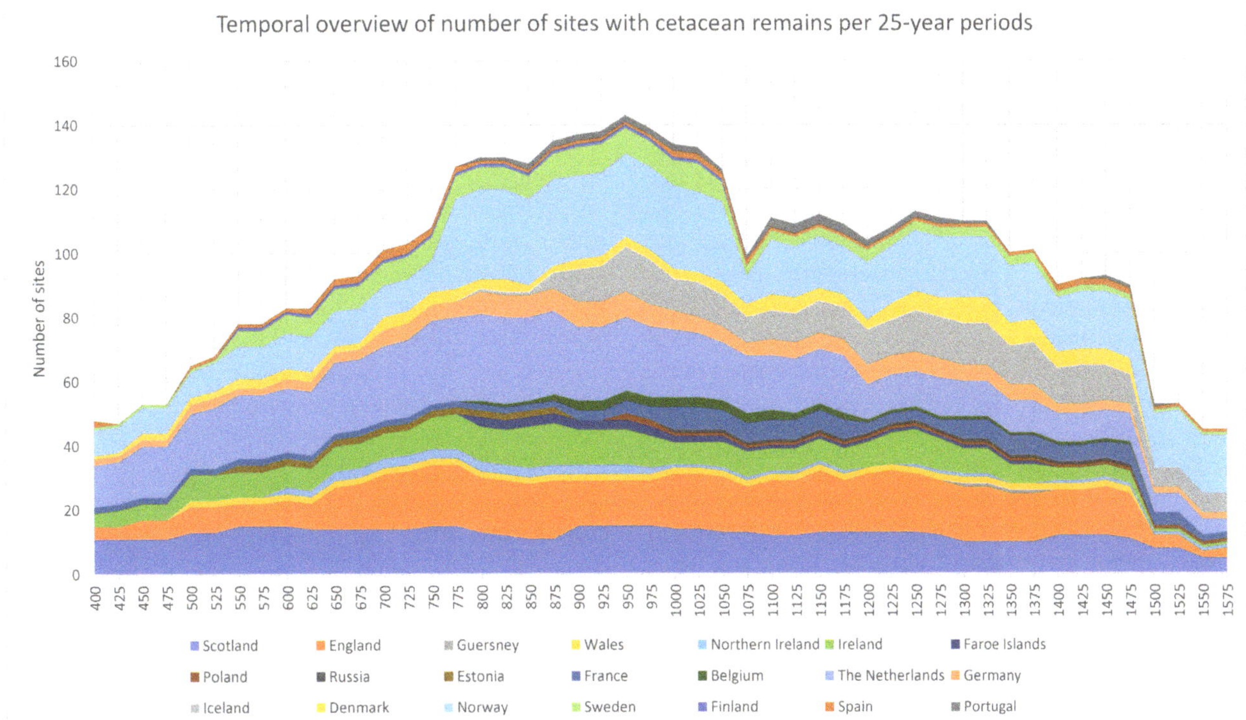

Figure 59. Temporal overview of number of sites with cetacean remains per 25-year periods, distribution based on countries. Countries listed in legend (from left to right, top to bottom) are displayed in the graph from bottom to top.

Another potential reason for an increased interest in cetacean exploitation is the Fish Event Horizon commencing around AD 850. Fishing increased over large parts of northern and western Europe and might have resulted in an increased exploitation of cetacean as well. However, an increased number of cetacean remains is

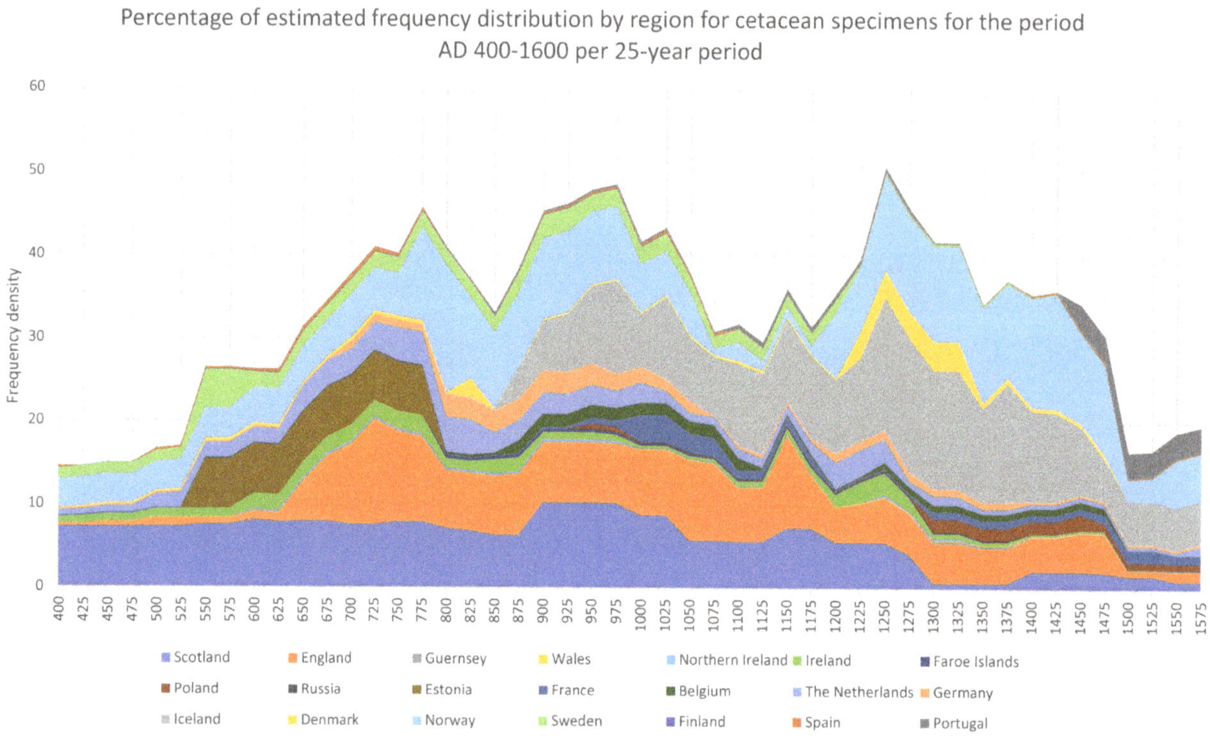

Figure 60. Temporal overview of frequency density per 25-year periods, distribution based on countries. Countries listed in legend (from left to right, top to bottom) are displayed in the graph from bottom to top.

only visible for about a century following the onset of the horizon. The increased exploitation of fish might even have triggered less cetacean exploitation being undertaken, as marine resources were now available in order forms (fish).

However, following AD 950/975, the number of sites gradually decreases after rapidly decreasing around AD 1050-1075. This can potentially be ascribed to the fact that AD 1066 is marked as the transition point from the Early Medieval Period to the High Medieval Period, and the dating of many sites ends or start there, creating an edge-effect. However, even ignoring the AD 1050-1075 dip, numbers of sites are still lower for the High Medieval Period in comparison to the ninth or tenth century data. With the spread of Christianity and the accompanied burial practices, grave good were no longer buried with the deceased in the Norse region. Many of the Norse grave goods made of whalebone, e.g. weaving swords, plaques and cleavers, disappeared, resulting in fewer cetacean remains ending up in the archaeological record and fewer cetacean remains being identified in "grave" contexts. Many of the cetacean remains deriving from "grave" contexts considered in this study are all indeed located in Norway and Sweden and date to the Norse period.

The numbers of sites for the High Medieval Period seem to be relatively stable, reaching its highest point at AD 1250. This is the period that is defined as the end of Medieval Warm period (potentially again creating another edge effect) and the commencement of the Little Ice Age. A similar pattern can be noted for the frequency density, though the frequency density numbers increased rapidly for the early thirteenth century, reaching its highest peak at AD 1250. Again, because of the Little Ice Age cetaceans might have stranded more frequently, leading to an increased exploitation, and an increased number of cetacean remains ending up in the archaeological record.

For the AD 1325-1349 period, numbers start to decrease again both for the numbers of sites as well as for the frequency density. This might have been caused by the Black Death spreading through Europe from AD 1347 to 1351. As a result, Eurasia's population lost between 75 to 200 million people. This decrease in human population, might have resulted in lower numbers of cetacean exploitation. Not only was less meat required to feed everyone, large parts of the land were uninhabited after the Black Death, which could have led to more cetacean strandings going unnoticed.

There is an additional dip following AD 1500. What caused this dip remains unclear. It might be that cetacean exploitation was less frequently undertaken during this period, or that less archaeological research has been conducted on post-medieval sites. It might also be that as part of this study, sites with cetacean remains dating to the sixteenth century were less frequently identified in comparison to medieval sites.

Assessing the contexts of all the sites the cetacean remains were found at on a temporal scale, several patterns emerge. First of all, grave context containing cetacean material primarily date to the early sixth to mid-eleventh century. They primarily derive from Norse contexts in Norway.

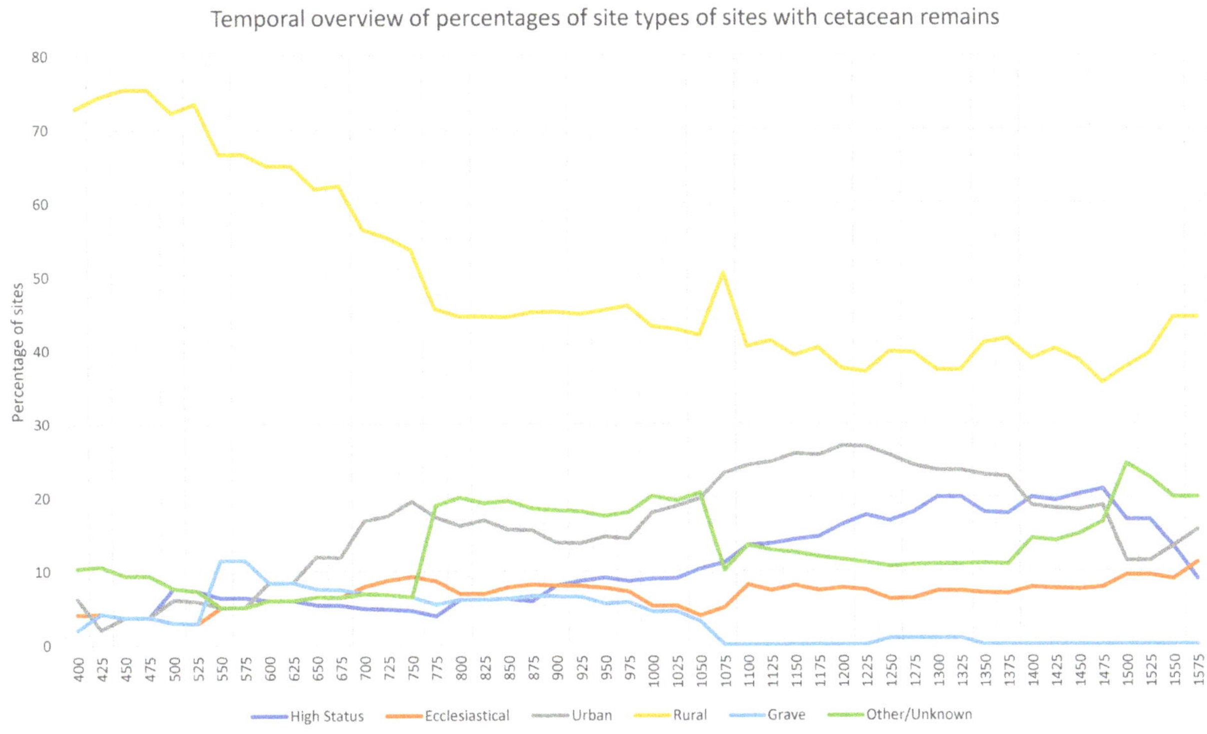

Figure 61. Temporal overview of percentages of site types of sites with cetacean remains per site type.

Figures 57 and 58 display high numbers of sites and frequency density of this country for this period as well.

"Rural" sites (small settlements or sites that housed only a couple of families) make up the largest portion of the site types for the entire temporal region (both for the number of sites and the frequency density). The highest proportion of rural sites is visible for the fifth century. After that point the proportion of rural sites decreases until the end of the eighth century. The proportion of rural sites remains relatively stable until the start of the High Medieval Period. After that point the proportion of sites decreases again. A different pattern emerges looking at the frequency density. For almost the entire Medieval Period rural sites make up around 50% of the total frequency density (figure 62). Only for the mid-thirteenth to the fourteenth century onwards this percentages increases to roughly 60%, after which it rapidly decreases again with the onset of the sixteenth century. The fourteenth century saw the European Great Famine and the Black Death hitting large parts of Europe, which might have triggered peasants to exploit any stranded cetacean they came across. They had to break the law to do this, but the numerous crises happening during the fourteenth century led to people exploiting any food source within their reach.

Looking at the number of sites, around the start of the High Medieval Period, the proportion of high-status and urban sites increases. Ecclesiastical sites also increase slightly, however the proportion of ecclesiastical sites is relatively stable from the mid-eighth century onwards. The increase in high-status can potentially be explained by an interest the European social elite got in cetacean meat around this period. Furthermore, cetacean meat was probably an expensive commodity during the Medieval Period, and only available to the social elite. This can explain the increase in urban sites with the onset of High Medieval Period, as the rich people of the cities tried to copy to diet of the social elite, in order to showcase their wealth.

The frequency density again shows a slightly different pattern (figure 58). Both the proportions of the frequency density for high status and ecclesiastical sites increases with the onset of the ninth century AD. For the eleventh century the frequency density of ecclesiastical sites increases strongly, but then decreases again with the onset of the thirteenth century. The frequency density of high-status site types displays the opposite. It decreases with the onset of the eleventh century and increases with the onset of the thirteenth century. The absolute numbers of the frequency density clearly indicate a rise for the high-status sites as well for especially the thirteenth century.

This confirms Gardiner's theory (1997) that the social elite increased the exploitation of cetaceans around the onset of the High Medieval Period, though both the frequency density and number of sites data suggest that an interest by the social elite in cetaceans can be traced back to at least the early tenth century, potentially even the early ninth century. It is interesting that around this time, the number of rural sites decreases (though this is less strongly pronounced looking at the frequency density data). This suggests that the social elite did try to monopolize the consumption of cetacean meat. They presumably did this by enforcing the

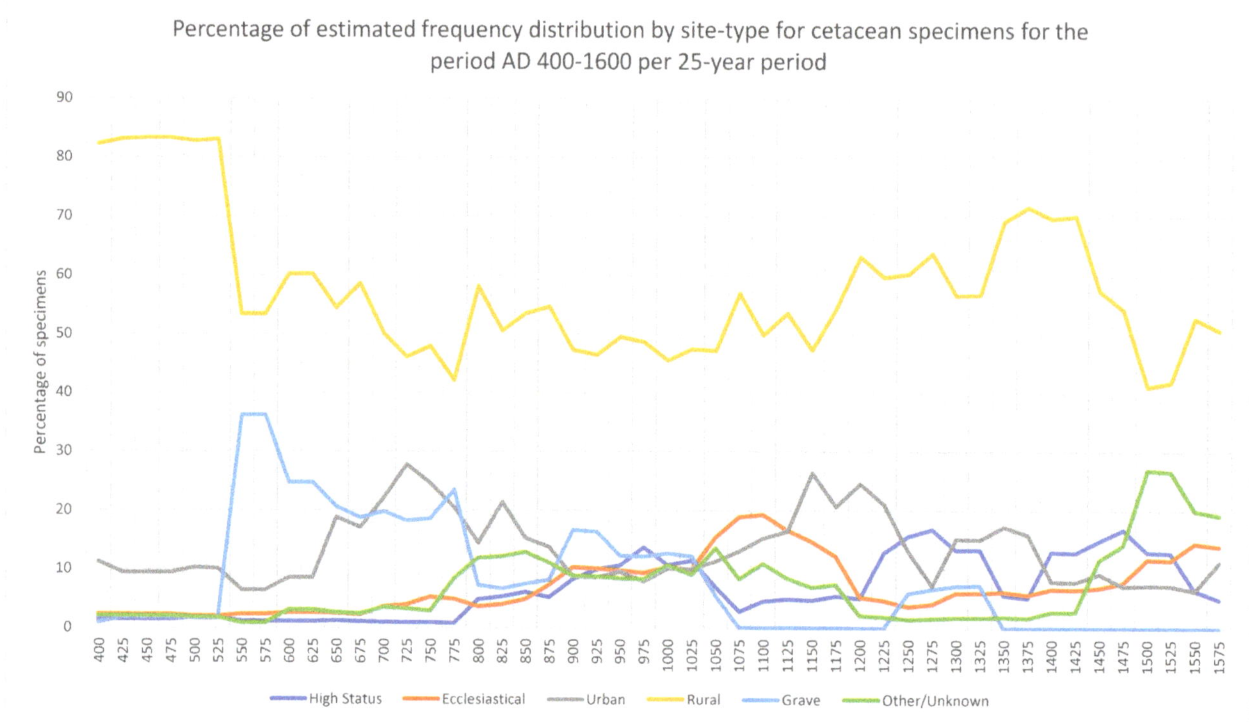

Figure 62. Temporal overview of percentages of frequency density per site type.

"wreck of sea" rights, in which any stranded cetacean was by law the property of the King, Queen, or to the social elite the King/Queen gave those rights to. As a result, less cetaceans were available to peasants, and eventually to less cetacean remains ending up in "rural" site types. This did however not stop peasants from exploiting cetaceans, and over the entire Medieval Period peasants are thought to have illegally exploited stranded cetaceans.

The increase in "urban" site types around the High Medieval Period, also suggests a certain form a commercialization of cetacean meat (as can be observed for the twelfth century frequency density). Indeed, several sources indicate that cetacean meat was available at the markets of several large European cities, including London, Calais, Boulogne, Damme, and Bayonne, though most of the zooarchaeological cetacean remains derive from London (Aguilar, 1986; De Smet, 1981; Gardiner, 1997). Cetacean meat was probably only available to the rich people of urban context, whom in this way copied the social elite's diet and showcased their wealth. The number of urban sites eventually dropped again at the end of the fourteenth century, possibly a direct result of the European Great Famine and the Black Death spreading through Europe with especially high numbers of casualties in urban contexts.

The patterns that emerge looking at the countries the cetacean specimens derived from are also interesting. The frequency density suggests that for the Early Medieval Period cetaceans were most frequently exploited in Scotland and Norway. For Scotland, cetacean appear to stay important until AD 1300, while for Norway cetaceans appear to stay important for the entire Medieval Period. This can be linked to the Norse and the fact that whale bone was frequently used for a variety of tools and artefacts in their regions. Between AD 600 and 1500, cetacean remains are also frequently recovered from England, and from *landnám* (AD 872) onwards also from Iceland. Cetacean remains are less abundant in the other countries considered as part of this study.

4.3. Conclusion

From this comprehensive assessment of a large number of zooarchaeological cetacean remains originating from a medieval context in northern and western Europe, it is clear that the exploitation of cetaceans was widespread, both on a geographical scale and a temporal scale. The most frequently encountered skeletal elements are the vertebrae. These are abundant in the skeleton of cetaceans, and therefore their abundance is not surprising. Additionally, a large portion of the identified remains appear to have been worked. Especially in the Norse region, worked cetacean remains are frequently encountered, and whale bone weaving swords, gaming pieces, plaques, and cleavers are examples of this. These are clear signs that whale bone was used for the creation of high-status artefacts. However, non-high-status artefacts are also known, most clearly the vertebral chopping blocks. These have been identified all over the research area, as well as outside of it, indicating that it was a widespread practice to use vertebrae of large cetaceans for this purpose. This might have happened on an *ad hoc* basis as well.

For the Early Medieval Period a gradual increase can be noted in regard to cetacean remains deriving from

Northern and Western Europe. This can partially be ascribed to the fact that many of these remains derive from Norway, where many artefacts were made of whale bone and were included in inhumation graves. Additionally, the onset of the Medieval Warm Period around AD 950 might have resulted in more strandings occurring around this period as well, resulting in an increase in the exploitation of stranded cetaceans. The Fish Event Horizon appears to not be associated with an increase in cetacean exploitation and might even resulted in the opposite. As more marine resources were available (fish), medieval people might have less frequently have exploited cetaceans.

For the High Medieval Period a decrease can be noted in the number of zooarchaeological remains and sites. This can be ascribed to the cessation of Norse burial practices with whale bone artefacts. However, an increased number of remains are recovered from the urban, high-status, and to a lesser extent, ecclesiastical sites from the early tenth century onwards already. This is probably the result of the social elite's attempt to monopolize cetacean consumption. However, it appears that cetacean meat was still available in some urban contexts, though it was probably only available to the rich, as they attempted to copy the social elite's diet. Furthermore, around AD 1250 another increase in sites was noted that could potentially be explained by another climate event, this time the Little Ice Age.

For the Late Medieval Period, numbers start to decrease, especially for the fourteenth century. This was caused by the Black Death and the European Great Famine, hitting large parts of Europe. Though lower numbers of sites can be observed for this period, higher numbers of cetacean remains are actually recovered from rural sites, suggesting that peasants were actively trying to get access to any food source within their reach. Furthermore, around AD 1500 numbers started to decrease even more, suggesting limited interest in cetacean meat for the sixteenth century.

Based on the species it is clear that the exploitation of the harbour porpoise, the common bottlenose dolphin, and the North Atlantic right whale was most frequently undertaken. These species are all coastal species, and especially the harbour porpoise and the North Atlantic right whale are frequently mentioned in medieval historical sources. From the data it is still clear that many remains are still only identified as "unknown cetacean", still hampering our understanding of medieval cetacean exploitation.

In order to partially overcome this, several case studies were undertaken, allowing to analyse the material more in depth. For several of these ZooMS was undertaken, and the ORCA-Manual (van den Hurk, forthcoming) was used as well for identification purposes. This allowed to see which species were exploited in particular regions and periods. Furthermore, these studies allowed to analyse cetacean exploitation of a smaller region on a larger temporal scale, thereby making it easier to identify any patterns in the exploitation of cetaceans.

The presence of zooarchaeological material does not automatically mean that active whaling was undertaken for a particular period or by a particular group or culture. To argue that active whaling was undertaken the zooarchaeological material needs to be assessed in comparison to the historical data analysed in chapter 2. This analysis will be undertaken in chapter 6.

It should be noted that no statistical analysis of the observed trends was undertaken, and the presented considerations are all based on the visual representation of figures 56-62. Statistical analysis of the trends in the chronological data might point to certain trends in cetacean exploitation. Statistical analysis should focus on each region separately as well as the region as a whole. This might indicate that for certain regions or periods, cetacean exploitation was restricted to the social elite, or that zooarchaeological data might be too scarce to make assumptions. Furthermore, statistical analysis regarding inland versus coastal sites, sites with worked whale bone remains versus sites with unworked remains, as well as representation of several species for particular regions or periods, should be undertaken to confirm discussed observed trends.

Furthermore, the trends visible might be a representation of archaeological excavations undertaken for the Medieval Period as a whole. It might be possible that more excavations have been undertaken on sites dating to AD 950 resulting in an overrepresentation of whale bone being recovered from that period. Future analysis should try to incorporate all archaeological excavations undertaken in the research area and look at the proportions of sites with cetacean remains for entire Medieval Period. This might reveal other trends in cetacean exploitation. However, even in that case, excavations in some areas might be over- or underrepresented giving a distorted overview of cetacean exploitation based on zooarchaeological remains and for some regions and periods will create considerable error ranges. The same is the case for site types, as archaeologists might have focused more on particular site types and have neglected others, again creating a bias in the data.

Moreover, this kind of analysis will be extremely time consuming to undertake, as thousands of archaeological excavations have been undertaken in northern and western Europe, and additionally a lot of archaeological data is only presented in grey literature. It was therefore not feasible to include such an analysis in this study, though a minor attempt was undertaken in the case study concerned with the Netherlands and Flanders. However, for the case studies similar error ranges obscure a complete understanding of medieval whaling practices and future attempts should focus on the statistical robustness of the analysed and presented data.

5

Case Studies

As the geographic as well as temporal scope of this study is considerable and covers a variety of cultures, only general trends in medieval cetacean exploitation can be observed. Focusing on a smaller region allows for in depth study and makes it possible to determine which species were exploited and allows a more detailed comparison to historical sources. This was undertaken for two regions as part of this study: The Netherlands and Flanders and England.

ZooMS was practiced on specimens from the Netherlands and Flanders, making it possible to identify which species were exploited and where. In the case of England, many historical sources are present that suggest that cetacean exploitation was restricted to the social elite, while for the Netherlands and Belgium these are rarer. These cases studies therefore provided a great opportunity to compare the two regions, to see whether cetacean exploitation was accompanied by any social implications, and in this manner answer the main question this study is concerned with, but then on a smaller scale.

In order to focus on one species more in depth, a species-specific case study was conducted as well. The species chosen was the grey whale. The grey whale is the only cetacean species to have gone extinct from European waters in historical times. This species is likely to have gone extinct on the European side of the North Atlantic during the Medieval Period, and as part of this study several new findings of grey whale specimens have been done. These specimens were compared to all other known grey whale specimens from both the European as well as the North American side of the North Atlantic, making it possible to reconstruct the demise of the species, and see whether anthropogenic factors played a role in it.

5.1. Netherlands and Flanders: Zooarchaeological Analysis of Medieval Cetacean Remains from the Netherlands and Flanders

An updated version of this case study is being published in Environmental Archaeology; The Journal of Human Palaeoecology: van den Hurk, Y., Spindler, L., McGrath, K. and Speller, C., forthcoming. Medieval Whalers in the Netherlands and Flanders: Zooarchaeological Analysis of Medieval Cetacean Remains. *Environmental Archaeology*.

Introduction

Zooarchaeological cetacean material regularly shows up in the Dutch and Flemish archaeological record (figure 63). Numerous remains date to the Medieval Period (roughly 400-1600 AD). It has often been suggested that zooarchaeological cetacean material merely derived from stranded individuals, however, several historical sources from the Netherlands and Flanders seem to suggest that active whaling was undertaken during the Medieval Period. Furthermore, historical sources seem to argue that cetacean exploitation was associated with the social elite and appeared to have been a precious fasting food consumed by the clergy and nobility.

As part of this case study, medieval zooarchaeological cetacean material was analysed in order to investigate whether cetacean exploitation was indeed associated with the social elite during the Medieval Period and which species were exploited. Cetacean material is however hard to identify to species or genus level, as the material is often too fragmented to allow more detailed morphological comparison. To overcome this problem, ZooMS (Zooarchaeology by Mass-Spectrometry) was performed on 38 cetacean specimens (37 from the Netherlands and 1 from Belgium) with a medieval date. As several species are unlikely to have been hunted during the Medieval Period and others could potentially have been hunted, the identification to species level can be a useful tool in determining whether active whaling was undertaken.

Furthermore, a comparison between the cetacean strandings that have occurred between 1969-2018 in the Netherlands and the zooarchaeological material identified as part of this study is conducted. This will both shed light on past cetacean distribution in comparison with modern distribution, and this data additionally also holds the potential to determine whether active whaling or opportunistic scavenging was conducted.

Historical context

For various medieval European regions cetaceans were claimed by the social elite. For England stranded cetaceans were by law the right to the King, unless those rights were given to local lord or ecclesiastical institution (Gardiner, 1997). Similar laws were set in place for *inter alia* France (from AD 850 onwards), Denmark (from AD 1200 onwards), and Norway (from AD 1100 onwards) (Schnall, 1993; Hybel and Poulsen, 2007, 55; Ardouin, Hadjouis and Arroyo, 2009; Laist, 2017, 89). This was also the case for the Netherlands. Though cetaceans were not explicitly mentioned, a letter from William II, Count of Holland and Zeeland (1234-1256) to Margaret of Constantinople, Countess of Flanders (1244-1278) specified the made arrangement between the two rulers regarding beach finds (De Groote, 1999). Based on this it appears that the social elite of the Netherlands and Flanders, tried to monopolize

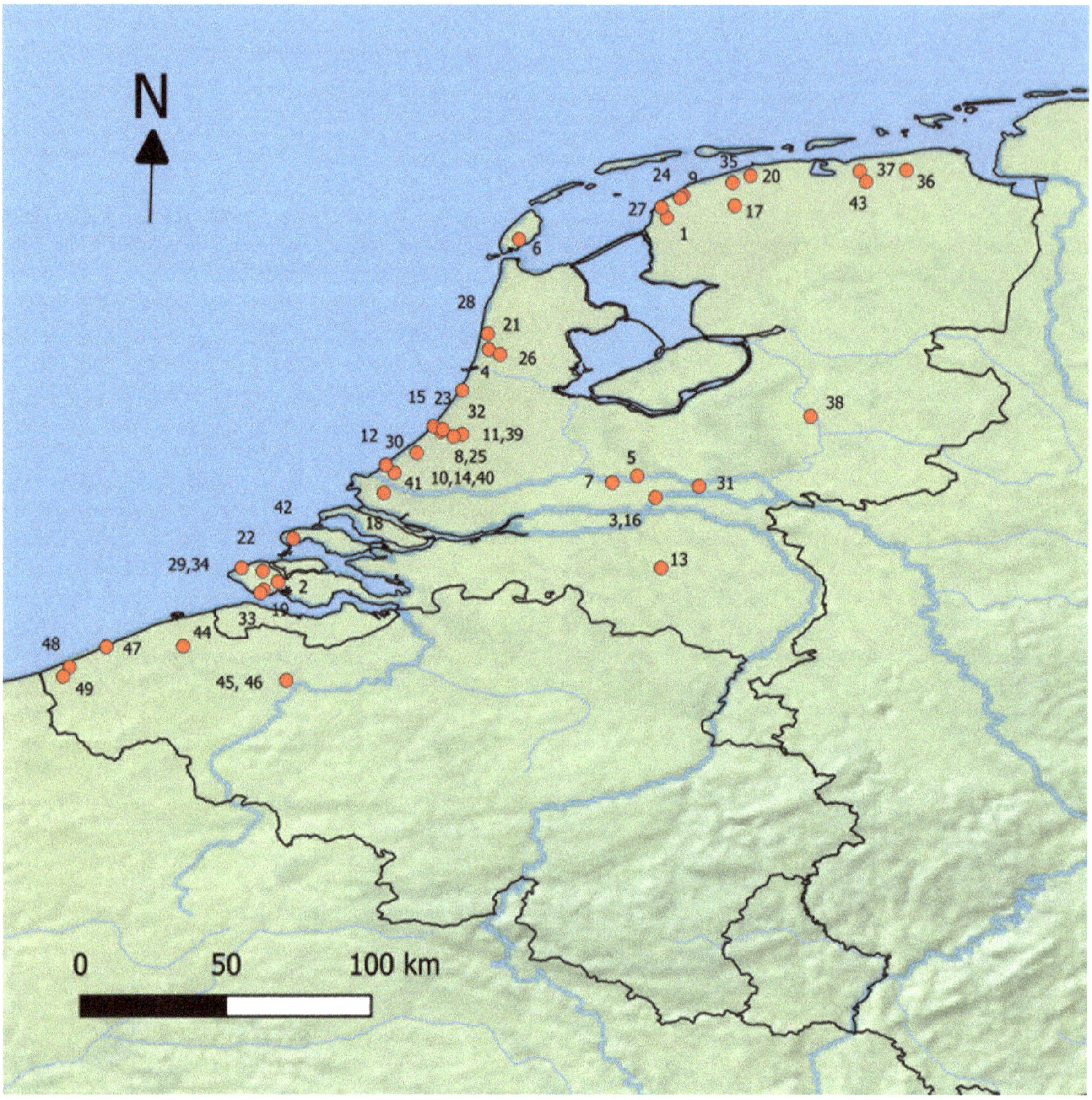

Figure 63. Medieval sites (AD 400-1600) located in the Netherlands and Belgium with zooarchaeological cetacean remains.

the exploitation of cetaceans, but to what extent this occurred remains unclear.

Several sources seem to suggest that the social elite in the Netherlands developed a taste for cetacean meat during the Early Medieval Period already, but especially during the High and Late Medieval Period this becomes clearer. Eventually laws were enforced in which the social elite made sure they received a portion of every whale caught or stranded. The St.-Maartenkerk (a church) in Utrecht, the Netherlands is known to have possessed right of wreck on their land from at least the eighth century until the first half of the tenth century AD (Moesker and Cavallo, 2016, 608). For England, these "wrecks" included stranded cetaceans as well and it is likely this was also the case for the Netherlands.

Ecclesiastical institutions, like the St.-Maartenkerk, gained interest in cetaceans during the Medieval Period. During fasting periods, the consumption of mammalian meat was not allowed, but as cetaceans were perceived as fish they were a welcome addition to the menu (this confusion still lives on in the Dutch language nowadays, as the Dutch word for "whale" is "walvis" (whale-fish)). However, some sources seem to suggest that fatty fish were still banned, this included especially the large marine mammals such as seals, whales, and dolphins, as their meat resembled that of terrestrial mammals. Other records seem to indicate that marine mammal meat was both considered a high-status food source as well as a food source consumed by the clergy (van Dam, 2003, 476). Especially the porpoise is relatively frequently mentioned and appears next to seal, swordfish, salmon, haddock, and sturgeon as the more prestigious varieties of "fish" (Van Dam, 2009, 310). Van Dam (2009, 310) noted that prestigious fish had one attribute in common – their size, as all these species can get large. Until at least AD 1300, fresh fish, marine fish and especially large fish (such as cetaceans) were seen as a delicacy in large parts of Europe (van Dam, 2003, 467).

One source that seems to confirm this for the Netherlands, is an account by the treasurer of the Count of Holland in the year 1395-96. He bought a seal and a porpoise that he gifted to a certain Duke Robrecht and furthermore gifted three seals to the Bishop of Liège (van Dam, 2003, 477). Another record indicates that during the siege of Utrecht (26 June to 27 July 1345), a large quantity of fish was shipped in for the troops participating in the siege. Besides, smoked and salted herring, salted eel, cod, haddock, and two porpoises were provided as well (Van Dam, 2009, 327). Duchess Catherine of Cleves, wife of Arnold, Duke of Guelders is known to have stayed at the Valkhof, Nijmegen for several years during the mid-fifteenth century. Her kitchen staff kept records of the food supplies they bought and received. These records indicated that the Duchess occasionally received porpoises from her family (Kruyff, 2009. The lords of Castle Doorwerth are also known to have consumed porpoises (Snoeren, 1981).

That whales were precious animals to medieval lords, is also indicated by the Siegfried II of Westerburg, who compared the invading John I, Duke of Brabant to a whale who swam too close to the shore and stranded itself and would make him and his mean rich. This ended however catastrophically for the archbishop as he was captured in 1288 (Stern, 2000).

None of these sources make it clear whether cetaceans were indeed actively caught or opportunistically exploited through strandings. It is often assumed that the latter was more frequently the case. However, Albertus Magnus, a German Dominican friar and Catholic bishop that lived from 1193 to 1280, clearly suggests that active whaling was undertaken in the northern parts of the Netherlands as well. Magnus visited Frisia and the Wadden Sea islands in northern Germany and the Netherlands where he states that he witnessed the catching of a whale by the Frisian locals (Albert the Great, 1987, 338-342). He describes how the Frisians worked in teams of several small boats, utilised music and noise to drive the animal in a specific direction and used harpoons and a powerful ballista to catch the animal. He noted that various species were exploited, though that the very large species were rarely exploited. When the hunt was successful, he stated that the Frisians conserved the oil, rendered the whale blubber and retrieved the baleen, meat, and bone (Albert the Great, 1987, 338-342). This clearly indicates that active whaling was occasionally undertaken, and that various species were exploited. Indeed, cetacean remains are frequently found in the terp sites (tell mound) in the northern part of the Netherlands, suggesting that these might have derived from actively caught whales.

By the end of the Medieval Period, in the sixteenth century, it appears that though cetacean meat was still highly prized, it was not exclusively available to the social elite. At the sixteenth century *Amsterdamse vismarkt* (fishmarket of Amsterdam) fresh and salted meat of seals (that were abundant in the Zuiderzee) was frequently sold. Additionally, meat of porpoises and swordfish were valuable goods sold at this market (Ypma, 1962, 30). The kitchen accounts of Kartuizerklooster (abbey) in Geertruidenberg record that a porpoise was bought for the members of the clergy present in the abbey during the early fifteenth century (Sanders, 1990, 92).

Moreover, the jaw of a sperm whale, that stranded between Katwijk and Scheveningen on the 2nd of February 1598, was gifted to Count Jan van Nassau (Landwehr, 1981, 102). This might have included the tongue which was perceived as a delicacy. The rest of the animal was sold for 126 guilders. Three other sperm whales had stranded near Heijde and the steward of the court of Holland brought the mandibles of the largest specimen to The Hague where they have stayed in the *Ridderzaal* until the late eighteenth century.

For Flanders even more sources concerned with cetacean exploitation are known. Though several cetacean strandings are recorded, such as the stranding of eight whales in Oostduinkerke in 1403 (Charlier, 2004) and the stranding of a sperm whale in the Scheldt nearby Antwerp and two others near Bieselinge in 1577, which were likely to have been exploited, a stronger case can be made to argue that active whaling was practiced. One of the oldest sources suggesting whaling was already undertaken during the Early Medieval Period is *The Life of St. Vedastus*, dating to around 875. Herein a group of Flemish fishermen from a monastery organized a contest with another group to hunt a whale. The story indicates that the hunt was communally organized and that the participants paid a fee into a "contubernium" (a co-operative society) and agreed on sharing the catch. Eventually, the group that prayed to the St. Vaast caught the whale (Chevallier, 2014).

Prayers to saints appear to have been frequently undertaken by whalers to ensure a successful hunt. A similar situation arose when Flemish fishermen tried to catch a whale in the tenth century AD and only by praying to St. Bavon the hunt ended successfully. Another case occurred in the twelfth century when fishermen prayed to St. Arnulf to ensure a successful hunt, as registered in *The life of St. Arnulf* (Chevallier, 2014).

For Flanders, several sources seem to indicate that cetacean meat was a valuable product as well, especially from the twelfth century onwards. In 1121, the Count of Flanders gifted "pinam de cetam" (tail of a whale) to the Abbey of Sint-Winoksbergen in Bergen (Steevens, 2014). From the accounts of the kitchen of the Sint-Pieters Abbey in Ghent dating to 1485, it is clear that porpoise meat was consumed there as well (Mortier, 2016, 224). Furthermore, in 1178, Count Philip of Alsace was gifted a monstrous beast that was hunted by whalers from Bruges. Bruges might have been a centre for whale meat as John II (John le Bon), King of France (1350-1364), during his imprisonment in London from 1357 to 1358, bought whale from Bruges as well (van Neer and Ervynck, 1993, 87). Whale meat was also available in Calais as in 1300 the Count of Artois bought 33 pieces of whale meat from the market there (De Smet, 1981).

The accounts by the bailiff of the County of Flanders also provide valuable information. These accounts indicate that the Count of Flanders regularly used his "wreck of sea" rights to claim stranded cetaceans, as well as large sharks and other fish. These accounts are often vague and not detailed in regard to what species is dealt with. In some instances, the accounts specify that the finders, people transporting the cetaceans or fish, as well as the people guarding the stranded cetaceans preventing poaching from happening, all got their share (De Groote, 1999). This is a clear sign that poaching of stranded whale carcasses occurred regularly.

One account from the *Brugse Vrije* dating to 1403 indicates that a large fish stranded on the island of Cadzand. Several people managed to ship the large fish to Monnikerede. The fish was eventually sold at Bruges for 36 Parisian pounds. Just over half of that went to the finders and the people who transported the animal to Bruges, while 15 pound 14 sous remained for the Count (De Groote, 1999).

In addition, in 1371, the Flemish Count Louis of Male sent whale meat to his daughter Margareta at the Burgundian Court. Whale meat appears to be prized at the Burgundian Court, as Charles the Bold, Duke of Burgundy, also served whale meat at his wedding with Margareta, Countess of Flanders in 1468 (De Haan and Oosterman, 1996, 51). The Duke of Burgundy, Count of Flanders is also known to have had a ship undertaking whaling in the North Sea in 1456 (De Smet, 1981).

These sources seem to suggest that whale meat was indeed frequently consumed by the social elite from the twelfth century onwards. Following the twelfth century, active whaling practices are also more frequently mentioned. Historical sources describe that four whale hunting ships (potentially fishing ships adapted to also hunt whales) had their homeport in Blankenberge in 1147 (Charlier, 2004). Other sources indicate that whaling was a specific activity which required the permission of nobility. The citizen of Blankenberge appear to have had ties with cetacean exploitation for a considerable time, as it was recorded in 1523 that the they presented a harbour porpoise to the councillors of Brugge (Viaena, 1971, 59).

In 1163, several towns were granted the rights to hunt cetacean in the Charter of Newport (De Gryse, 1940-1945; van Neer and Ervynck, 1993, 86; Charlier, 2004). Additionally, in 1340 Wenduine was granted the right to hunt cetaceans, more specifically the harbour porpoise. A picture of a harpooned porpoise was also present at Wenduine's coat of arms (Charlier, 2004).

A taste for especially harbour porpoise meat persisted until at least the sixteenth century. In 1534 the city of Oostende gifted a harbour porpoise to Anton van Croÿ, Lord of Sempy, who resided in Brussels. Furthermore, in 1568, the dune-abbey of Newport bought a fresh harbour porpoise (De Baets, 2013).

It appears that cetacean meat was not exclusively restricted to the nobility and clergy. Sources indicate that in 1024 taxes had to be paid for every hundredth part of whale meat at the city of Arras (Steevens, 2014). Other cities, such as Boulogne, Calais and Damme are also known to have sold whale meat at the local markets between the eleventh and twelfth century (De Smet, 1981). In the city accounts of Bruges records indicate that porpoises were sold to the city by several fishermen and merchants for the period of 1360-1372 (Espeel, 2016). This indicates that cetacean meat was widely available at medieval Flemish markets, though probably was only available to the rich and in this way can therefore still be associated with social elite. A Dutch cookbook dating to the second half of the fifteenth century contains a recipe that contains porpoise meat (van Winter, 2013). This was however probably still only available to the social elite.

Historical sources from the Netherlands and Flanders seem to indicate that cetacean exploitation was already practiced during the Early Medieval Period, was widespread, in some regions well organized, and indeed associated with the social elite. From these historical sources it is however not known which species were actually exploited. Zooarchaeological cetacean material is frequently found in medieval contexts in the Netherlands and Flanders and offers the possibility to compare the historical data with the archaeological data and allows to assess which species were exploited.

Methodology

As part of this study 38 zooarchaeological specimens were analysed using ZooMS. Samples were collected from the Groningen Institute of Archaeology in Groningen, the archaeological depots of Groningen, Friesland, and Drenthe; North Holland; South Holland; North Brabant, and Zeeland. Furthermore, samples were acquired from RAAP Oost, ADC ArchaeoProjecten, ArchaeoBone, and ArcheoPlan Eco, covering a large geographical region. Samples of 0.03 gram were taken by cutting a small piece of the bone using a Dremel, damaging the bone as little as possible. These were subsequently taken to BioArCh, York, UK where they went through acid demineralisation for two weeks. The samples were subsequently put through buffer extraction, gelatine extraction, trypsin digestion, and peptide extraction. Following this, the samples were spotted on a 384 MALDI plate, after which the actual mass-spectrometry was undertaken.

The majority of the specimens analysed were originally merely identified as "whale" or "large whale". These specimens were often large chunks of cancellous bone material. However, five specimens were identified to species as well, these include two killer whale, one humpback whale, one North Atlantic right whale, and one common bottlenose dolphin. Additionally, one specimen was identified as either bowhead whale or sperm whale. ZooMS allowed to check whether these identifications based on morphology were accurate.

For those specimens that were complete or partially complete, morphological and osteometric analysis was undertaken as well. This was based on the Osteological Reference for Cetaceans in Archaeology-Manual (ORCA-Manual) (van den Hurk, forthcoming). This allowed for some specimens to be more precisely identified, as for some of the specimens ZooMS was only able to identify to the genus or sub-family level. Additionally, for some of the vertebral remains the ORCA-manual (van den Hurk, forthcoming) allowed to identify those to a particular region of the vertebral column (cervical, thoracic, lumbar, or caudal). This was based on osteometric analysis performed on the samples and compared to osteometric data from modern cetacean specimens held at the Smithsonian Institution, Washington DC, USA and the Naturhistoriske Museum in Copenhagen, Denmark.

Results ZooMS

Of the 38 samples analysed using ZooMS, 35 were successful (figure 64 and 64). Of these 35 specimens, one specimen was identified as a sheep (*Ovis aries*), red deer (*Cervus elaphus*) or fallow deer (*Dama dama*). More detailed analysis of the specimen confirmed it was a piece of antler, most likely from red deer. In addition, another specimen was previously morphologically identified as a scapula of large whale. This specimen came from the early medieval site of Plantage in Leiderdorp (AD 800-850; Moesker and Cavallo, 2016). ZooMS analysis identified it as elephant. Further analysis indicated that the bone was sub-fossilised and probably was the proximal posterior portion of a tibia of a mammoth. Interestingly a large hole was drilled in the bone. However, it is not clear whether this was done during the Medieval Period or earlier.

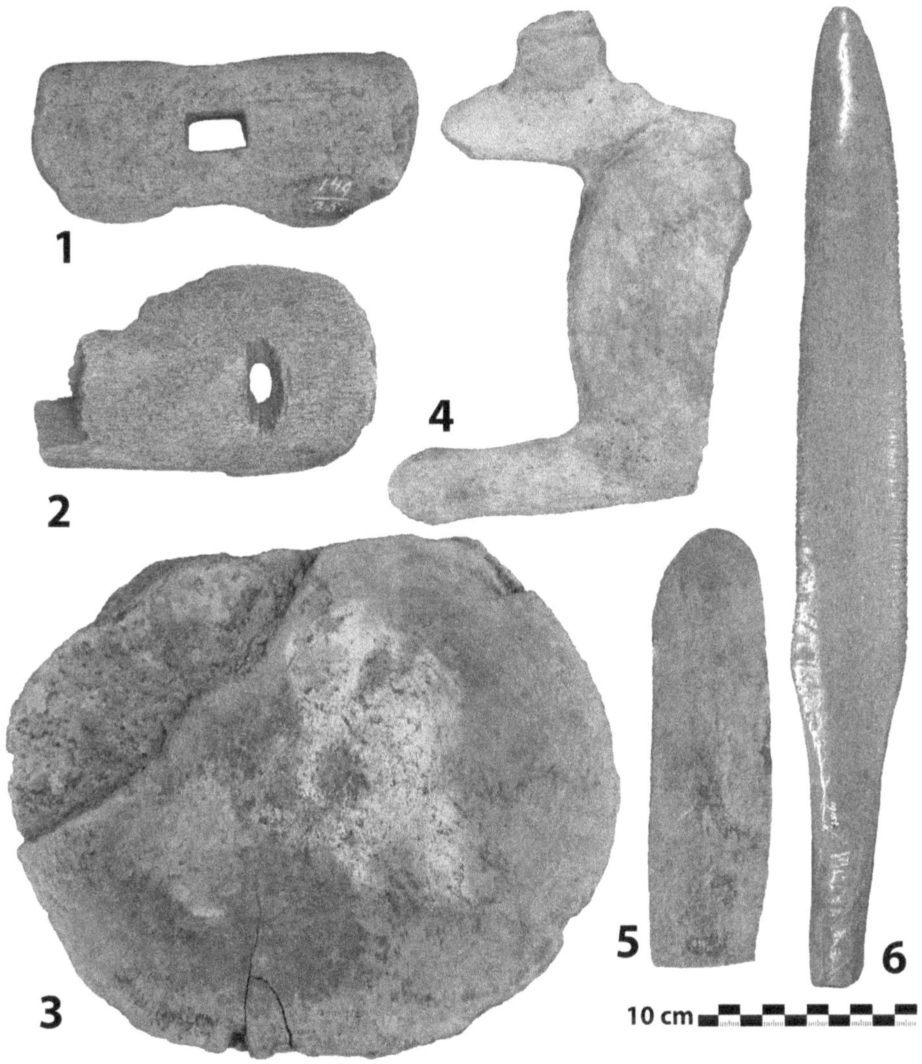

Figure 64. A selection of cetacean material analysed using ZooMS. 1. Worked piece of bone from Achlum, Friesland (identified as sperm whale), 2. Worked piece of bone from Tzummarum, Friesland (identified as fin whale), 3. Vertebral body from Achlum, Friesland (identified as northern bottlenose whale), 4. Cervical vertebra from Hallum, Friesland (identified as grey whale), 5 Weaving sword from Leens, Groningen (identified as North Atlantic right whale), 6. Weaving sword from Rottum, Friesland (identified as grey whale).

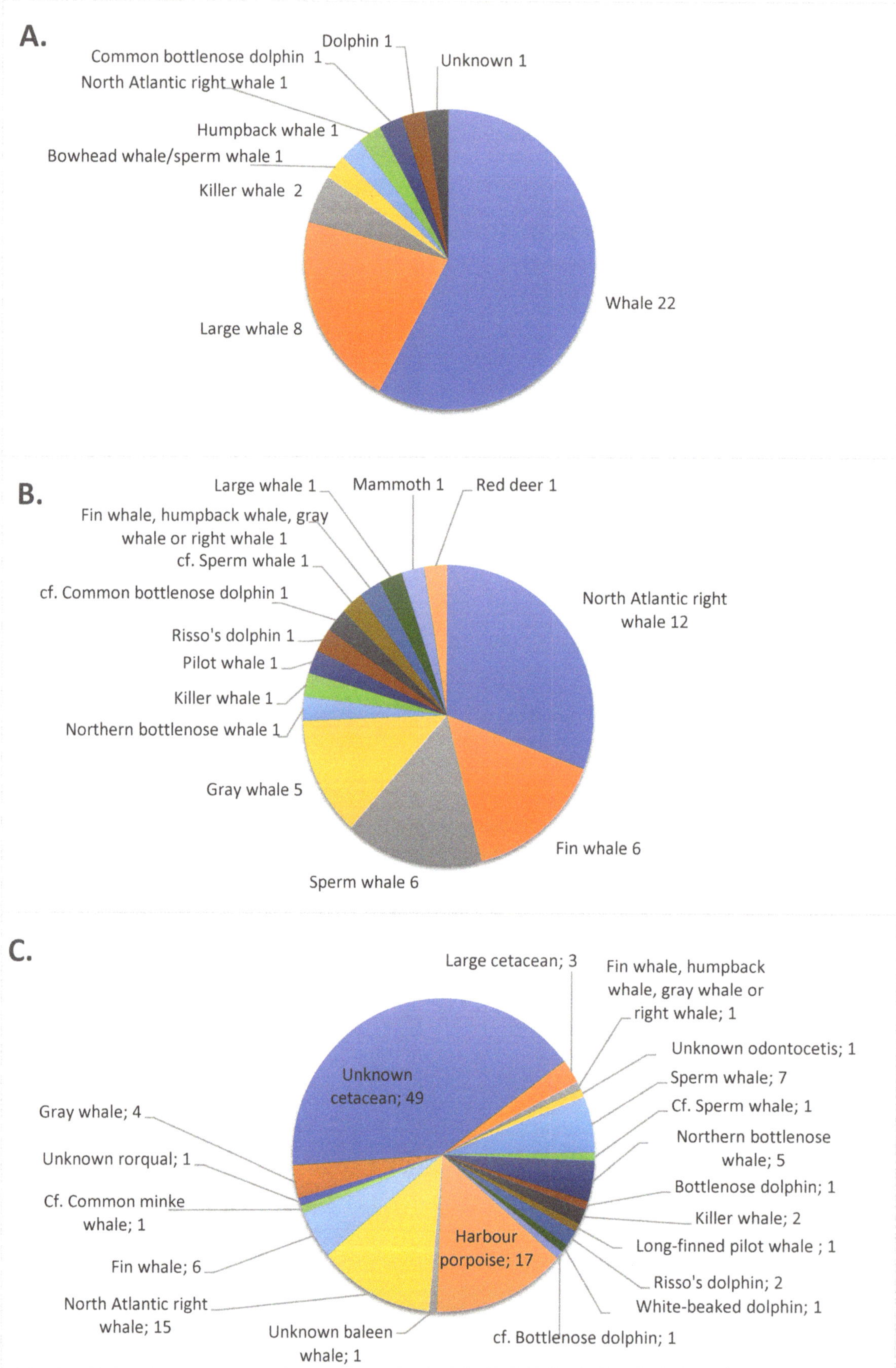

Figure 65. Identification of the cetacean material analysed as part of this study. A. Original identification of the material (n=38), B. Identification based on ZooMS, morphological analysis and osteometric analysis (n=38), C. All medieval (400-1600 AD) zooarchaeological cetacean specimens from the Netherlands and Belgium, including the specimens analysed as part of this study (n=120).

The remaining 33 specimens were all identified as being cetacean, for which the majority could be identified to species, however several specimens could only be identified as belonging to a particular group of species. It can furthermore be noted that of the five specimens that were previously morphologically identified to species level, ZooMS determined that the original identification for four of these were wrong. The fifth bone was one of the three samples for which not enough collagen was left to allow ZooMS identification.

Results of morphological and osteometric analysis

Based on morphological and osteometric analysis and comparison to data that was incorporated in the ORCA-Manual (van den Hurk, forthcoming), several of the specimens could be more precisely identified to species and element. While there is substantial variation in size within a species, it is assumed that the general proportions of the osteological features are generally the same for each individual of a particular species and can aid identification. All of the specimens identified to species level, were vertebral remains. The dimensions (length, breadth, and height) of the vertebral bodies are useful for identification purposes, especially the length varies for some species, making it an ideal measurement to aid identification purposes. ZooMS was already undertaken for all these specimens, but not a clear identification could be reached as ZooMS is not able to distinguish between several species. Therefore, for all the specimens considered here identification could be narrowed down from four, three, or sometimes two species, to just one. Osteometric was performed on the specimens, following the instructions provided in the ORCA-Manual (van den Hurk, forthcoming) and the dimensions of the vertebral specimens are provided in table 5. These were then compared with the vertebral osteometric data from the ORCA-Manual (van den Hurk, forthcoming).

Three specimens were identified using ZooMS as Globicephalinae, a group of six dolphin species frequently referred to as "blackfish". Three of these species are relatively frequently sighted in the European Atlantic: the long-finned pilot whale (*Globicephala melas*), Risso's dolphin (*Grampus griseus*), and false killer whale (*Pseudorca crassidens*). All three specimens were vertebrae for which the central body was partially complete and allowed osteometric comparison to individuals of the three species from the Smithsonian Institution (figure 66).

As can be noted from the three graphs, there is considerable difference in especially the length of the vertebral bodies for the three species, which allowed the identification to species level. The two specimens from Egmond aan den Hoef (Slot op den Hoef; find numbers 713 and 714) were based on the length in comparison to the height and breadth of the vertebrae identified as Risso's dolphin, most probably one of the last lumbar vertebrae or one of the first caudal vertebrae, and likely to have been from one individual. This was accomplished as the vertebral remains did not possess the length of false killer whale vertebrae, and there was considerable difference between the length and the breadth and height of the vertebral bodies.

The specimens from Santpoort-Zuid (Castle Brederode; find number 1531-2), previously identified as killer whale, likely derived from one of the last thoracic or one of the first lumbar vertebrae of a long-finned pilot whale. This was based again on the length of the specimen, as it was too short to be from a false killer whale, and too long to be from a Risso's dolphin.

Interestingly all three specimens were original mis-identified (table 5). The two specimens from Slot op den Hoef were originally identified as common bottlenose dolphin, while specimen 1513-2 was identified as killer whale. The latter was found next to another vertebrae (1513-1), which was originally identified as killer whale as well. This identification was confirmed through ZooMS, and morphological and osteometric analysis indicated that it was a lumbar vertebra (figure 67). This indicates that remains of a killer whale as well as a long-finned pilot whale were found at the site.

The specimen from Molenslag was identified by ZooMS as common bottlenose dolphin, white beaked dolphin, common dolphin, or striped dolphin. For this specimen (a vertebra, both cranially and caudally unfused), it was determined that based on the position of the laminae (located more to the cranial side and pointing in the cranial direction as well), it was one of the last lumbar or first couple of caudal vertebrae. The length of the vertebral body (29 mm) was helpful in determining the species as well. The length of all the vertebrae of the vertebral bodies for the four species are plotted in figure 68. The lumbar vertebrae are the middle section of the vertebral column and as can be observed are much less long than the thoracic or the later caudal vertebrae. Based on the length, and the fact that the epiphyses of the specimen was still unfused (meaning that the length of the vertebral body would be longer if the specimen would have been fused), the common bottlenose dolphin is the most probable option. While variation for different populations within a species are apparent for most species, and while this might also be the case for the four species discussed here, the common bottlenose dolphin remains the most likely candidate.

Based on the height to breadth ratio of the vertebral body of the vertebra from Dokkershaven (identified through ZooMS as either a bowhead whale or a North Atlantic right whale), this specimen is probably one of the last lumbar or one of the first caudal vertebrae of a North Atlantic right whale (figure 69). This species was probably more abundant than the bowhead whale in the southern North Sea as well, making it the more likely candidate. Furthermore, the vertebra displayed chopping marks, indicating it was used as a chopping block.

Specimen 159-35 from Achlum was identified as a beaked whale through ZooMS. The size of the vertebral body

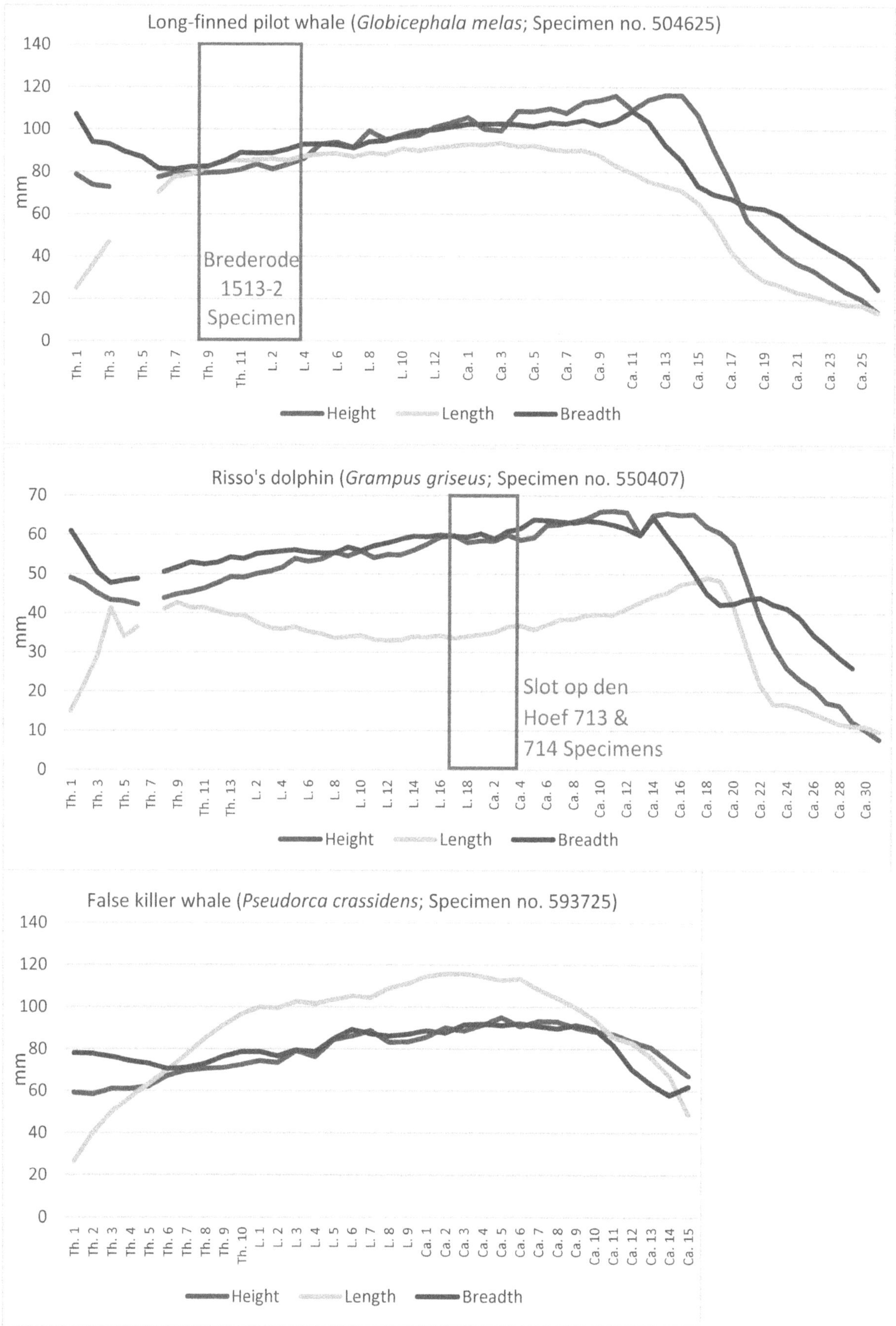

Figure 66. Height, length, and breadth of the vertebral bodies of the thoracic (Th.), lumbar (L.), and caudal (Ca.) of the long-finned pilot whale (specimen 504625), Risso's dolphin (specimen 550407), and false killer whale (specimen 593725), all part of the Smithsonian Institution, in comparison to zooarchaeological remains of Brederode (1513-2) and Slot op den Hoef (713 and 714). Measurements of the archaeological specimens are provided in table 5.

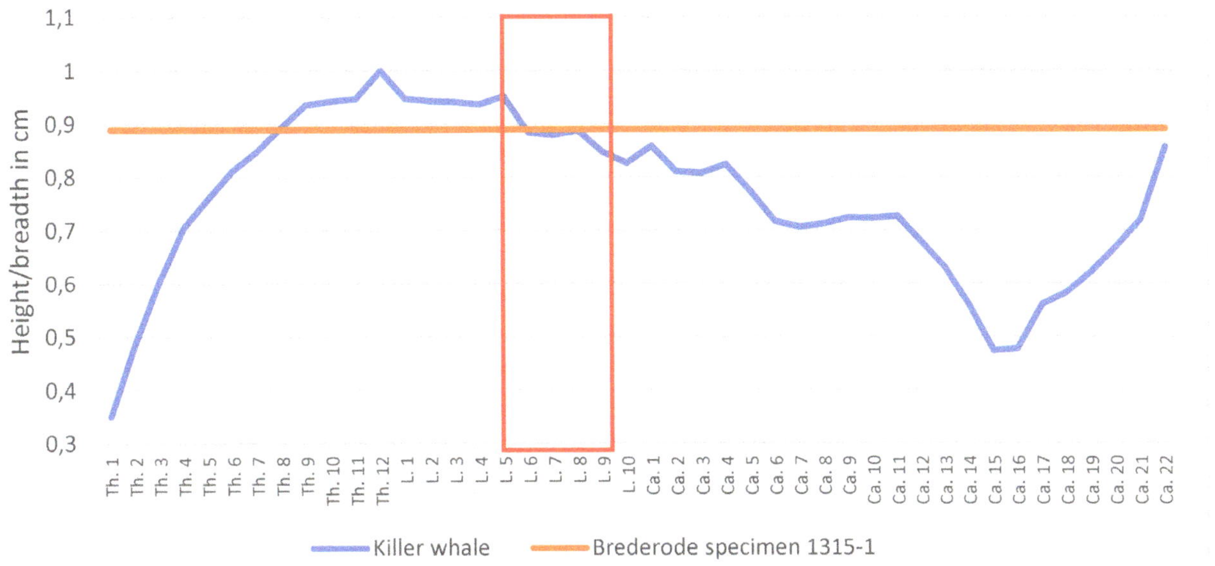

Figure 67. Height/breadth of the vertebral body of the thoracic (Th.), lumbar (L.), and caudal (Ca.) of the killer whale (specimen 23004), part of the Smithsonian Institution, in comparison to the zooarchaeological specimen of Brederode (1513-1). Measurements of the archaeological specimens are provided in table 5.

made it clear that the northern bottlenose whale was the only likely option, as the other beaked whale species in the North Atlantic are considerably smaller. Osteometric analysis of the specimen with a northern bottlenose whale from the Smithsonian Institution made it clear that it was one of the last thoracic vertebrae (figure 70).

These various cases demonstrate the difficulties that arise when identifying cetacean remains to species based on morphology. The ORCA-Manual (van den Hurk, forthcoming) has the potential to optimise the identification of zooarchaeological cetacean material, not only to species level, but also holds the potential to identify vertebral remains to a segment of the vertebral column (e.g. cervical, thoracic, lumbar, or caudal). Savelle and Friesen (1996, 713-721) performed a section on a sub-adult harbour porpoise to create a meat utility index and in this way found out what parts of the harbour porpoise contained the most flesh. By

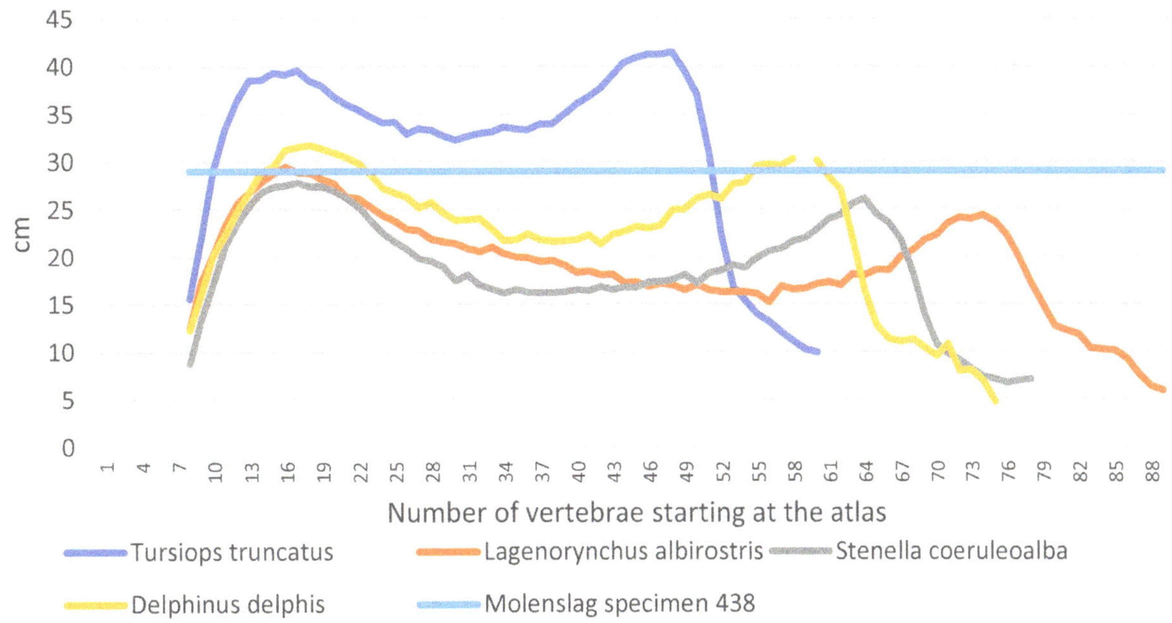

Figure 68. Length of the vertebral body of the vertebral column of the common bottlenose dolphin (specimen 593772), white-beaked dolphin (specimen 267573), striped dolphin (specimen 504350), and the common dolphin (specimen 571620), part of the Smithsonian Institution, in comparison to the zooarchaeological specimen of Molenslag (438). Measurements of the archaeological specimens are provided in table 5.

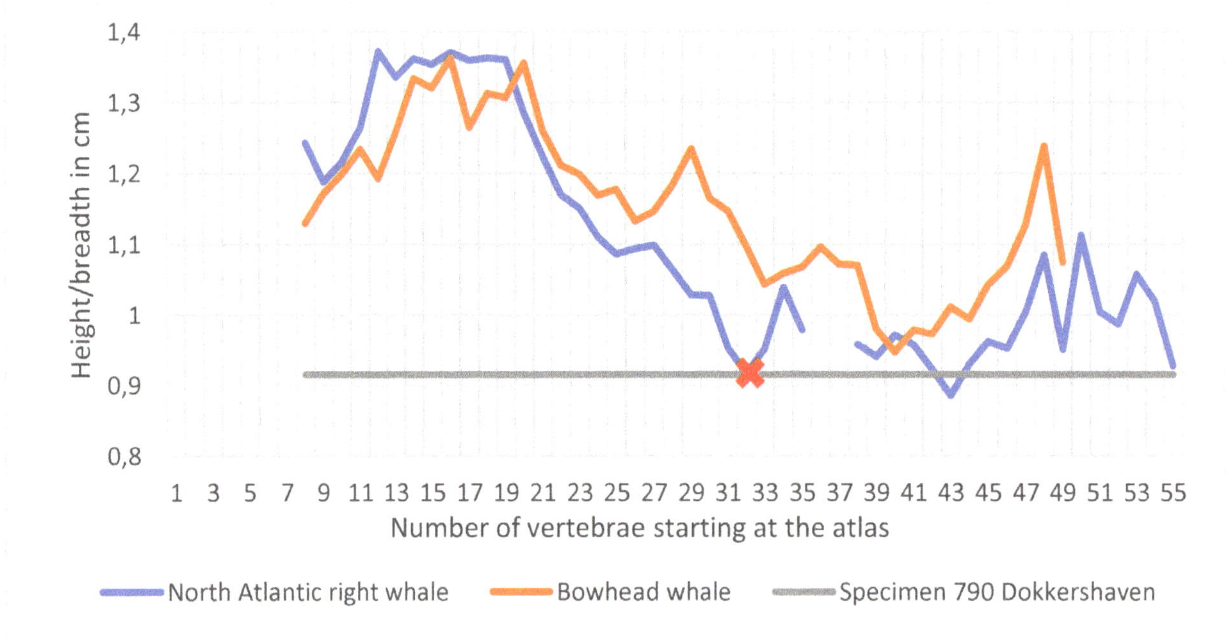

Figure 69. Height/breadth ratio of the vertebral column of the North Atlantic right whale (specimen 593893; part of the Smithsonian Institution) and the bowhead whale (specimen 1596; part of the Naturhistorisk Museum) in comparison to the same ratio for specimen 790 from Dokkershaven. The red X indicates that the specimen from Dokkershaven is most likely the 32nd vertebra of a North Atlantic right whale, which is the last lumbar vertebra. Measurements of the archaeological specimens are provided in table 5.

identifying the zooarchaeological remains to element as well it might therefore be possible to indicate whether the animal was exploited for its meat, or for other purposes (oil, ivory, baleen, etc.). However, in order to do this a large sample size is needed from one site, which is unfortunately not the case for the specimens analysed as part of this study.

Comparison to modern stranding data

Comparison between the zooarchaeological material identified for the Medieval Period and modern stranding data from the Netherlands for the period of 1969-2018 as recorded by Walvisstrandingen (2019) was undertaken to

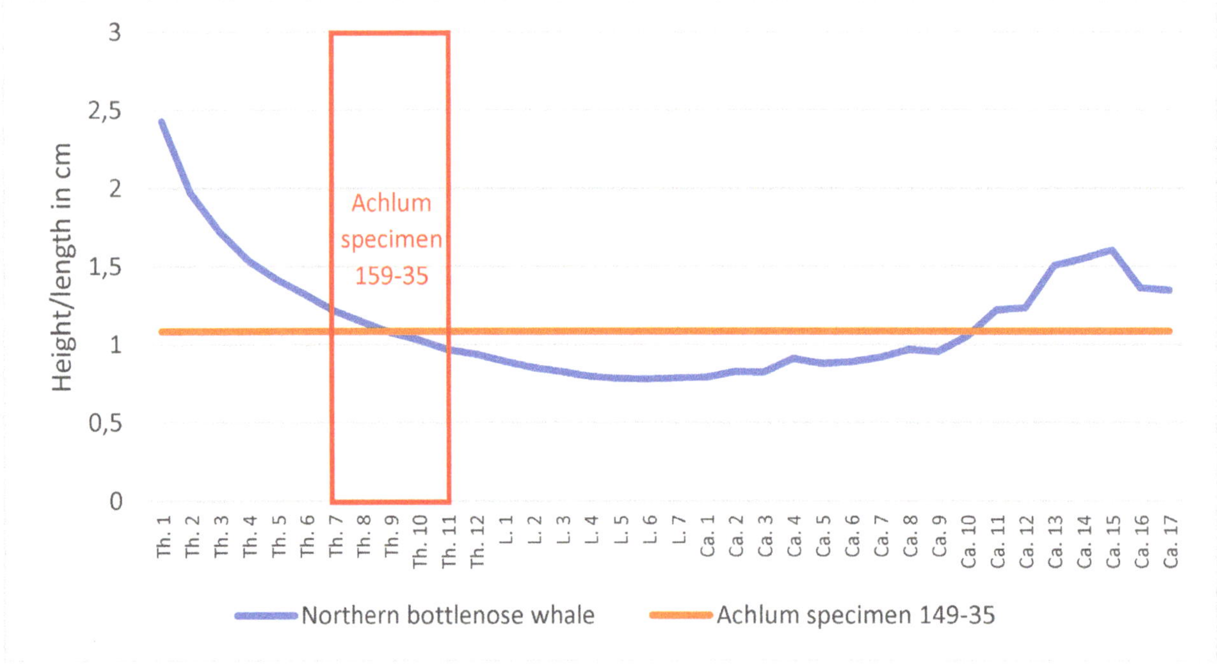

Figure 70. Height/breadth ratio of the vertebral column of the northern bottlenose whale (specimen 14447; part of the Smithsonian Institution) in comparison to the same ratio for specimen 149-35 from Achlum. Measurements of the archaeological specimens are provided in table 5.

Table 5. Measurements performed on (partially) complete vertebral remains (all in mm). Measurements undertaken include the maximum height, length and breadth of the vertebral bodies, according to the instructions provided in the ORCA-Manual (van den Hurk, forthcoming).

Site	Specimen number	Height	Length	Breadth	ZooMS and osteometric identification	Original identification
Slot op den Hoef	713	63,92	37,08	69,43	Risso's dolphin	Common bottlenose dolphin
Slot op den Hoef	714	62,65	38,61	69,8	Risso's dolphin	Common bottlenose dolphin
Brederode	1513-2	92,83	104,32	106,16	Long-finned pilot whale	Killer whale
Brederode	1513-1	122,6	-	137,65	Killer whale	Killer whale
Molenslag	438	53	42	29	Common bottlenose dolphin	Dolphin
Achlum	149-35	218,9	184,74	170,85	Northern bottlenose whale	Sperm whale
Dokkers-haven	790	355	325	255	North Atlantic right whale	Sperm whale or bowhead whale

see whether species distribution has changed between the Medieval Period and the modern period (table 6). However modern stranding data should be treated with caution as they are often the result by anthropogenic factors such as ship strikes or the swallowing of plastic.

From this data it appears that a number of species which were identified in the archaeological record are absent in modern stranding data, these species include the North Atlantic right whale, grey whale, and killer whale. This first species is close to extinction and is rarely sighted on the European side of the North Atlantic anymore. The Atlantic population of the grey whale is completely extinct. The killer whale has not stranded in the Netherlands for the past 50 years, though a sick individual was rescued from Dutch waters in 2010 (Walvisstrandingen, 2019).

Other species are only represented in modern strandings and have not been recorded in the medieval archaeological record. This might be the results of the relatively small sample size of medieval zooarchaeological cetacean material and an even smaller number of these been identified to species level. A large number of species is however both identified in the archaeological record and are known through modern strandings as well.

Table 6. Strandings to have occurred in the Netherlands over the 50 years period of 1969-2018 compared with the zooarchaeological cetacean material dating to AD 400-1600 from both the Netherlands and Belgium. Numbers in "()" are cf. identifications.

	Species	Strandings 1969-2018*	Archaeological material dating to 400-1600
Both represented in modern strandings and archaeological record	Harbour porpoise	9315	17
	White beaked dolphin	202	1
	Sperm whale	28	8 (1)
	Common bottlenose dolphin	24	1 (1)
	Fin whale	14	6
	Long-finned pilot whale	14	1
	Northern Bottlenose whale	3	5
	Risso's dolphin	1	2
Represented in modern strandings - not represented in archaeological record	Common minke whale	21	0
	Sowerby's beaked whale	12	0
	White sided dolphin	11	0
	Striped dolphin	11	0
	Humpback whale	6	0
	Common dolphin	2	0
	Sei whale	3	0
	Blainville's beaked whale	1	0
Not represented in modern strandings - represented in archaeological record	Grey whale	0	4
	Killer whale	0	2
	North Atlantic right whale	0	15

Chronological results

Zooarchaeological data from archaeological sites in the Netherlands from each time period have been collected by the Rijksdienst voor het Cultureel Erfgoed (the Cultural Heritage Agency of the Netherlands) in the database "BoneInfo". This database contains site information from archaeological reports, dissertations, theses, articles, and grey literature and is freely accessible to anyone with an interest in zooarchaeology (Rijksdienst voor het Cultureel Erfgoed, 2019). The database also contains information regarding medieval sites. For this study this database offers the unique opportunity to examine all sites with cetacean remains in comparison to those without cetacean remains. By doing this, it will be possible to see for which periods cetaceans were relatively more frequently exploited than for others.

For this study, all Dutch medieval sites were accessed and information regarding site type, dates of occupation and the location was collected. Furthermore, all sites were grouped into six categories, including: settlements (rural sites, small settlements, farms, etc.), terps (terp/wierde, tell sites in the coastal areas of the Netherlands; especially predominant in Friesland and Groningen), urban (middle to large sized settlements), high status sites (castles and other settlements with clear high status occupation), ecclesiastical (monasteries, churches, etc.), and other (cemeteries, graveyards, tanneries, etc.). This was undertaken for all 869 medieval sites that were available through BoneInfo and all the medieval sites from which cetacean remains were uncovered (46 sites in total for which 31 sites were also recorded in BoneInfo, bringing the total number of medieval sites analysed here to 884).

When comparing the Dutch sites with cetacean remains with those without cetacean remains, some interesting patterns emerge (figure 71 and 72). All the sites were plotted over a temporal range of AD 500-1500, with 25-year intervals. A site was counted as "1" as its temporal range fell within a 25-year period interval, creating a temporal overview for which all sites are equal. The sites were divided into two categories: sites with cetacean remains and sites without cetacean remains.

It appears that the number of sites with cetacean remains do not correspond with the number of overall sites in the Netherlands. The number of medieval sites overall goes up for the High and Late Medieval Period, but the number of sites with cetacean remains visibly goes down. As a result, there is a lower percentage of sites with cetacean remains during the High and Late Medieval Period, in comparison to the Early Medieval Period. The highest point of 13.4% in AD 850 and the lowest point in AD 1400-1500 3.0%.

Looking exclusively at the sites with cetacean remains (for both the Netherlands and Belgium) and when they are plotted chronologically (figure 73), there appears to have been a peak in cetacean exploitation around the ninth and tenth century AD, after which the number of sites with cetacean remains dropped gradually. There is sharp decline in number of sites for the beginning of the thirteenth century AD. This is the result of multiple terp-sites not being dated precisely and therefore ending roughly at the end of the twelfth century AD, at which point dykes were created allowing people to leave the terps, this is a so-called edge-effect. The historical sources suggested that there would be a peak in cetacean exploitation for the twelfth century, though the zooarchaeological remains do not confirm this.

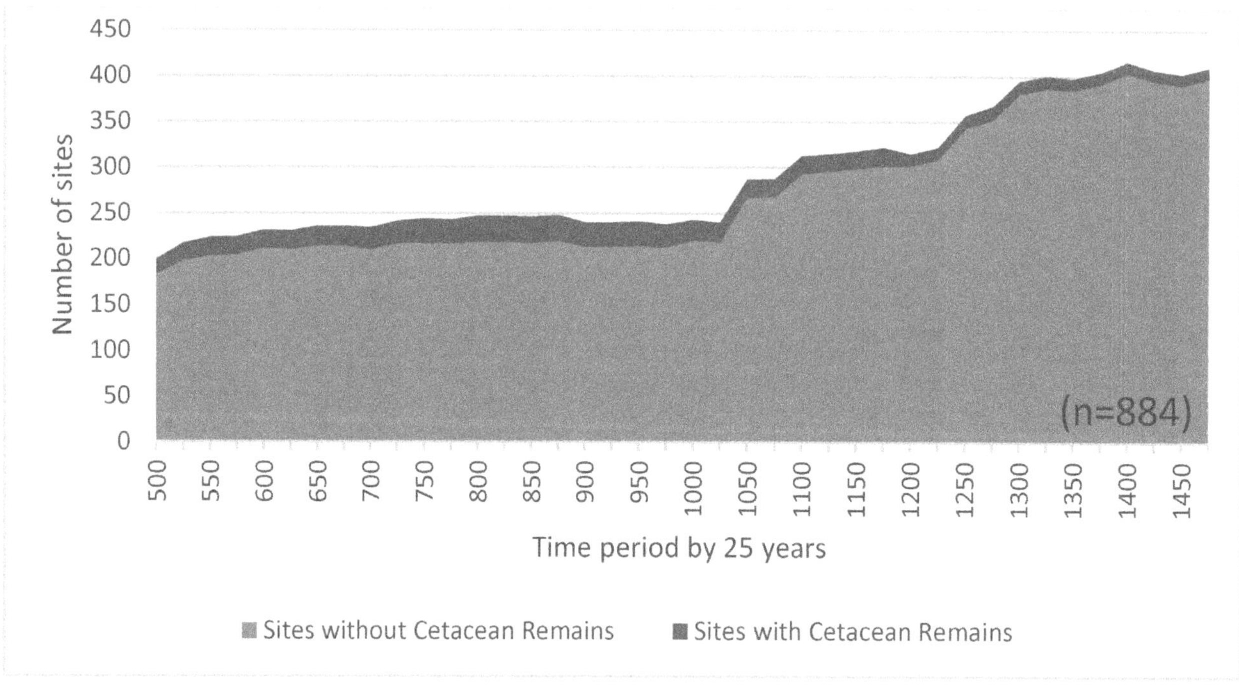

Figure 71. Comparison between medieval sites with and without cetacean remains. Data available through BoneInfo (Rijksdienst voor het cultureel Erfgoed, 2019).

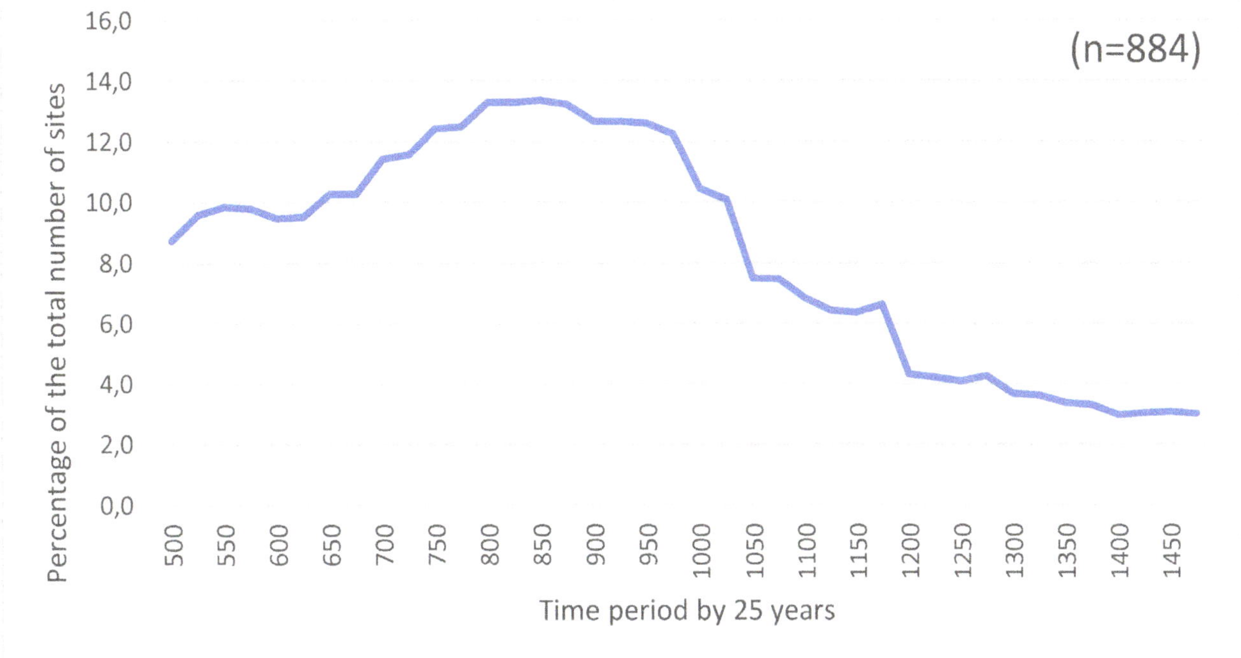

Figure 72. Percentage of medieval sites with cetacean as a part of all medieval sites in the Netherlands with zooarchaeological remains. Data available through BoneInfo (Rijksdienst voor het cultureel Erfgoed, 2019).

The decline in sites suggests that following this period, cetaceans were less frequently exploited.

A frequency density graph (figure 74) based on site-types and plotted for the period of AD 400-1600 with 25-year period intervals, in a comparable way as figure 58 was created. The frequency density graph is comparable to the number of sites graph, though the increase of urban site-types for the frequency density can be explained by the presence of seven northern bottlenose whale remains deriving from Aalmarkschool, Leiden dating to AD 1200-1275 (Esser, Beerenhout and Kootker, 2010).

Based on the site types, the social elite (both ecclesiastical and high-status sites) seems to get an interest in cetacean exploitation from the beginning of the eighth century AD, and at least continuing to the mid-thirteenth century AD. After this no ecclesiastical sites with cetacean remains are known, but high-status sites still display some interest in cetacean exploitation for the remainder of the Medieval

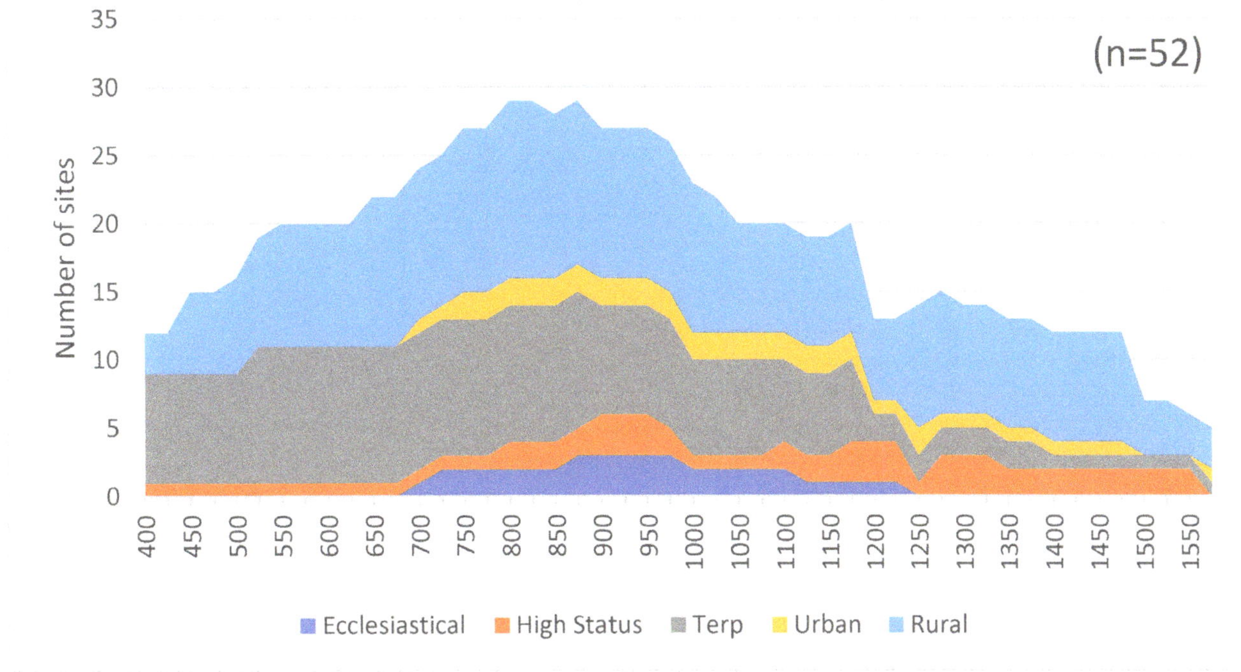

Figure 73. Temporal overview of sites with cetacean remains based on site-types 25-year period.

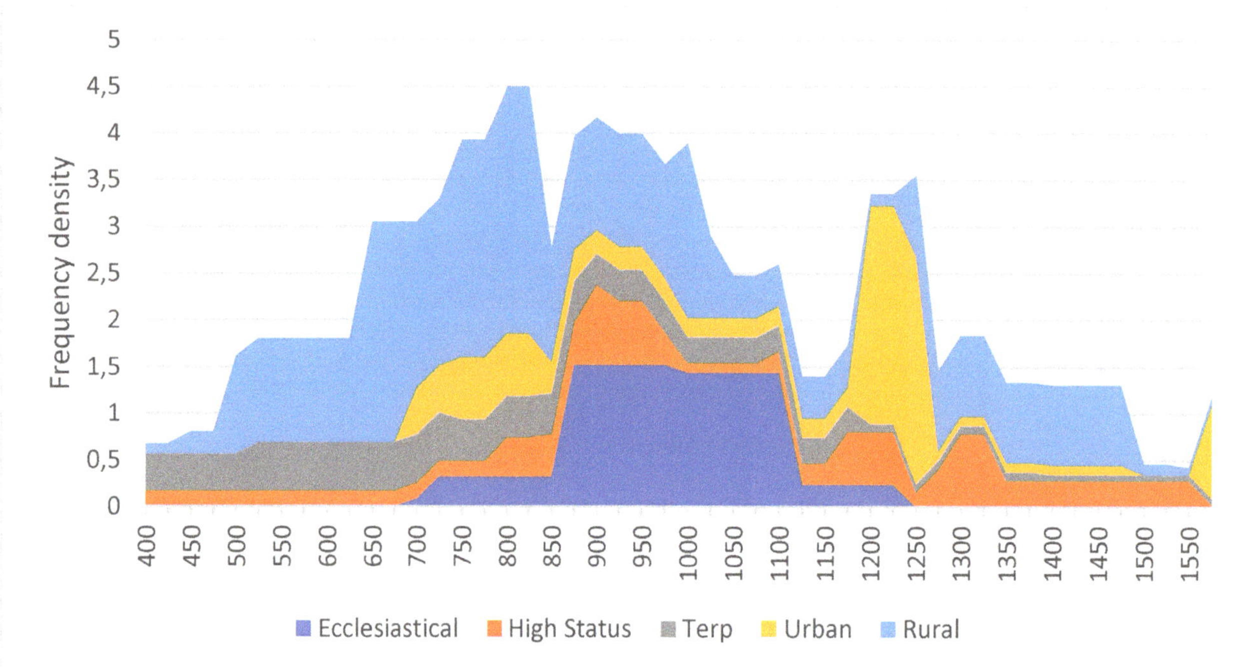

Figure 74. Temporal overview of estimated frequency density based on site-types per 25-year period.

Period up until the sixteenth century AD. Urban sites also display an interest in cetacean exploitation from the eighth century onwards, suggesting a certain commercialisation of cetacean exploitation already from this period onwards.

Discussion

This case study suggests that the exploitation of cetaceans was relatively more frequently undertaken during the Early Medieval Period. There can be several reasons why the number of sites with cetacean remains dropped for the High and Late Medieval Period. It might be that whaling was more frequently undertaken during the Early Medieval Period, but historical sources in general are scarce for this period, making it impossible to prove this. Historical sources for the High and Late Medieval Period do seem to imply that cetaceans were still sought after, and that whaling was at least occasionally undertaken. This suggests that a lack of interest in cetacean meat during the latter half of the Medieval Period is not the reason for the drop in numbers of sites with cetacean remains. Another potential reason is an increased interest in marine fish exploitation. The onset of the Fish Event Horizon around AD 950 (Barrett, Locker and Roberts, 2004a), might have eliminated the need to exploit stranded cetaceans as ample marine fish was now available.

The identification of cetacean material to species level based on morphology is hard as can be observed from the ZooMS results in comparison to the specimens identified through morphology. This clearly indicates that caution should be paid when identifying cetacean remains to species or even genus level. The results of this study also clearly demonstrate that several species dominate the archaeological assemblages.

Though not identified through the ZooMS analysis performed as part of this study, the harbour porpoise is still the best represented species in the archaeological material. This species is abundant in the North Sea, which is the most important area for the species with an estimated population of around 250,000 individuals in 2016 and the best represented species in modern stranding data (Walvisstrandingen, 2019). The historical sources suggest that it was the most widely exploited cetacean species for the Netherlands, which the zooarchaeological material seems to confirm. Indicating that the species has been hunted for centuries.

Of the species identified through ZooMS the North Atlantic right whale is the best represented species. This species could likely have been caught through active whaling. The species is a migratory species that often moves close to the coast. Furthermore, it a slow swimmer with a maximum speed of fifteen km/h (Jefferson, Webber and Pitman, 2008, 28-30) and tends to float after being killed, therefore making them the "right" whale to hunt. The Basques in northern Spain and southwestern France are known to have targeted the species during the Medieval Period (Aguilar, 1981). It appears that at least occasionally right whales were hunted in the Low Countries as well, though if this happened, not to the same extent as the whaling practices that occurred in the Basque area. However, as this species is known to have been more abundant in the southern North Sea in the past, natural stranded probably occurred more frequently as well, making it possible that these remains derived from those events.

The sperm whale is also relatively strongly represented. This species frequently strands along the shores of the southern North Sea. Barthelmess (1997) has recorded at least 12 sperm whale strandings on the Dutch and Flemish shores between 1519 and 1617, suggesting that strandings happened regularly for the Medieval Period as well. Albertus Magnus (1193-1280) saw the locals of Friesland butchering a sperm whale, piercing the animals eye resulting in the spermaceti flowing out of the hole. The locals filled eleven large flagons with it and also stripped the blubber from the animal (Albert the Great, 1987, 338-342). One interesting finding is a partial sperm whale skeleton displaying multiple chop- and cutmarks found on the beach in Walraversijde, Belgium dating to the Late Medieval Period. This young individual probably stranded along the coast and was subsequently butchered on the beach (van Neer and Ervynck, 1993, 87). It is however unlikely that sperm whales were actively hunted in the Medieval Period in the Low Countries, as the species is the necessary technology only appears several centuries following the Medieval Period.

The fin whale is the fourth most strongly represented species. Walvisstrandingen (2019) only records 39 records of this species stranding in the Netherlands, of which the oldest one dates to 1306. Strandings were not regularly recorded in the past as they are currently, but it is striking that fourteen strandings date to the period of 1998-2017, while there is a large gap between 1998 and 1956 with no strandings. At the end of the nineteenth and early twentieth century there are also more records for this species. This might suggest that the species stranded more frequently during the Medieval Period as well. It is likely that this species was only opportunistically exploited through strandings, as it is a large and fast species.

Grey whale remains are frequently found in the North Sea. The five new specimens identified through this case-study, suggests that this species, of which the North Atlantic population is now extinct, was once far more abundant along the Dutch and Belgian coast. This species, like the North Atlantic right whale, is relatively slow and tends to stay close to the shore and might have shown a similar migration route as the North Atlantic right whale. The remains identified in this study are the most recent grey whale specimens from the Netherlands and amongst the most recent findings in the whole of Europe. Only the grey whale remains coming from Babbacombe Bay in the UK dating to 1300-1800 AD are younger (Alter *et al.*, 2015). It appears that the species was hunted in the Low Countries during the Early and High Medieval Period, though it remains unclear whether whaling led to the depletion of the North Atlantic population.

The killer whale has not stranded in the Netherlands over the period of 1969-2018, however older strandings are known to have happened occasionally (Walvisstrandingen, 2019). The zooarchaeological material resembling this species might have been derived from such a stranding event, though also this species could have been actively hunted.

Strandings of northern bottlenose whale, pilot whale, Risso's dolphin, common bottlenose dolphin, and white beaked dolphin occasionally occur in the Netherlands. These species, with the exception of the bottlenose dolphin, do not frequently venture into the southern North Sea region and it is therefore likely that these species were not frequently actively hunted and therefore these specimens were likely acquired through scavenging of strandings or where opportunistically hunted.

The bottlenose dolphin however, had an established population in the western Wadden Sea area, hunting the herring present in the region. This population however disappeared after the construction of the Afsluitdijk, closing off the Zuiderzee from the Wadden Sea. Strandings of bottlenose dolphins still occur in the Netherlands (the most recent one in 2013), however it appears that in the past this happened more frequently with several individuals each year (Walvisstrandingen.nl). If this population was present in the area during the Medieval Period as well, it seems likely that it has been targeted as well.

Conclusion

Historical sources argue that whaling was occasionally undertaken, especially for Flanders and the Frisian region of the Netherlands. The abundance of harbour porpoise, North Atlantic right whale, and grey whale argues that active whaling was indeed practiced, from the beginning of the High Medieval Period already for the Flemish and potentially the Frisians as well, as these species were probably within the reach of medieval hunters. The abundance of zooarchaeological material from these species is an indication of a higher abundance of these species during the Medieval Period as well. This might indicate that natural strandings for these species occurred more frequently as well for the Medieval Period, and that the exploitation of these stranded individuals occurred as well, though there is clear evidence for active whaling as well.

Zooarchaeological cetacean remains appear to be relatively frequently identified at high-status sites such as castles, royal strongholds, or *stins* (stronghold or villa in province of Friesland). Furthermore, three specimens were identified at ecclesiastical sites. This suggests that the social elite indeed did have an interest in cetacean meat as the historical sources suggested as well. Cetacean remains however also derive from rural sites. This is clear evidence for peasant efforts to undermine elite control of stranded cetaceans. The accounts by the bailiff of the County of Flanders indicate that whenever a whale stranded, guards were assigned, ensuring that peasant poachers stayed away.

However, it should be noted that the presence of osteological remains of cetaceans on a site does not necessarily indicate that their meat was consumed but can also suggest that their osteological remains were merely used as a raw source for the production of various tools and artefacts such as chopping blocks. Moreover, as cetaceans are so large, their osteological remains might have been left on shore, while their meat was taken to site. This is probably what happened to the remains found in Walraversijde, Belgium (van Neer and Ervynck, 1993, 87).

The combination of both historical sources and zooarchaeological sources seem to confirm that cetaceans were indeed hunted during the Medieval Period, however it is not possible to argue this on a case by case method, and it can only be stated that it happened. The extent to which it was undertaken will remain unclear and will always be part of the enigma of cetacean material in archaeological contexts.

This case study made it also clear that the identification of zooarchaeological cetacean material faces a lot of problems. It is however vital to identify the remains to genus or species level in order to understand the complexities of early whaling practices. ZooMS, as well as the ORCA-Manual (van den Hurk, forthcoming), hold the potential to unravel the early beginnings of cetacean exploitation. Though the last decade has seen an interest in the zooarchaeology of cetaceans, numerous cetacean remains from various time periods in the Low Countries, Europe, and elsewhere still remain unanalysed.

5.2. England: "The Exploitation of Sea-Mammals in Medieval England: Bones and their Social Context": >20 years on

Introduction

In 1997, Mark Gardiner published "The Exploitation of Sea-Mammals in England: Bones and the Social Context". This extensive study, primarily focused on historical sources to reconstruct patterns in medieval cetacean exploitation, though several zooarchaeological sources were incorporated as well. This has led to Gardiner (1997) proposing a three-phase system for cetacean exploitation in medieval England.

Gardiner proposed that from the Anglo-Saxon period until the eleventh century (phase one), cetacean exploitation was limited to coastal communities, and they primarily relied on the opportunistic scavenging of stranded cetaceans. Cetacean meat did not travel far inland, and it was not restricted to the social elite.

This changed during the second phase, that started in the early eleventh century and lasted until 1300. During this phase the King, nobility, and clergy were interested in cetacean consumption and attempted to monopolize it. Stranded cetaceans were from this period onwards a royal and seigneurial right, and cetacean meat was perceived as a "high-status" food source. Porpoises were occasionally exploited, though active hunting on other species was still rarely undertaken. This was different for the other side of the English Channel, and whale meat was imported to England from France.

The third phase started around AD 1300 and Gardiner argued that the whale population (most likely the North Atlantic right whale population) was in decline from this period onwards. This led to less whale meat being available to the social elite and it fell out of favour. Porpoise meat continued to be sold as high-status food.

This system was set out over twenty years ago and was based on historical sources and limited zooarchaeological material. A lot of new cetacean findings have been done over the past twenty years, making it possible to assess the accurateness of this phase system and see whether the historical sources and the zooarchaeological sources confirm each other.

Methodology

As part of this case study, an interdisciplinary analysis based on both zooarchaeological and historical data was conducted. This was done in a similar manner as undertaken by Gardiner (1997), though in opposition to this study, the main emphasis will be placed on zooarchaeological data. The acquired data will be the framework against which the historical data will be compared. Archaeological sites dating to the Medieval Period from which cetacean remains have been uncovered were assessed, in order to create a temporal overview of cetacean exploitation in medieval England. The Medieval Period is considered to be the period between AD 400-1600 and was split into three periods: the Early Medieval Period/Anglo-Saxon period (AD 400-1066), the High Medieval Period/Norman period (AD 1066-1216), and the Late Medieval Period (AD 1216-1600).

Zooarchaeological data was examined, and the number of identified specimens (NISP) for all cetacean remains were assessed. This provided a general idea of the representation of each species within the archaeological record. The majority of the remains were merely identified based on morphology, though several specimens deriving from London were analysed using ZooMS (van den Hurk, Rielly and Buckley, forthcoming))

A large number of remains were not identified to the species level and were merely identified as "unknown cetacean", "large cetacean, "medium cetacean", or "small cetacean". As part of this study large cetaceans are considered to be the sperm whale and all the baleen whales with the exception of the common minke whale. These species are all generally larger than ten meters. The common minke whale, beaked whales, killer whale, long-finned pilot whale, short-finned pilot whale, false killer whale, beluga, and the narwhal are considered medium sized cetaceans and generally range between four to ten meters in length. The small cetaceans are the species generally shorter than four meters and include the majority of the Delphinidae as well as the pygmy sperm whale, the dwarf sperm whale, and the harbour porpoise.

Special emphasis was placed on the species represented, the location of the sites, and the contexts the material derived from. This information was subsequently compared to the three-phase system set out by Gardiner (1997) and was additionally compared to records of the *Calendar of Patent Rolls* concerned with cetaceans. The *Calendar of Patent Rolls* dating to 1216 and 1452 were analysed, using the webpage created by Boynton (2016). This provided data regarding whaling endeavours in England, the stranding of cetaceans, as well as to who exploited and consumed cetaceans.

The combination of the zooarchaeological and historical sources allowed for the possibility to examine whether the exploitation and consumption of cetaceans was restricted to the social elite (nobility and clergy), as proposed by Gardiner (1997).

Historical Sources

Historical sources can provide valuable information regarding status and diet. For Anglo-Saxon England, beef, pork, poultry, and wild animals were considered to be feasting foods, while mutton and lamb were rarely mentioned (Hagen, 1994). Cetaceans are also rarely mentioned in Anglo-Saxon historical records but Bede's Historia Ecclesiastica (AD 731), and Ælfrcic's Colloquy (AD 955-1010), appear to suggest that cetacean exploitation was at least occasionally undertaken (Swanton, 1975, 110-111; Gardiner, 1997).

Just prior to the Norman Conquest, cetacean exploitation started to get restricted to the social elite. Although, enforcing these restrictions was hard. Many coastal areas of medieval England were not densely populated making it relatively easy for a cetacean stranding to go unnoticed or for peasants to conceal them and exploit them themselves. Peasants appear to have illegally exploited stranded cetaceans relatively frequently. However, this was a risky business. Whenever they were caught, punishments and fines were enforced.

A historical source in which whales are frequently discussed in England are the *Calendar of Patent Rolls*. These are a series of administrative records that comprise a register of the letters patent issued by the Crown. They cover grants of official positions, commissions, pardons and privileges. The rolls date back to 1201 to the reign of King John (Boynton, 2016).

As part of this study, the Patent Rolls between AD 1216 and 1452 were analysed, using the webpage created by Boynton (2016) concerned with the Patent Rolls which allows to search for specific words or phrases. For this study the words: "whale", "whal", "porpoise", "cete", "grampus", "balena", "porpoise", and "grapays" were searched for. The texts that contained these words were further analysed in order to see whether a stranding occurred, or active hunting was undertaken.

A simple search for "whale" in the *Patent Rolls* provided 64 results, of which a total of 45 are actual records that are concerned with whales (the remaining 19 are hits in the indexes). In comparison, a search (including index hits) for "deer" provides 1277 hits, "hare" 803 hits, "partridge" 387 hits, "pheasant" 349 hits, and "rabbit" 330 hits, indicating that the terrestrial animals were more frequently the topic of the administrative records. Although, as these animals are considerably more common in comparison to whales, the number of hits for whales is relatively high. The instances that a whale stranded along the coast were still probably rare, but whenever this happened, the local elite appears to have tried to claim them and punish the commoners who illegally exploited the carcass.

A total of 52 texts concerned with cetaceans have been identified (based on all the words searched for) of which most date to the period 1325-1349 (during the final years of the reign of Edward II and the first two decades of the reign of Edward III; figure 75; table 7). Cetaceans were claimed several times for the King, but the texts also indicate that a large number of nobility members (e.g. the Earls of Surrey, Angos, and Lincoln, and some other lords) and clerical members (e.g. the Bishops of London, Ely, Lincoln, Exeter, and Durham, and the Abbot of Ramsey) claimed cetaceans. These nobility and clerical members used their "wreck of sea" rights to claim stranded cetaceans on their land. The cetaceans had often already been butchered or taken away by peasants. The Patent Rolls recorded claims, and commissions (*commission of oyer and terminer*) were subsequently issued to investigate and potentially fine the people who illegally exploited the stranded cetaceans.

In another record, King Edward III confirms the rights the church of St. Stephen, Caen had to whales and other fish cast ashore on their land, which rights had previously been given to them by King Henry II. Additionally, one record dating to the 4th of March 1399, deals with John German who bought a whale for 10 marks from William Godmanston of Frynton. William however, illegally exploited this stranded whale and was ordered to pay 10 marks to the King as a fine. Yet another source recorded a group of people that stole the boats, nets and engines of the Bishop of Lincoln near his manor of Newerk (Newark-on-Trent), which they used to illegally catch a porpoise in the river Trent.

For the third quarter of the fourteenth century a decrease in texts concerned with cetaceans occurred. This drop can probably be ascribed to the Black Death that reached England in June 1348 and spread rapidly after that. As a result, the population of England dropped dramatically and potentially led to fewer claims of stranded cetaceans. Alternatively, the desire to exploit cetaceans by the social elite had ceded, but zooarchaeological data might be able to answer that question.

The geographical distribution of the records from the *Patent Rolls* (figure 76), indicates that cases concerned with cetaceans were distributed along the entire eastern and southern coast of England. Records are predominantly concerned with cases coming from The Wash region, a bay and estuary at the north-western corner of East Anglia.

Figure 75. Number of entries in the Calendar of Patent Rolls concerned with cetaceans per 25 years.

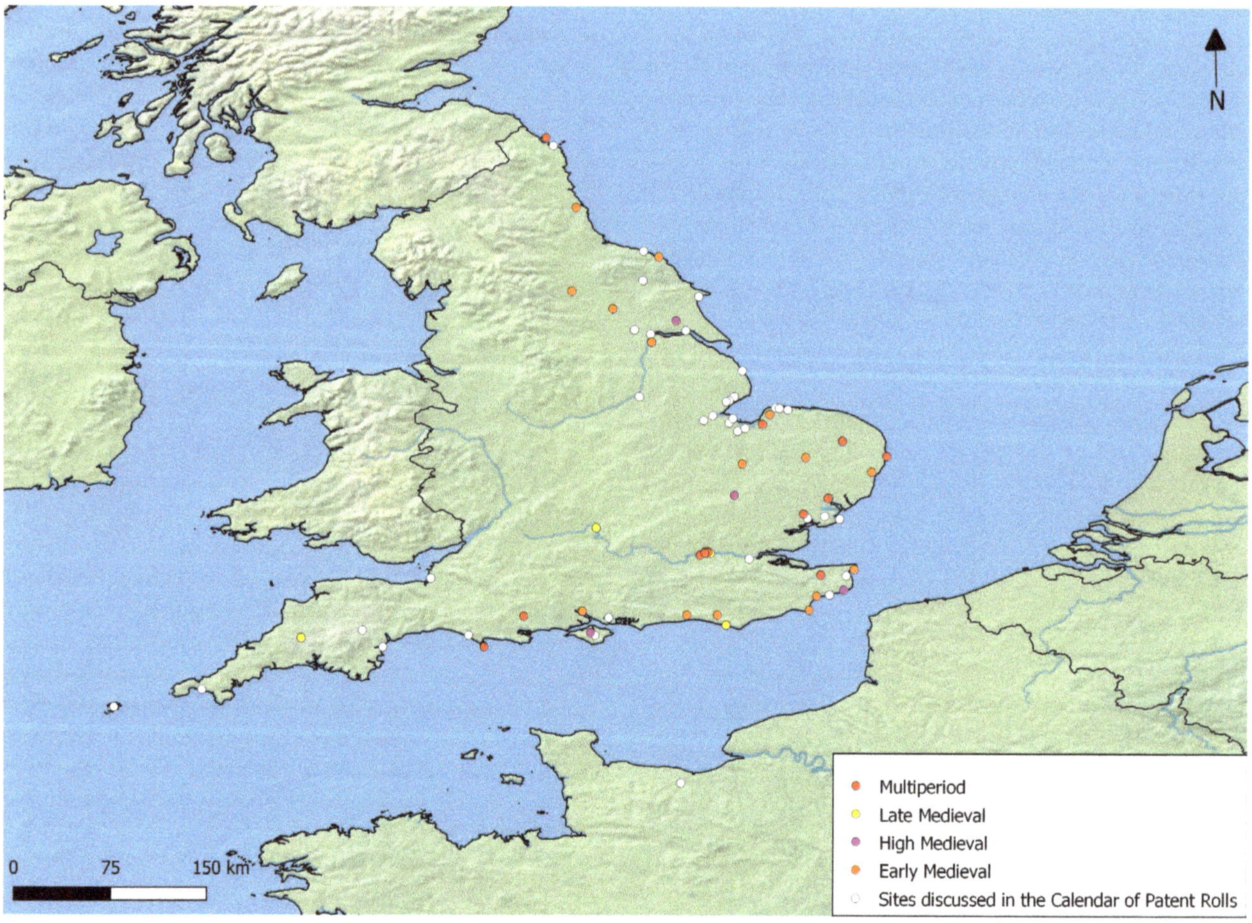

Figure 76. Location of archaeological sites and records in the Calendar of Patent Rolls assessed as part of this study.

Table 7. Calendar of Patent Rolls records concerned with cetaceans

Key word	Name	Vol.	P.	Day	Month	Year	Claimant	Location (as written in Rolls)	Activity
Whale	Edward I	1	468	15	January	1281	Isabella de Albiniaco, Countess of Arundel.	Thornham and Tychewell, co. Norfolk	Peasants carried away a stranded whale
Whale	Edward I	1	469	17	January	1281	Abbot of Ramsey	Brauncestre, co. Norfolk	Peasants carried away a stranded whale
Whale	Edward I	2	445	10	September	1291	Peter, Bishop of Exeter	Teynkton, co. Devon	Peasants carried away a stranded whale
Whale	Edward I	2	513	24	January	1292	Master Andrew de Kilkenny, Master Peter de Insula and Master Robert de Veteri Terra, executors of the will of Peter, sometime bishop of Exeter	Teynton, co. Devon	Peasants carried away a stranded whale
Porpoise	Edward I	2	520	22	August	1292	Bishop of Lincoln	Manor of Newerk	Peasants carried away a stranded whale
Whale	Edward I	3	16	15	May	1293	King	Sandwich	Peasants carried away a stranded whale
Whale	Edward I	3	547	29	March	1300	Bailiffs and Commonalty of Colecestre	Colecestre	The people of the town caught a whale, intended to hand it over to the King, but the prior of Mereseye and other men took away the whale
Whale	Edward I	4	405	26	October	1305	Henry de Lacy, Earl of Lincoln	Button in Holand, co. Lincoln.	Peasants carried away a stranded whale
Whale	Edward I	4	480	22	November	1306	Ramilph de Albo Monasterio	Isle of Sullye, co. Cornwall. (Isle of Scilly)	Peasants carried away a stranded whale
Whale	Edward II	1	129	16	March	1309	Ranulph de Albo Monasterio	Scilly Islands, county Cornwall	Peasants carried away a stranded whale
Whale	Edward II	1	248	11	December	1309	John de Warenna, Earl of Surrey	Brunham within the precincts of his hundreds of Galehowe and Brothercross, co. Norfolk.	Peasants carried away a stranded whale
Whale	Edward II	1	365	6	March	1311	John de Warenna, Earl of Surrey	Brunham, within the precincts of his hundreds of Galehowe and Brothercross, co. Norfolk	Peasants carried away a stranded whale
Whale	Edward II	1	359	2	July	1311	Earl John de Warenna	Brunham	Peasants carried away a stranded whale
Whale	Edward II	2	134	20	January	1314	Abbot of Rameseye	Brauncestre, co. Norfolk	Peasants carried away a stranded whale
Whale	Edward II	2	317	16	April	1315	Abbot of Abbodesbury	Abbodesbury	Abbot of Abbodesbury claims to have wreck of sea of all cast ashore of Abbodesbury, sheriff took it away for the King
Whale	Edward II	2	299	11	June	1315	Benedict, Abbot of Abbodesbury	Abbodesbury	John de Erie, sheriff of the county of Dorset, took two tuns of a whale carcass stranded on the abbots land, but the abbot claims he has the rights to it
porpais	Edward II	5	143	14	May	1325	King	Pykeryngg	Peasants carried away a stranded whale
Grapays (great fish)	Edward II	5	143	14	May	1325	King	Manor of Faxflet, co. York	Peasants carried away a stranded whale

Key word	Name	Vol.	P.	Day	Month	Year	Claimant	Location (as written in Rolls)	Activity
Cete	Edward II	5	283	18	February	1326	Stephen, Bishop of London, and the dean and chapter of the church of St. Paul, London	Manor of Walton, co. Essex.	Peasants carried away a stranded whale
Cete	Edward II	5	290	12	April	1326	Stephen, Bishop of London, and the dean and chapter of the church of St. Paul, London	Manor of Walton, co. Essex.	Peasants carried away a stranded whale
Whale/ Cete	Edward III	1	481	20	December	1329	John, Bishop of Ely	Walpol in Merskland, co. Norfolk	Peasants carried away a stranded whale
Whale	Edward III	1	561	20	April	1330	John, Bishop of Ely (manor and lordship of Walpol in the parts of Merskland, co. Norfolk)	Walpol in the parts of Merskland, co. Norfolk	Peasants carried away a stranded whale
Whale	Edward III	1	572	16	Augustus	1330	John, Bishop of Ely (with a manor in Tiryngton)	Tiryngton	Peasants carried away a stranded whale
Whale	Edward III	2	203	15	October	1331	King	Porcestre, County of Southampton	Peasants carried away a stranded whale
Whale	Edward III	3	64	20	September	1334	Ebulo Lestraunge, Lord of the manor of Fryskencye, co. Lincoln	Fryskencye	Peasants carried away a stranded whale
Whale	Edward III	3	137	20	February	1335	Henry de Lancastre, Lord of the manor of Pykering	Pykering, co. York	Peasants carried away a stranded whale
Whale	Edward III	3	141	26	March	1335	Ebulo Lestraunge and Alice his wife (Lord of the port of the town of Friskeney)	Friskeneye	Peasants carried away a stranded whale
Whale	Edward III	3	141	1	April	1335	Henry, Earl of Lancaster (with manor at Pickering)	Fyfle, Yorkshire	Peasants carried away a stranded whale
Whale/ Grampuses/ Balenas	Edward III	3	287	14	April	1336	King	Isle of Wight	Peasants carried away a stranded whale
Whale	Edward III	3	374	4	December	1336	King	Fossedyk in the parts of Holand, co. Lincoln	Peasants carried away a stranded whale
Whale	Edward III	3	375	10	December	1336	King?	Coast of the Thames between Grenewich and Northflete, co. Kent	Peasants carried away a stranded whale
Whale	Edward III	4	77	26	April	1338	King	Gedeneye (Gedney), co. Lincoln	Peasants carried away a stranded whale
Whale	Edward III	4	558	8	June	1340	Richard, bishop of Durham (lordship of his manor of Houeden, co. York)	Houeden, Yorkshire	Peasants carried away a stranded whale
Whale/ Grampus	Edward III	5	362	8	November	1341	Gilbert de Umframvill, Earl of Angos - lord of the manor of Malberthorpe, co. Lincoln	Malberthorpe, co. Lincoln	Peasants carried away a stranded whale
Whale	Edward III	5	373	26	January	1342		Church of St. Stephen, Caen	Confirmation of the Grant by Henry II (1154-1189), that they should have whale and other fish Grant by Henry II, that they should have whale and other fish cast ashore in their land.
Whale	Edward III	6	28	20	May	1343		Kyngeston upon Hull	Peasants carried away a stranded whale
Whale	Edward III	6	93	20	May	1343	Richard, bishop of Durham, Lord of the manor of Houeden	Houeden, Yorkshire	Peasants carried away a stranded whale

Key word	Name	Vol.	P.	Day	Month	Year	Claimant	Location (as written in Rolls)	Activity
Whale	Edward III	6	581	18	January	1345	Edward, prince of Wales, Duke of Cornwall, and Earl of Chester	Portlegh by Kenegy	Peasants carried away a stranded whale
Whale/ porpais	Edward III	7	304	10	February	1347	Thomas de Thwenge, Katharine late the wife of William de Thwenge, John Faucumberge of Skelton and Bartholomew de Fanacourt	Hildrewell, Kyrkelythum, Eston in Clyveland, Ormesby and aclum on Tepe	Peasants carried away a stranded whale
Whale	Edward III	8	83	20	April	1348	King	Folkeston, co. Kent	Peasants carried away a stranded whale
Whale	Edward III	8	316	10	April	1349	Henry, earl of Lancaster	Wrangle	Peasants carried away a stranded whale
Whale	Edward III	8	319	18	May	1349	Queen Philippa	Lek, co. Lincoln	Peasants carried away a stranded whale
Whale	Edward III	8	589	12	July	1350	Simon Semeon, Lord of the manor of Surflete, co. Lincoln,	Surflete, co. Lincoln	Peasants carried away a stranded whale
Whale	Edward III	11	202	20	May	1359	Walter de Byntree	the port of Ipswich and all ports and places thence to Boston	Purveyance of several fish and whales
Porpoise	Edward III	11	475	25	October	1360	William, provost of the church of St. John, Beverley	Great and Little Monkwyk	The provost claimed 2 porpoises stranded there, but disputed as might be the King's right, but he is eventually acquitted of all charges
Whale	Edward III	12	288	26	October	1362	Queen Philippa	Ipswich and the vicinage	Peasants carried away a stranded whale
Whale	Edward III	12	445	10	July	1363	Queen Philippa	Ipswich, co. Suffolk	Peasants carried away a stranded whale
Whale	Edward III	12	361	26	October	1363	Queen Philippa	Ipswich and in the neighbourhood, co. Suffolk	Peasants carried away a stranded whale
Whal	Edward III	16	59	27	November	1374	King	King's lordship of Baumburgh	Peasants carried away a stranded whale
Whale	Richard II	2	417	20	January	1384	Crown?	Devon	Peasants carried away a stranded whale
Whale	Richard II	6	481	4	March	1399	John German	The hundred of Tendryng, co. Essex	John German bought a whale for 10 marks from William Godmanston of Frynton, who sold it to him for 10 marks. William has to pay the 10 marks to the King.
Whale	Henry VI	5	46	13	May	1447	Gilbert de Gaunt and Richard Malebyse	Port of Fysle, Yorkshire	Peasants carried away a stranded whale

The Calendar of Patent Rolls clearly indicate that cetaceans were a highly prized and prestigious resource the social elite attempted to monopolize. These records therefore provide a great medium to assess the complex social aspects that are associated with cetacean exploitation during the Medieval Period in England. However, from these sources it is often not possible to determine which species are dealt with. The term "porpoise" was probably also used to describe dolphins, and the term "whale" for a wide variety of baleen whales, the sperm whale and potentially even some of the beaked whales or large dolphin species. Zooarchaeology however holds the potential to unravel which species were exploited.

Zooarchaeology

Gardiner (1997) stated that few medieval sites in England have yielded cetacean remains, which indeed was the case twenty years ago, but numerous new findings have been done in the meantime. Although cetaceans are indeed not frequently found (in comparison to terrestrial mammals), they are not rare among medieval zooarchaeological assemblages either. As part of this study 64 sites were identified at which cetacean remains have been found.

Considering a temporal overview of the 64 sites, numbers of sites are low for the period after the Roman period and the migration period (figure 77). From the mid seventh century onwards however, an increase can be noted. This is around the time that Anglo-Saxon England became Christianised. Cetacean remains appear to have been present at ecclesiastical sites from the beginning of the Medieval Period, but from AD 700 the number of sites increased, remaining relatively stable during the remainder of the Medieval Period. The high number of urban sites, mostly located in London, additionally suggests that cetacean meat was transported to these regions as it was perceived as a high-status food source from the seventh century onwards.

During the period of Viking influence the number of sites with cetacean remains went gradually down, only to increase right before the Norman Conquest of 1066. Following the Norman Conquest, the number of sites rises again, and the number of sites is situated between 15 and 20 sites until the mid-fourteenth century. From that point a decrease in number of sites can be noted. This can probably be ascribed to the Black Death, though the total number of sites remains at or above 14.

The number of high-status sites increased right after the Norman Conquest of AD 1066, suggesting that only from this point onwards cetacean consumption was associated with nobility, as has also been suggested by Gardiner (1997). It is furthermore interesting to note that the number of rural as well as urban sites went down from this point onwards. As Gardiner (1997) suggested, this might be the result of the attempt to monopolize the consumption of cetaceans by the nobility. Stranded animals were from this point onwards perceived as a "wreck of sea" right and were the exclusive right of the King or other members of the nobility of clergy he granted those rights to. The number of ecclesiastical sites also increased slightly from the Norman Conquest onwards, suggesting that the clergy also took an increased interest in cetacean meat from this point onwards.

A frequency density graph (figure 78) based on site-types and plotted for the period of AD 400-1500 with 25-year period intervals, in a comparable way as for figure

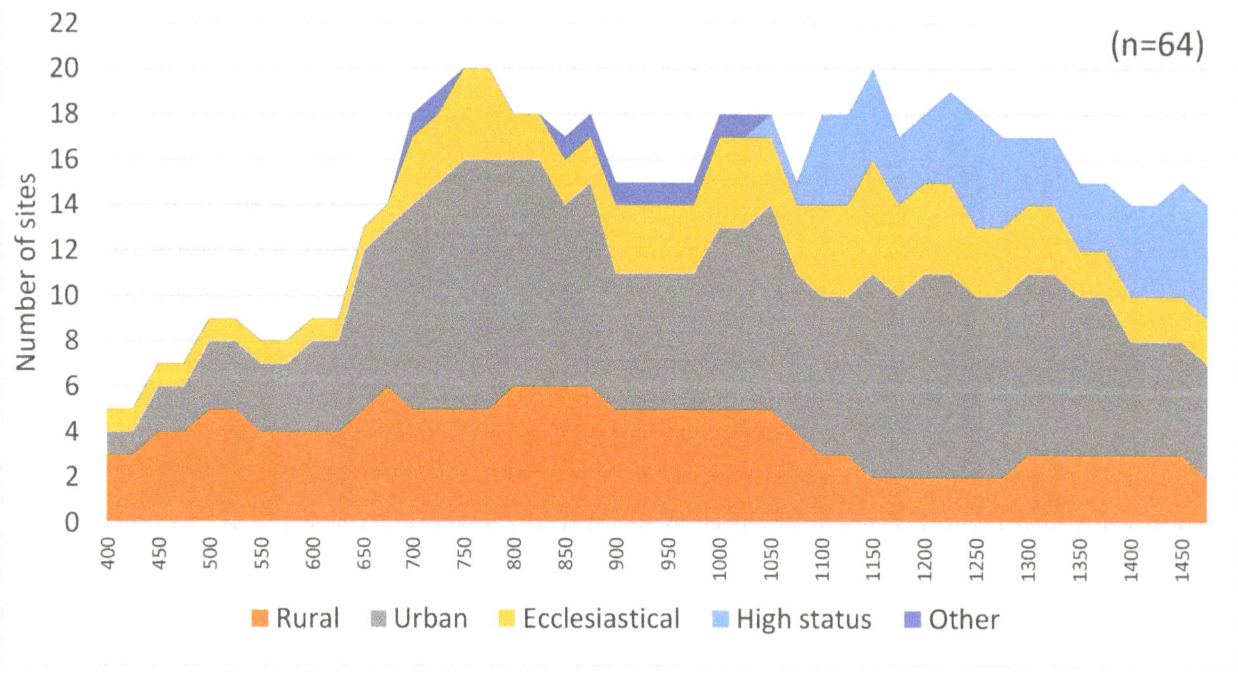

Figure 77. Temporal overview of English medieval sites with cetacean remains per 25 years.

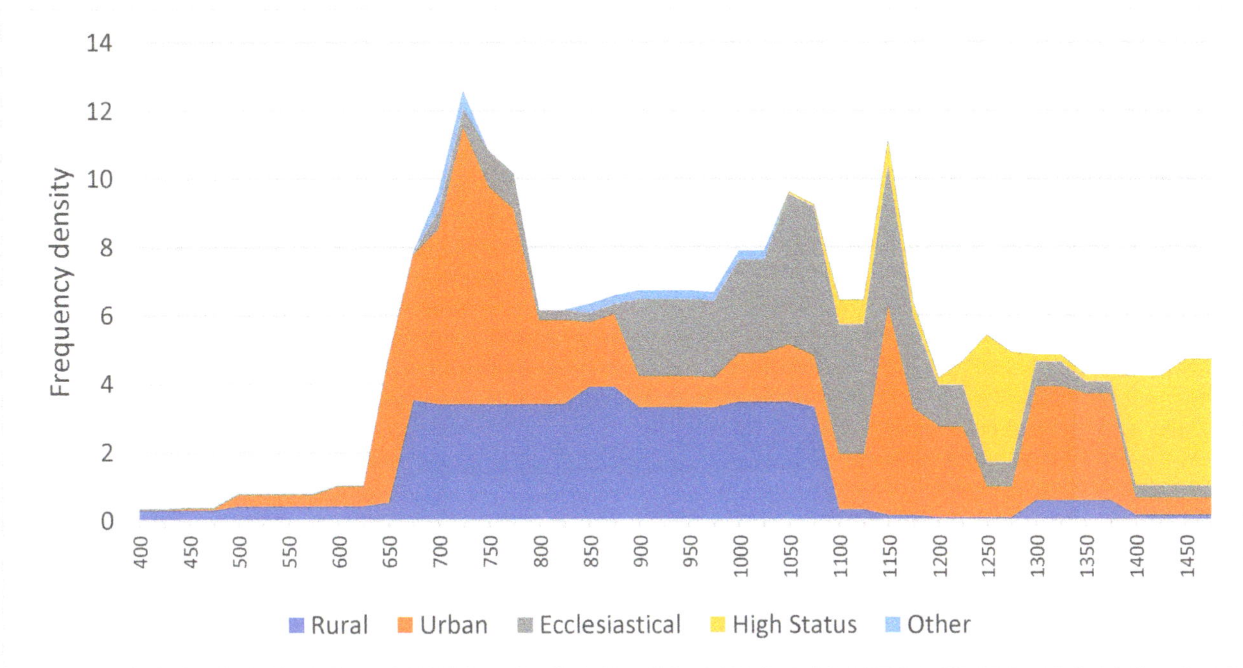

Figure 78. Temporal overview of frequency density based on site-types per 25-year period for England.

58, was created. A similar pattern as the number of sites graph in figure 77 can be noted. Limited remains predate the mid-seventh century, after which a sudden increase can be noted. A high number of remains derived from urban contexts in London for the seventh to the ninth century, after which the numbers decline again and remain relatively stable up until the onset of the High-Medieval Period, for which an increase can be noted again. From this period onwards limited cetacean remains derive from rural contexts, and the majority now originates from the ecclesiastical, high-status, or urban contexts. This is again comparable to figure 77 and can be ascribed to the social elite trying to monopolize cetacean consumption, though some form of commercialisation and transportation of cetacean meat to urban markets, also occurred.

Considering the species exploited, it can be noted that a considerable number of remains could not be assigned to any species, and for 323 specimens (of which 240 belong to the site of Hamwic, Ascupart Street) not even an indication of the size of the animal was provided, resulting in them being labelled merely as "unknown cetacean" (figure 79).

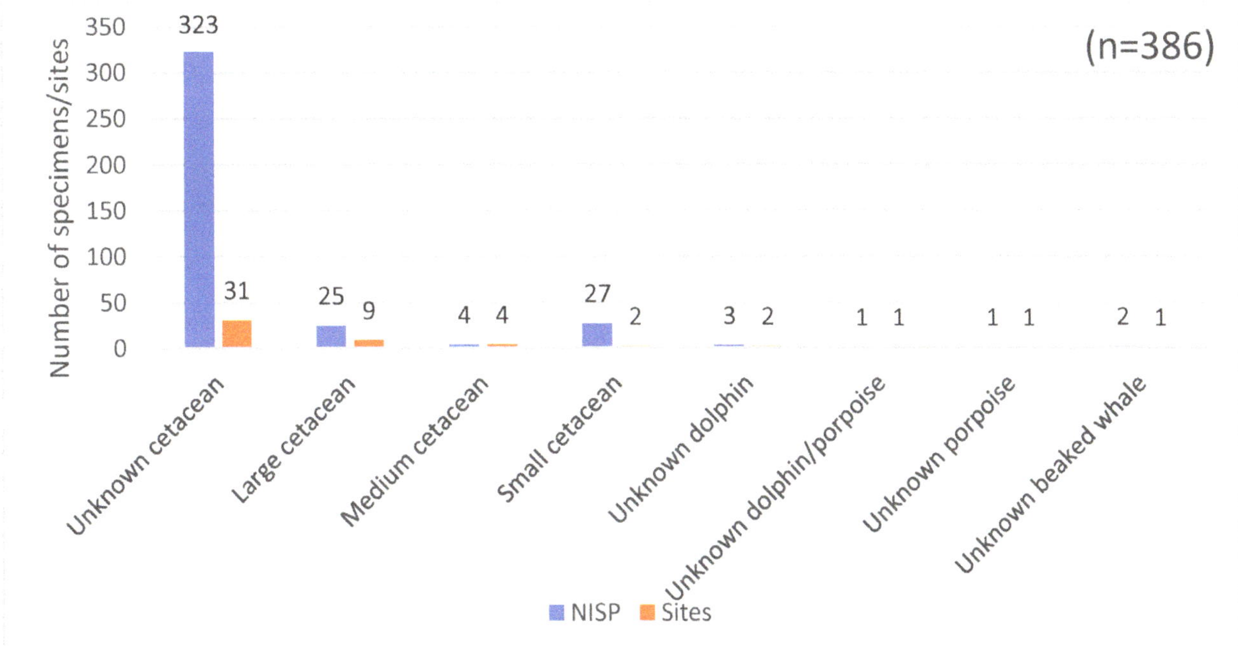

Figure 79. Number of cetacean remains which could not be identified to species level.

Examining the remains that could be identified to species level (figure 80), it is clear that the common bottlenose dolphin (*Tursiops truncatus*) is the best represented species, although it should be noted that of the 116 specimens, 115 derived from Flixborough (Dobney *et al.*, 2007). Furthermore all 50 specimens identified as common bottlenose dolphin cf. also all derived from Flixborough. Besides common bottlenose dolphin remains, killer whale and common minke whale remains have also been identified at this site. The remains from this site have been dated to AD 677-1099 and comprise several phases, suggesting a long-standing tradition of cetacean exploitation in the region. It has been suggested that a common bottlenose dolphin population was present in the area during the medieval times. This population went extinct at least 100 years ago (Nichols *et al.*, 2007).

Dolphins are known to occasionally swim up rivers, and this has also been reported for the Humber (on which Flixborough is situated) and these therefore might have been exploited in the river. The high number of remains clearly indicate that there was a steady supply of these animals, suggesting that active hunting was undertaken in the region (Loveluck, 2001, 93-94). It has been argued that the presence of the dolphins on the site indicates high-status dietary practices. Moreover, the presence of other animals in the zooarchaeological assemblage, including crane, and high proportions of fish and fowl, has also been used to suggest this. However, the material derived from Flixborough predates the phase set out by Gardiner (1997) as the period for which cetacean consumption was a high-status practice. It might therefore have been a locally undertaken practice, not associated to social status.

The second site in England with a high number of cetacean remains is the site of Hamwic in Southampton. Several excavations have been carried out at this site, but particularly from the sites at Ascupart Street and Six Dials have whale bones been recovered. Over 240 fragments have been recovered from these sites and the majority

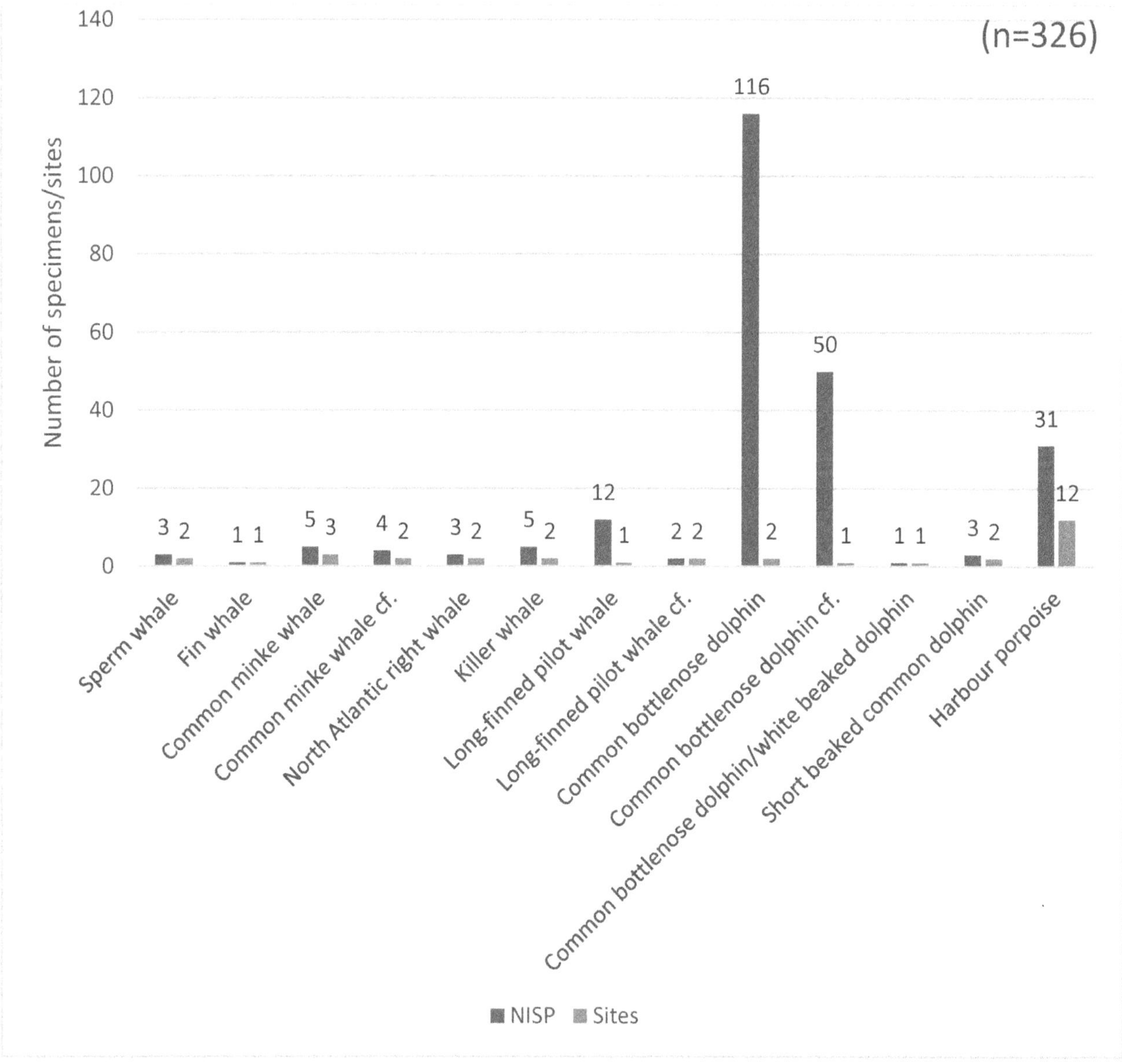

Figure 80. Number of cetacean remains which could be identified to species level.

appears to have been used for the creation of tools and artefacts, including combs, spindle whorls, and needles. Whale bone is thought to have been a replacement for antler, which was rare in the eight and first half of the ninth century (Riddler and Trzaska-Nartowski, 2014).

The harbour porpoise is the second-best represented species in the English medieval archaeological record and has been identified at eleven sites. The majority of these sites are ecclesiastical sites, suggesting the clergy developed an interest in porpoise meat. Indeed, the harbour porpoise is relatively frequently mentioned in medieval historical sources and frequently associated with an ecclesiastical diet (Gardiner, 1997).

Several large whales are also represented in the zooarchaeological data, including the sperm whale, fin whale, common minke whale, and North Atlantic right whale. Of these four, the former three are not likely to have been actively hunted, as these were too fast or aggressive for medieval whalers in England to have been exploited. The North Atlantic right whale is however considered to have been in the reach of medieval whalers and is known to have been frequently exploited by Basque whalers (Aguilar, 1981). Two carcasses of this species were recovered from the site of Dengemarsh, in the south-eastern part of England, and have been dated to AD 840-1043. The carcasses clearly showed butchery signs, indicating that they were stripped of their valuable resources, including meat and blubber (Gardiner, Stewart and Priestley-Bell, 1998).

Since processing of stranded or caught cetaceans most likely took place on the foreshore, these remains are rarely encountered. Therefore, the two North Atlantic right whale carcasses represent an exceptional finding. However, even for these excellently preserved specimens, it is not clear whether they were actively caught or opportunistically stranded, highlighting the difficulties arising trying to reconstruct past cetacean exploitation.

Discussion

The temporal pattern of cetacean exploitation observed for medieval England is considerably different to that of the Netherlands and Belgium, as can be seen in the case study concerned with that geographical region. Cetacean exploitation for the Netherlands and Belgium showed a peak between AD 800-850, after which the number of sites gradually, but steadily decreased.

This appears to not have been the case for England. Right after the Roman period cetacean exploitation seems to have been rare, but from the mid-seventh century AD there appears to have been an increase in cetacean exploitation, peaking at the eighth century. It remains unclear whether active whaling was practiced, although Bede's *Historia Ecclesiastica,* dating to AD 731, indicates that whaling was already undertaken in Britain during the eighth century (Gardiner, 1997). The material from Flixborough suggests that at least the common bottlenose dolphin was actively caught by this point in time.

Furthermore, the site of Hamwic also poses an interesting case, as the high number of remains might be indicative of active whaling practiced in the area. However, as the material is worked and fragmented it is unclear which species the Hamwic material represents. A selection of material from Hamwic is currently being analysed using ZooMS in order to establish which species the material derived from. This might be helpful in determining whether active whaling was undertaken, as several species are more likely to have been caught than others. Whaling is known to have occurred in the English Channel from an early date onwards already and it is likely that the material found at Hamwic represents this.

Moreover, the hunting on porpoises might have also regularly been undertaken or might have been a by-product of regular fishing activities.

From the mid-seventh century AD, cetacean remains are also more frequently encountered at ecclesiastical sites. This pattern might be explained by the Christianising of the Anglo-Saxons and the adoption by the clergy of fasting practices, which excluded the meat of terrestrial mammals for particular days of the year but allowing the consumption of fish and (semi-)aquatic mammals (Fagan, 2006, 15-57).

Following the Norman Conquest, the nobility also appears to get an interest in cetacean consumption, and cetacean remains are more frequently found in high-status contexts. This is in line with historical documents, which suggest that the interest in cetaceans by the clergy predates that of the nobility by several centuries. This interest by the nobility in cetaceans appears to have lived on until the end of the Medieval Period, and the records of the *Calendar of Patent Rolls* seem to imply that the social elite indeed did try to maintain their rights over stranded cetaceans until at least the mid-fourteenth century. Following the Great Famine and the Black Death, historical sources are becoming rarer, but zooarchaeological sources seem to suggest that interest in cetacean meat lived on during fifteenth century.

Conclusion

The zooarchaeological data to some extent validates Gardiner's (1997) three-phase system, though it appears that the clergy and nobility developed a taste for cetacean meat several hundred years apart, with the former starting to develop a taste for cetacean meat during the seventh century and the latter at the end for the eleventh century. This signifies a cultural interest in cetacean meat which was eventually enforced by laws that ensured the social elite receiving at least a portion of every whale caught or stranded. From the Norman Conquest onwards, cetacean meat appears to have been less available to peasantry, which is probably the result of the attempt by the nobility to monopolize the exploitation of stranded cetaceans. Although in some cases peasants appear to have broken the laws and exploited stranded cetaceans, as attested by the records of the *Calendar of Patent Rolls* and the recovery of cetacean bones from several rural and low-status contexts. This is clear evidence of peasants undermining the social elite's power.

The harbour porpoise is relatively frequently recovered from various sites all over England, as was the case for the Netherlands and Flanders as well. The combination of zooarchaeological and historical sources clearly indicate that whaling was undertaken in England from the Anglo-Saxon period onwards already. The exploitation of large species at Hamwic, and of smaller species at Flixborough, indicates that no uniform whaling technique was practiced in England however. New zooarchaeological findings of cetacean material, as well as the identification of those remains to the species level, will optimize our understanding of medieval cetacean exploitation. If one species dominates the archaeological assemblages it might be argued that the species was actively hunted, but even in those cases other factors could have led to the abundance of one species in a specific region or period, making it almost impossible to identify active whaling practices based solely on zooarchaeological evidence. ZooMS has already been undertaken on several remains, but more in-depth research using ZooMS, as well as aDNA analysis, will continue revolutionizing zooarchaeological research on cetaceans.

5.3. Grey whale (Eschrichtius robustus): Greys in Grey Literature: The Demise of the North Atlantic Grey Whale (*Eschrichtius robustus*)

Introduction

The grey whale (*Eschrichtius robustus*) is the one extant species of the family Eschrichtiidae. Adult individuals normally are between twelve and fifteen meters long (figure 81) and can weigh up to 35 ton (Jefferson, Webber and Pitman, 2008, 70-73). They primarily feed on benthic organisms, but their diet also includes planktonic and nektonic organisms at the sea surface or midwater. Foraging activities normally take place in waters not deeper than 120 meters, primarily restricting the species to coastal waters. The whales concentrate their feeding during approximately five months from May/June to October/November in high latitude waters. During the remaining six to seven months they migrate to more southern waters. During this migration they fast and rely on stored lipid (Jones and Swartz, 2009, 503-511).

There are currently two populations. The first one presumably migrates between the coastal area of southern Korea and the Sea of Okhotsk and numbers about 130 individuals. This population is endangered, though shows signs of recovery thanks to heavy conservation action. The second one numbers between 20,000 and 22,000 individuals and migrates between northern Alaska and Baja California Sur (Jones and Swartz, 2009, 503-511). Both these populations are located in the North Pacific, but in the past another population was present in the North Atlantic. This population has now been extirpated, but appears to have survived in the North Atlantic until the seventeenth- or eighteenth-century AD. Among baleen whales, grey whales are the only species to have been extirpated from an entire ocean basin during historical times (Alter *et al.*, 2015).

Several historical sources presumably discuss the grey whale's presence in the North Atlantic, including an account from sixteenth century Iceland described as the "sandloegja" (Fraser, 1970). Lindqvist (2000) analysed numerous Icelandic historical sources mentioning cetaceans dating from the twelfth to eighteenth century AD and has suggested that the grey whale was present in Icelandic waters until presumably the eighteenth-century AD. A short paper by Dudley (1725) describing the "scrag whale" off New England, and a commission from the Muscovy Merchants to Thomas Edge in 1611 describing the "otta sotta", have also been interpreted as Atlantic grey whale (Mead and Mitchell, 1984).

The descriptions of whales in historical sources are often vague and hard to ascribe to a particular species. Findings of sub-fossil remains are more clear evidence of the grey whale's presence in the North Atlantic. Findings have been made on both sides of the North Atlantic dating to several hundred years to 50,000 years ago. A multitude of studies have been undertaken on these remains, and though it has often been suggested that the demise of the Atlantic grey whale population was the result of premodern whaling enterprises, a study by Alter *et al.* (2015) suggested that low genetic diversity, a reduction in haplotype diversity, and a loss of suitable habitat underlies their extirpation. They furthermore suggested that premodern whaling might have played a role as well, but with few grey whale specimens originating from the archaeological record, they proposed that this might have only been a minor factor.

Recently however, several grey whale bones have been identified from the archaeological record. This can partly be ascribed to an increase in interest in whales in the field of zooarchaeology, but also to development of Zooarchaeology by Mass Spectrometry (ZooMS), which aids the identification of archaeological and subfossil cetacean remains. As part of this case study, five new grey whale specimens were identified, all originating from archaeological contexts from the Netherlands. Moreover, grey whale specimens from archaeological contexts have also been identified by Rodrigues *et al.* (2018) in Spain and Morocco, Hufthammer *et al.* (2018) in Norway, and Szabo and her team have been able to identify grey whale remains at Greenland, Iceland, and the Orkney Islands (personal communication Vicki Szabo, 20-05-2019).

Dispersal into the Atlantic

Grey whales spend a considerable portion of their life in shallow shelf habitats in high-latitude waters. They do

Figure 81. Grey whale skull. Photo created by the United States National Museum Photographic Laboratory. Smithsonian Institution Archives, Acc. 11-007, Box 013, Image No. MNH-2910. Photo courtesy of the Smithsonian Institution. CC0 License. Available at: <https://www.si.edu/object/gray-whale-skull:siris_arc_388654>.

not live close to heavy sea ice like the bowhead. Because of the massive ice sheets in the Arctic Ocean, migration from the Pacific to the Atlantic Ocean was last possible during the last interglacial period (130-115 thousand year ago). Furthermore, because of the low sea level during the period of 70-10 thousand years ago the Bering Strait was closed off, making migration impossible.

Alter *et al.* (2015) indicated that there appear to have been at least two temporal groups for Atlantic grey whale specimens. The first group ranges from ~40,000 to >50,000 cal BP, while the second one ranges from <250 to ~10,000 cal BP. There are no specimens dating to the period in between (~11,000 to ~40,000 cal BP). During this time the sea-level was lower. If the grey whale persisted in the North Atlantic during this period, their habitat would have been much smaller (only 61% of suitable habitat during the height of the last glaciation). More specimens from this period of time might turn up, but Alter *et al.* (2015) performed aDNA research on the available specimens in the North Atlantic, compared those to modern Pacific grey whale specimens and suggested that there were several migration events from the Pacific to the Atlantic over the past ~100,000 years. Alter *et al.* (2015) suggested that there was low genetic diversity and a reduction in haplotype diversity during the mid-Holocene. This might indicate that grey whales were extirpated from the Atlantic at least once before during the mid-Holocene.

A potential new colonization of the grey whale occurred about 3000 cal BP. The Northwest passage of the Arctic was open probably from 1-12 Ka BP. During this time the whales could have migrated from the Beaufort Sea to the Hudson Sea and subsequently enter the North Atlantic Ocean (Jones and Swartz, 2009, 503). The Northwest passage became closed off during the Little Ice Age when temperature cooled enough for ice to block the way (1400-1850 AD).

aDNA research on late-Holocene specimens indicated an even lower haplotype diversity than was present in the mid-Holocene (Alter *et al.*, 2015). Low genetic diversity, loss of habitat resulting from climate change, as well as premodern exploitation of grey whales have all been listed as possible causes of the demise of the Atlantic population of the grey whale. The disappearance of the grey whale from the Atlantic Ocean has frequently been ascribed to whaling endeavours, although Alter *et al.* (2015) state that very few specimens originate from archaeological context, but more findings have been done since 2015.

Material and methods

As part of concerning medieval exploitation of cetaceans in the Low Countries, zooarchaeological cetacean remains were assessed using ZooMS. Samples were acquired through visits to the provincial archaeology depots of the Netherlands, including the archaeological depots of the three northern provinces (Groningen, Friesland, and Drenthe), North Holland, and Zeeland. There the specimens were sampled for 30 mg of bone. These samples were then taken to BioArCh, the University of York, UK. Here the samples were demineralized in 0.6 M hydrochloric acid, gelatinized, digested with trypsin and purified using a C18 resin ZipTipw pipette tip (EMD Millipore). A Bruker ultraflex III MALDITOF/TOF mass spectrometer was used to run the samples in triplicate. The mass spectra were subsequently assigned to species.

Furthermore, sixteen bones labelled as "grey whale" belonging to three or four individuals were found at the Groningen Institute for Archaeology (GIA), part of the University of Groningen, and were examined based on morphology and osteometry, comparing the bones to the grey whale skeleton at the Smithsonian Institution, Washington DC, USA (specimen NMNH USNM 593558, subadult male with a total length of 1125 cm). The sixteen specimens were collected by Robert Kosters (former employee of the GIA) in late 1970s/early 1980s from the depot of the "Rijksdienst voor de IJsselmeerpolders" (IJsselmeerpolders Agency) in Ketelhaven. The agency stored material found during the creation of the various Dutch polders, including bone material. The agency wanted to discard the remains, but the GIA showed interest in the material, and besides several moose (*Alces alces*) and steppe wisent (*Bison priscus*) the sixteen grey whale specimens were taken. Until this point no one analysed the material and its existence is not known to the zooarchaeological community.

Furthermore, a reanalysis of published data regarding sub-fossil and zooarchaeological grey whale specimens was undertaken. As many of the remains come from the Netherlands a lot of the data is still hidden in grey literature. Moreover, several publications appear to have published erroneous data, obscuring our understanding of the demise of the Atlantic grey whale. The zooarchaeological data will then be compared to historical sources, which will allow one to assess the possibility that grey whales were actively hunted from the Medieval Period to the eighteenth century on both sides of the North Atlantic.

Results

ZooMS analysis indicated that four of five specimens are grey whale specimens (figure 82). For the fifth specimen the ZooMS provided no results. However, morphological comparison indicated that this fifth specimen was grey whale as well. The Netherlands has by far produced the most grey whale remains of any of the regions in the North Atlantic region. The five specimens identified as part of this study all originate from archaeological context and are therefore unique findings. The five specimens all originate from different provinces of the Netherlands. None of these have been radiocarbon dated but have been dated based on the stratigraphy of the various sites, and range between 600 BC and 1200 AD.

The specimen from the terp (artificial dwelling mounds in northern Netherlands and Germany) Hallum, Friesland is half a cervical vertebra, displaying three chopmarks, two on the vertebral body and one on the transverse process. It has been interpreted to resemble a cat by IJssennagger-van der Pluijm (2018), but besides the chopmarks it displays no signs of working. It does not have a precise date, but as it is from the terp it most likely dates between the first century BC to the twelfth century AD.

The second specimen is from the terp Rottum, Groningen and also does not have a precise date, but is estimated to be from somewhere in between 600 BC and AD 1200

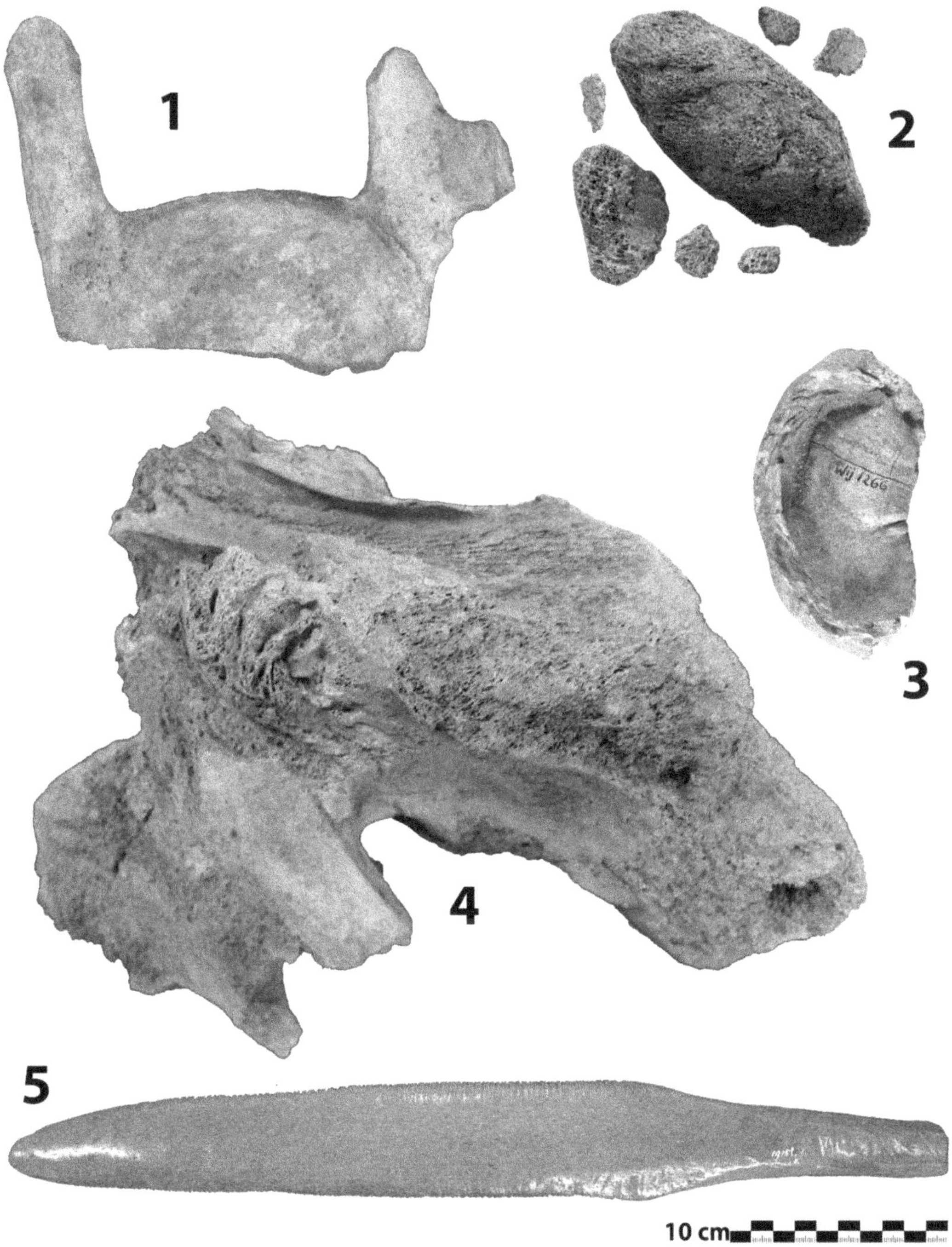

Figure 82. Grey whale specimens analysed using ZooMS as part of this study. 1. Hallum, 2. Den Burg, 3. Wijster, 4. Domburg, 5. Rottum. Photos by Y. van den Hurk.

(NADNuis, 2018). It is a weaving sword with markings all over the edge of the blade. A rune was recorded on the blade as well but does not match any of the runes described by Phillippa and Quak (1994).

The third bone is from Den Burg, located on the island of Texel, North Holland. It is dated to AD 750 to 1000, at what time Texel was still attached to the mainland of Holland (Krauwer, 1982). The bone is extremely fragmented, and it cannot be determined what skeletal element it is from. From the site a radius of a North Atlantic right whale (*Eubalaena glacialis*) and a bone fragment of a fin whale (*Balaenoptera physalus*) have also been identified, but it remains unclear whether these were acquired through active whaling or opportunistic scavenging.

The fourth bone is a cranial fragment found at the excavation of a circular rampart at Domburg, Zeeland from a site of Badstraat 1-3/Motel 't Groentje. The skull is dated to 850 to 1000 AD and displays numerous chopmarks, potentially suggesting that oil was extracted from the bone (Buitenhuis, 2011). This is the first finding of grey whale from the province of Zeeland, although a whale barnacle of the species *Cryptolepas rhachianecti*, that is associated with the grey whale, has been recovered from Zoutelande, Zeeland as well, suggesting that this barnacle species was also present on Atlantic grey whale individuals (Bosselaers and Collareta, 2016).

The fifth specimen was identified as grey whale based on the morphology (Prummel *et al.*, 2018). ZooMS was conducted on the specimen as well, but there was not enough collagen left in the bone and the ZooMS unfortunately provided no results. The specimen is a tympanic bulla which contains only 14% collagen (Reitz and Wing, 2008, 18), which might have problematized the ZooMS analysis. The bulla was recovered from an animal grave for a cow (*Bos taurus*) and a horse (*Equus caballus*) in Wijster, Drenthe and is dated to the third or fourth century AD (Clason, 1967; van Es, 2018). Why the bulla was located in the grave remains unclear, but it was potentially brought to the site as a curiosa.

These are not the first grey whale specimens coming from archaeological contexts. Three other grey whale specimens identified by Rodrigues *et al.* (2018) are also from archaeological contexts. Rodrigues *et al.* (2018) indicated that grey whale was present in the far western corner of the Mediterranean, where they identified three specimens using ZooMS and aDNA analysis. Identification of grey whale specimens based on morphology can lead to errors. Macé (2003) has previously identified grey whale specimens from southern France but ZooMS and aDNA analysis conducted by Speller *et al.* (2016), indicated that these were wrongly identified.

Besides the material analysed using ZooMS, sixteen bones identified as grey whale stored at the GIA were also analysed. The identification of the sixteen bones, was originally done by the IJsselmeerpolders Agency, but unfortunately the provenance of the bones is unknown (besides the fact that the bones were retrieved from one of the polders). Fifteen of the sixteen bones material have been numbered by the IJsselmeerpolders Agency and all start with either "Z" or "ZW". The meaning of these abbreviations is not clear. Additionally, the year the bones were found at are also incorporated, which is useful in determining at least from what region the bones derived.

Two of the bones have been found in 1935 (specimens ZW1935/V 244-2 (a cranially and caudally unfused lumbar vertebra) and ZW1935/V 344-1 (a cranially and caudally unfused thoracic vertebra)), suggesting that these have not been collected from the IJsselmeerpolder or the Noordoostpolder as work had not commenced on these polders by that year. This suggests that it might have been recovered from the Wieringermeerpolder from which a partial skeleton was retrieved as well in 1935 (Van Deinse and Junge, 1937; Bryant, 1995). The two bones might have belonged to this skeleton as well, however the partial skeleton is currently stored in the Naturalis which is being renovated, not allowing to check whether the two bones are part of this skeleton.

The vertebral bodies are unfused allowing limited osteometric analysis. The breadth and height of the vertebral bodies were measured using the instructions specified in the ORCA-Manual (van den Hurk, forthcoming). The breadth to height ratio of the vertebral bodies were compared to those of the grey whale skeleton specimen NMNH USNM 593558 (figure 83). The results indicate that ZW1935/5 244-2 was a 2nd thoracic vertebra,

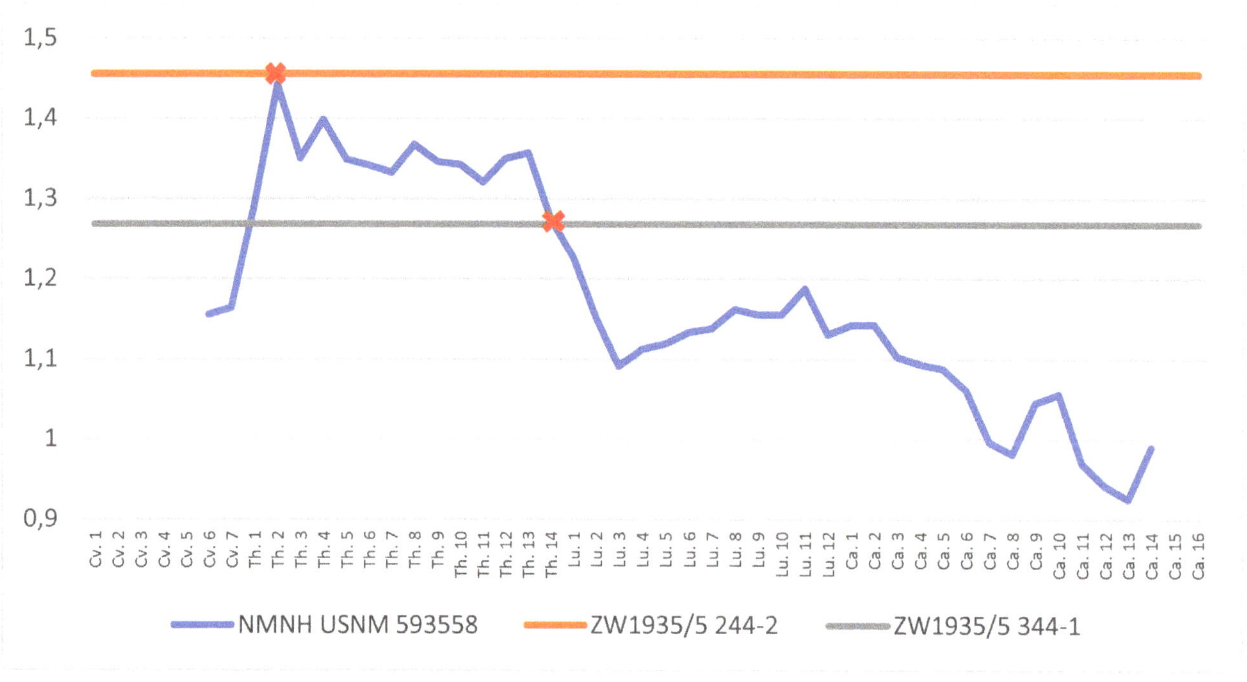

Figure 83. Breadth/height for grey whale specimen NMNH USNM 593558 compared to the two specimens ZW1935/5 244-2 and ZW1935/5 344-1. The red crosses indicate the intersection between the samples and grey whale specimen NMNH USNM 593558, indicating that ZW1935/5 244-2 most likely is a 2nd thoracic vertebrae and ZW1935/5 344-1 a 14th thoracic vertebra.

while specimen ZW1935/5 344-1 most likely was the 14th thoracic vertebra.

Furthermore, thirteen bones have been labelled as "Z1956/IV". Ten of these bones are vertebral epiphyses (62 and 63 are complete; 60, 64, 67, and 68 are almost complete; and 69, 70, 71, and 72 are fragments; figure 85) The other three specimens are all partial unfused lumbar of caudal vertebrae (57, 75, and 76). It is likely that these specimens belonged to one individual. It is however unknown what happened to numbers 58, 59, 61, 65, 66, 73, and 74, and whether these were grey whale bones belonging to the same individual as well. These might have been the bones discarded by the IJsselmeerpolders Agency. The fact that these bones were collected in 1956 suggests that the bones were recovered from the eastern part of the Flevoland polder, as work on that polder commenced in 1950 and finished in 1956.

The three vertebrae were too badly damaged to allow osteometric analysis, although the breadth to height ratio of four epiphyses were again compared to grey whale skeleton specimen NMNH USNM 593558 (figure 84). Based on the comparison it is most likely that the epiphyses belonged to one of the last couples of lumbar or one of first couple of caudal vertebrae. Indeed, the morphology of the three vertebrae also seem to indicate that these are the last lumbar or first caudal vertebrae, indicating that the bones are likely from one individual.

The sixteenth bone part of the collection present at the GIA does not have a "Z" or "ZW" number. This specimen is a left mandible and is 141 cm long labelled as number "8216". The bone appears to have been fragmented and glued back together. Additionally, on both the medial and lateral side there are two metal strips keeping the bone together. Considering this size of the bone this is juvenile individual. The morphology (the curvature of the mandible, the foramina, the processus coronoideus, and the condyle) clearly indicates that this mandible is from a grey whale (figure 86).

Another left mandible, originally identified as common minke whale (*Balaenoptera acutorostrata),* present in the Poldermuseum in Heerhugowaard (figure 87), also analysed as part of this study, equally displays a morphology typical for grey whales. Likewise, this bone originated from a juvenile individual.

Additionally, a reanalysis of published data on the Atlantic grey whale has revealed some errors. As some of the grey whale remains are old and have been published in multiple publications, some erroneous data duplication has been done. Garrison *et al.* (2019) published a list of specimens, but duplicated that data of IJmuiden, the Netherlands. Furthermore, a specimen currently stored in Oostduinkerke-Koksijde was not found at that location but is merely stored there in the local Fishery Museum. It was found near the mouth of Thames river by Belgian fishermen (Asselberg, 1981).

Furthermore, several grey whale specimens have only been published in grey literature, making them inaccessible to foreign zooarchaeologists or palaeontologists. An updated list of known specimens is provided in table 8 and geographical distribution of the specimens displayed in figure 88 and 89. At the time of writing, the total number of identified grey whale specimens for the eastern side

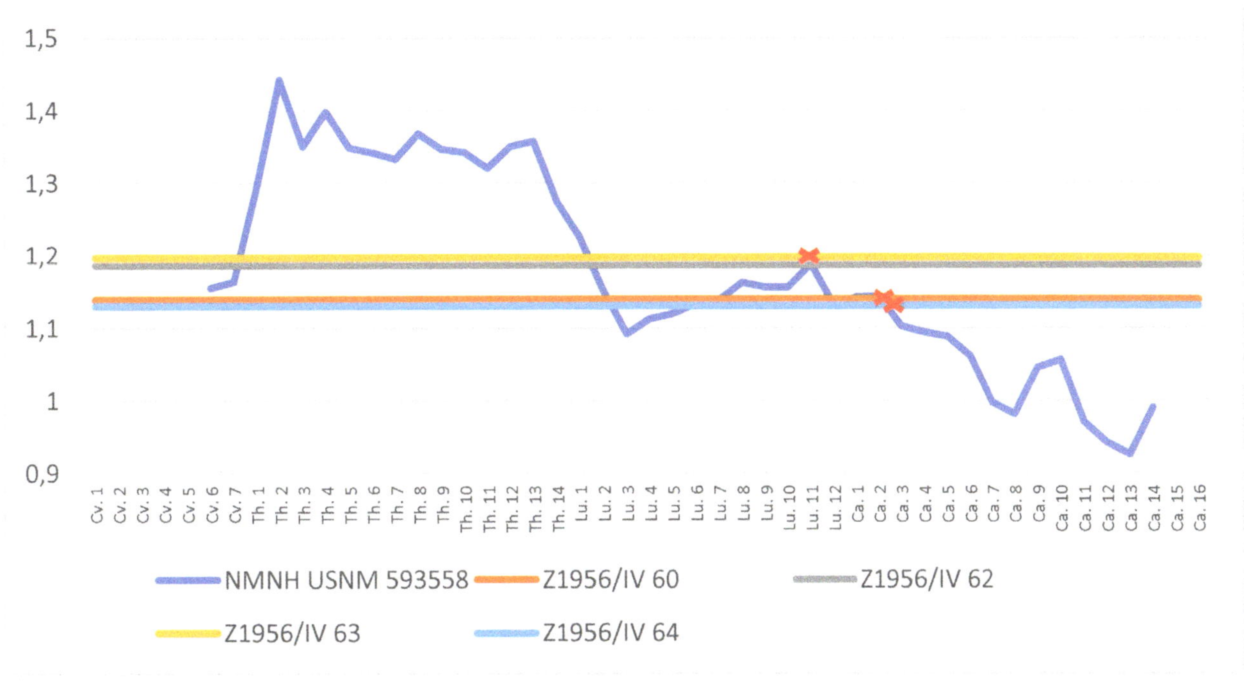

Figure 84. Breadth/height for grey whale specimen NMNH USNM 593558 compared to epiphyses Z1956/IV 60, 62, 63, and 64. The red crosses indicate the intersection between the samples and grey whale specimen NMNH USNM 593558, indicating that the specimens belonged to one of the last couple of lumbar or one of first couple of caudal vertebrae.

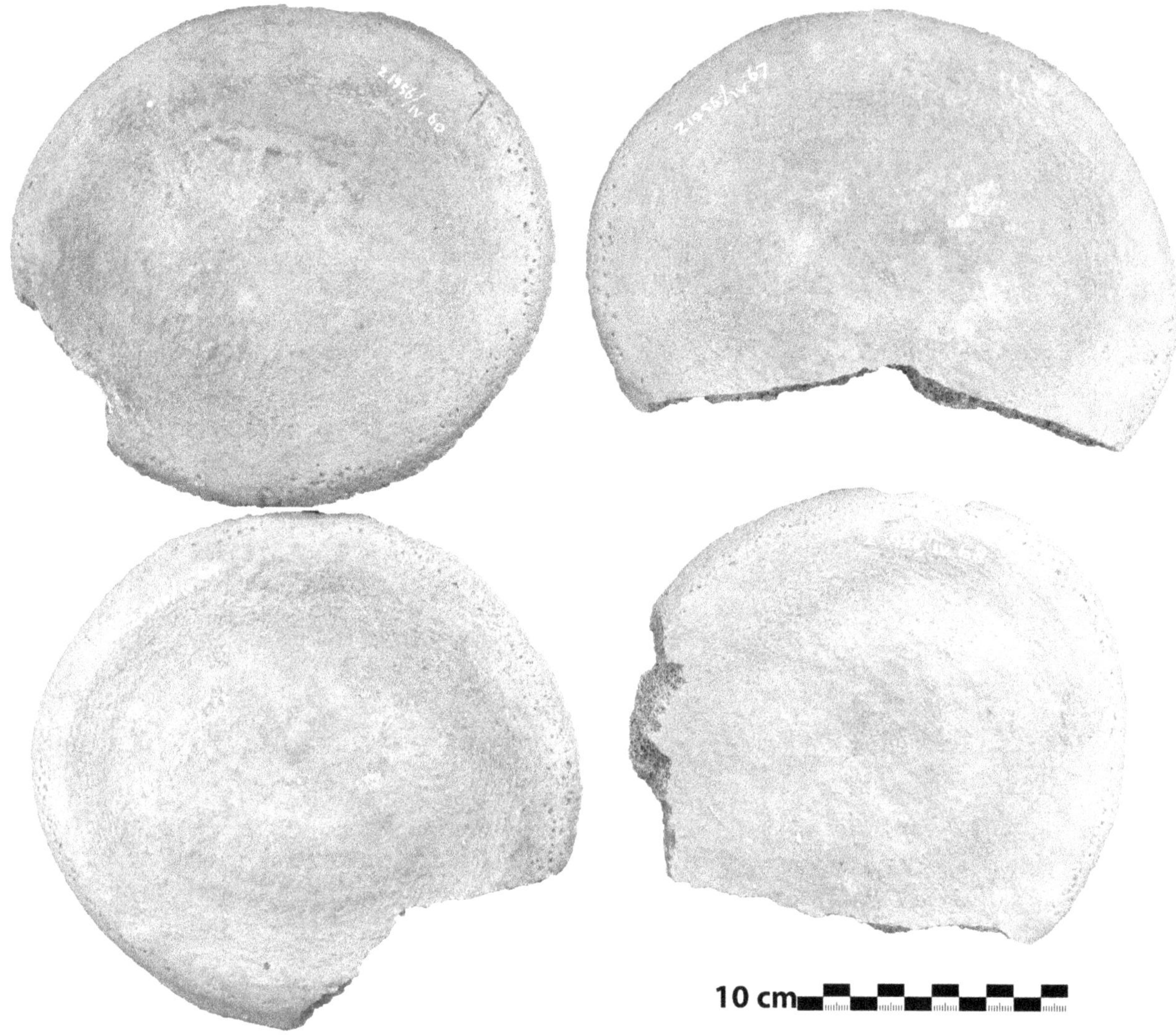

Figure 85. Vertebral epiphyses Z1956/IV 60, 67, 64, and 68. Photo by author.

of the North Atlantic (including Iceland) was 56 and for the western side (including Greenland) was 18, bringing the total to 74 grey whale specimens. This study has contributed to the identification of 10 of these specimens, all originating from the eastern side of the North Atlantic.

Discussion

- Distribution

Grey whale specimens have now been discovered in the Netherlands, England, Spain, Norway, Sweden, Scotland, Iceland, Greenland, Morocco, and the USA. The majority of these are from palaeontological contexts, but recently archaeological specimens have been identified more frequently. If the Atlantic population displayed a similar migration pattern as the Pacific population, an attempt can be made to reconstruct the distribution. The east Pacific herd breeds and calves in January and February off Baja California Sur, migrates northward from March to May, feeds from June to October in the Bering and Chukchi Seas, and migrates back southwards in November and December (Pike, 1962).

If this migration pattern is projected on the Atlantic herd, this would mean that the Atlantic herd would visit foraging ground in high latitude locations during the summer months and migrated south to breeding and calving grounds during the winter (figure 86 and 87). Rodrigues *et al.* (2018) suggested that the eastern North Atlantic grey whales might have used the western Mediterranean as a calving ground to which they migrated during wintertime. They additionally showed a reference by Pliny the Elder who made note of whales coming to Gibraltar and the Cadiz region where they according to him "conceal themselves in some calm capacious bay, in which they take a delight in bringing forth".

Rodrigues *et al.* (2018) suggested that grey whale might have shown a similar migration pattern as the North Atlantic right whale, which calved off the western Sahara, but also potentially in the Mediterranean for which two historical records of North Atlantic right whale sightings exist (one near Alger and the other in the Gulf of Taranto. The area between Corsica and Sardinia could potentially also have been used by the North Atlantic right whale as a calving ground. Two of the three specimens identified by

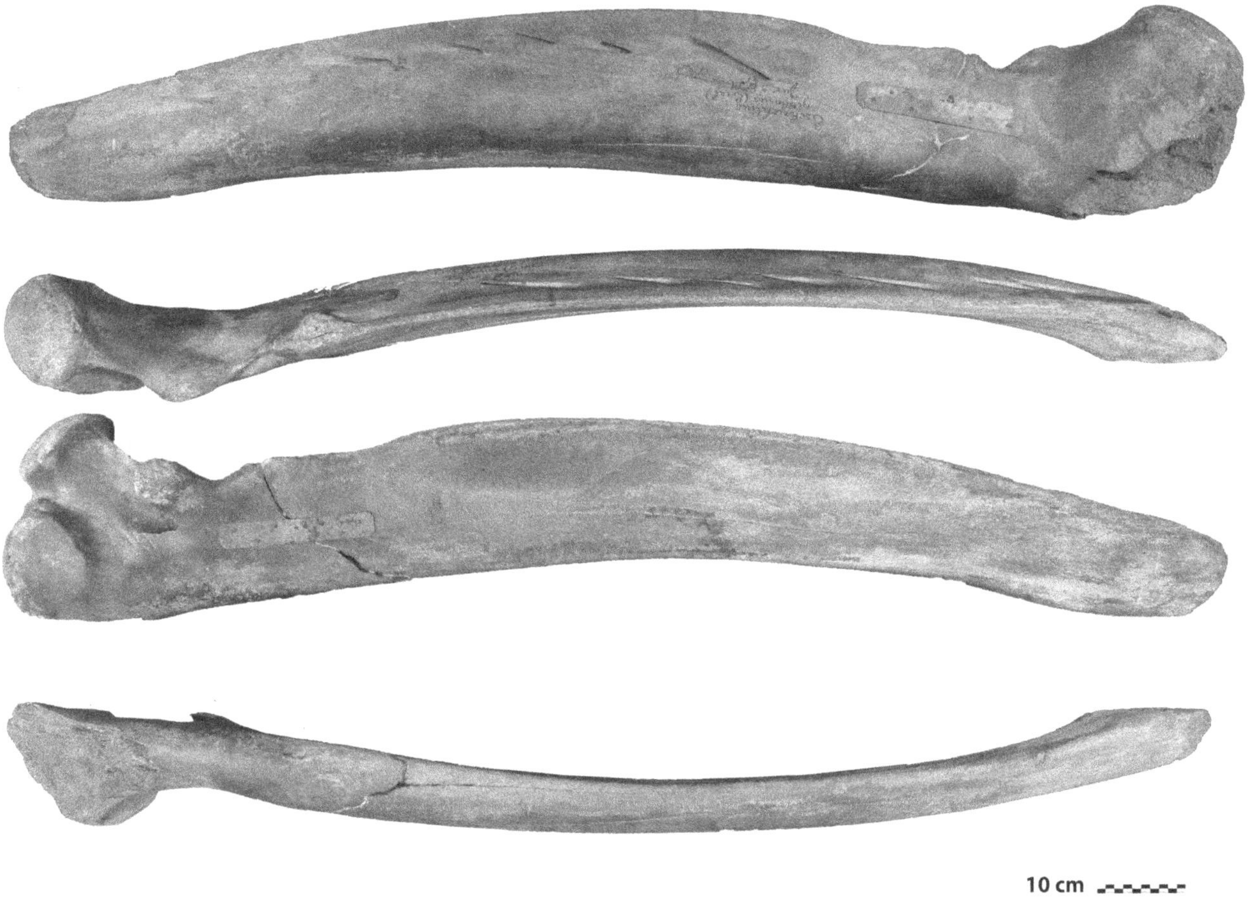

Figure 86. Mandible 8216 Lateral, dorsal, medial, and ventral view. Photos by author.

Figure 87. Grey whale mandible on display at the Poldermuseum, Heerhugowaard, the Netherlands. Photo courtesy of H. Aandewiel.

Rodrigues *et al.* (2018) come from the eastern side of the Strait of Gibraltar, suggesting that the grey whales indeed might have entered the Mediterranean to calve there.

De Smet (1981, 307), however, has suggested that grey whales might have used the shallows and estuaries along the coast of the North Sea and the English Channel as a calving ground. De Smet stated that the area is comparable to the calving grounds the eastern Pacific population calving ground in Baja California. The coastal area of the Netherlands might have been a suitable area for this as well. The delta of the rivers Rhine, Meuse, and Scheldt could have provided suitable lagoons and bays.

Helpful for migration reconstruction, is the location of osteological specimens of juvenile grey whale individuals. A higher proportion of these individuals is expected in the breeding and calving regions. On the Eastern side of the Atlantic, juvenile individuals are frequently found in the southern North Sea area (these juvenile age estimations are based on the size of palaeontological specimens found), especially in the Netherlands, suggesting that de Smet's (1981) theory might be accurate.

However, the region is on a higher latitude than expected for a calving ground. Instead, as the area of the southern North Sea is relatively shallow and supports a rich benthic biodiversity, it might have been a suitable foraging ground for grey whales. Moreover, since grey whale remains have also been found further to the south in Spain and Morocco, suggests that the Atlantic population migrated to those regions or even further south for calving or breeding purposes. The southern North Sea might therefore have been a favoured foraging ground for grey whales to which mothers took their calves.

Regarding the western North Atlantic population, a migration pattern between Georgia and Florida in the south and the Bay of Fundy, Gulf of St. Lawrence, Labrador Sea, and potentially even Baffin Bay and the Hudson Bay in the north is likely. The presence of a skull fragment of a new born grey whale calf at Jupiter Island along the south-eastern coast of Florida has been used to argue that the area might have been used as a calving ground (Garrison *et al.,* 2019). There are numerous bays and protected lagoons in the area which might have been used by female grey whales to deliver their young.

- *Killer whale predation on Atlantic grey whale?*

Rodrigues *et al.* (2018) noted that Pliny the Elder wrote about killer whale predation on whales during the winter (calving season) in the Gibraltar region. This might indicate that killer whales attacked new born calves of grey whales (and North Atlantic right whale), much like they still do in the Pacific Ocean, near Unimak Island, the Chukotka

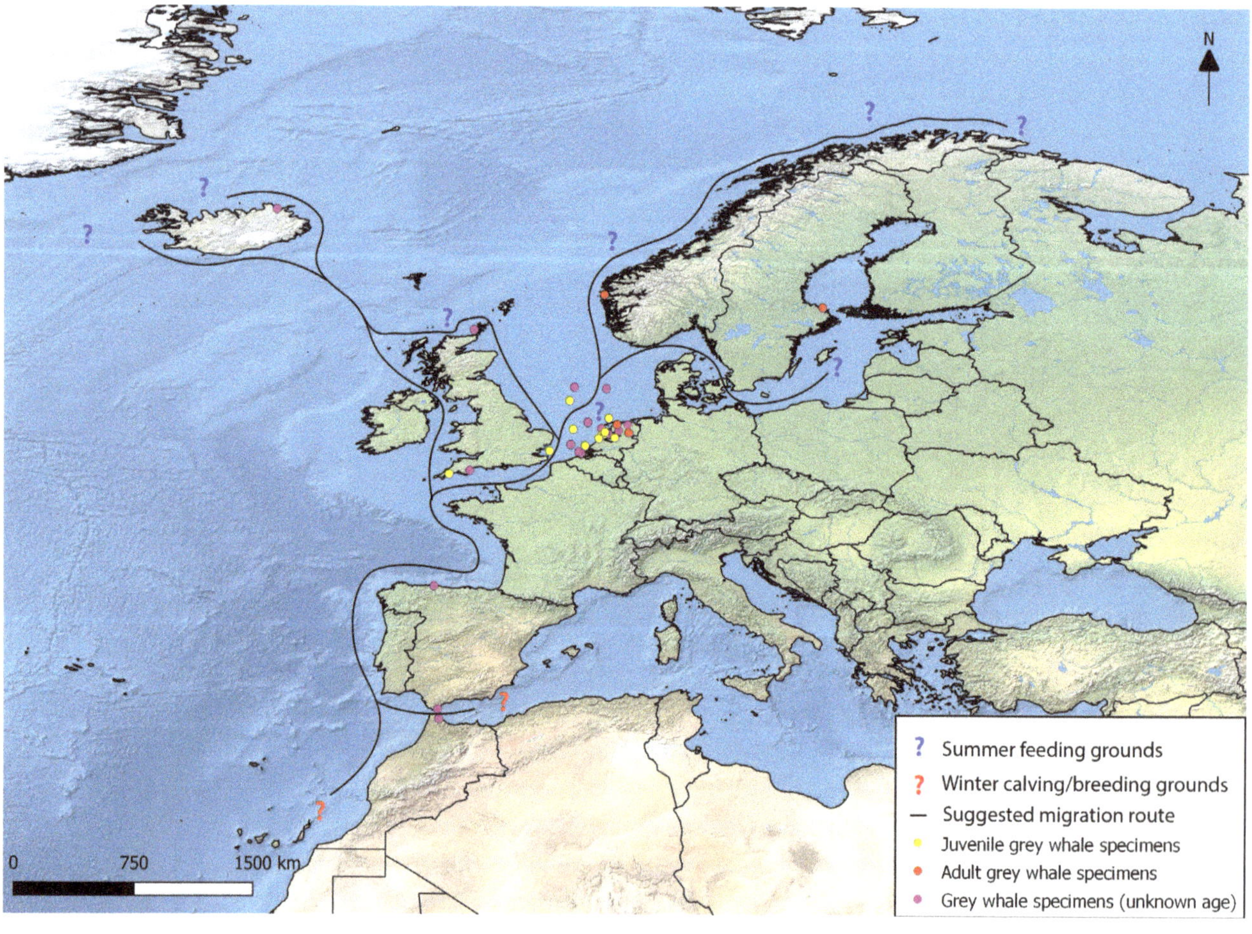

Figure 88. Grey whale findings from the eastern North Atlantic and suggested migration route. Created by author.

Table 8. North Atlantic grey whale specimens. Toponym, country (Neth. = Netherlands, specimen number (if known), Identification technique aDNA, ZooMS, or Morphology ((X) = undertaken but failed), Element, estimated age (based on fusion or size), find date, date, and references.

Toponym	Country	Specimen	aDNA	ZooMS	Morph.	Element	Estimated age	Find Date	Date	References
Hallum	Neth.	26C-122		X		Cervical vertebra	Adult	2018	100 BC - AD 1200	NADNuis, 2019; This study
Rottum	Neth.	1915-VI.1		X		Mandible?	Unknown	2018	600 BC - AD 1200	NADNuis, 2019; This study
Wijster	Neth.	Find no. 1266			X	Tympanic bulla	Adult	2018	AD 200-400	Prummel *et al.*, 2018
Den Burg (Texel)	Neth.	75.17		X		Bone fragments	Unknown	2018	AD 750-1000	Groenman van Waateringe and Wijngaarden-Bakker, n.d.; This study
Domburg (Badstraat; Groenje)	Neth.	Find no. 29173		X		Skull	Unknown	2011	AD 850-1000	Buitenhuis, 2011; This study
Heerhugowaard	Neth.	?			X	Mandible (left)	Juvenile	1975	4000 BP	Personal communications Henk Aandewiel, 20-05-2019; This study
Oudelandertocht (Lambertschaag)	Neth.	Find no. 103			X	13 vertebral epiphyses and 3 ribs	Juvenile	2008	>4810±40 cal BP	Buitenhuis, 2008
Wieringermeer-Polder	Neth.	Leiden NML 20350	X			Skull (Skull, mandible, hyoid several vertebrae, ribs and limb elements)	Juvenile (8 m)	1935	>4810±40 cal BP	Van Deinse and Junge, 1937; Bryant, 1995; Post, 2005
North Sea	Neth.	GrN 28549			X	Vertebra		2005	42800 ± 4100-2700 cal BP	Post and Bosselaers, 2005; Post, 2005
North Sea	Neth.	DM 47 (GrA 22182)			X	Vertebra		2005	>45200 cal BP	Post, 2005; Mol *et al.*, 2006
9 Miles north of Terschelling	Neth.	NMR999100001877	X			Mandible	Juvenile	2001	1150-1270 cal BP	Alter *et al.*, 2015
North Sea	Neth.	?	X			n/a		n/a	1350-1500 cal BP	Alter *et al.*, 2015
Witte Bank	Neth.	NMR999100001783	X			Mandible	Juvenile	1997	1350-1500 cal BP	Alter *et al.*, 2015
Eurogeul	Neth.	NMR999100001780	X			Cervical vertebra		2003	2650-2730 cal BP	Alter *et al.*, 2015
Eurogeul	Neth.	?	X			Vertebra		n/a	>48000 cal BP	Alter *et al.*, 2015
Dutch coast south	Neth.	NMR999100001938	X			Large vertebra		2005	>48000 cal BP	Alter *et al.*, 2015
Eurogeul	Neth.	NMR999100001781	X			Atlas (half)	Juvenile	2003	42500-43300 cal BP	Alter *et al.*, 2015
Ijmuiden	Neth.	Leiden NML 13130	X			Skull	Juvenile	1879	9470-9550 cal BP	Alter *et al.*, 2015
Ijmuiden	Neth.	Leiden NML 630	X			Skull	Adult	1916	1600-1800 cal BP	Alter *et al.*, 2015
North Sea	Neth.	NMR9990-00082	X			Skull		n/a	4950-5250 cal BP	Alter *et al.*, 2015
Domburger Rassen	Neth.	Brabant CollNr42002	X			Skull		1954	3830-3960 cal BP	Alter *et al.*, 2015
SL27 53'N 3'E	Neth.	NMR999100001790	X			Thoracic vertebra	Juvenile	1994	960-1120 cal BP	Alter *et al.*, 2015
Southern Bight	Neth.	NMR999100001785	X			Lumbar vertebra		1995	4230-4420 cal BP	Alter *et al.*, 2015

Toponym	Country	Specimen	aDNA	ZooMS	Morph.	Element	Estimated age	Find Date	Date	References
Southern Bight	Neth.	NMR999100001788	X			Axis		1996	>48000 cal BP	Alter *et al.*, 2015
Southern Bight	Neth.	NMR999100001789	X			Axis		2005	1820-1950 cal BP	Alter *et al.*, 2015
Southern Bight	Neth.	NMR999100001786	X			Caudal vertebra		2005	>50000 cal BP	Alter *et al.*, 2015
Southern Bight	Neth.	NMR999100001791	X			Lumbar vertebra		2005	3480-3630 cal BP	Alter *et al.*, 2015
55'45"N 5'10"E	Neth.	NMR999100001784	X			Radius		2005	10000-10180 cal BP	Alter *et al.*, 2015
North Sea	Neth.	NMR999100002102	X			Thoracic vertebra		2005	5280-5430 cal BP	Alter *et al.*, 2015
North Sea	Neth.	NMR999100001999	X			Ulna		2005	6620-6700 cal BP	Alter *et al.*, 2015
Gaatje van Ellen	Neth.	B2-1494 (Collection Vonk)	X			Mandible		1980s	5320-5470 cal BP	Alter *et al.*, 2015
North Sea	Neth.	B2-1493 (Collection Vonk)	X			Mandible		1980s	3470-3620 cal BP	Alter *et al.*, 2015
North Sea, Southern Bight	Neth.	NMR999100001998	X			Ulna		n/a	40200-41400 cal BP	Alter *et al.*, 2015
North Sea, Southern Bight	Neth.	NMR999100001994	X			Mandible		n/a	1680-1800 cal BP	Alter *et al.*, 2015
North Sea, Southern Bight	Neth.	NMR999100001996	X			Thoracic vertebra		n/a	42400-43600 cal BP	Alter *et al.*, 2015
North Sea	Neth.	NMR999100003115	X			Mandible		2007	3930-4070 cal BP	Alter *et al.*, 2015
Andijk	Neth.	?	X			Complete Skeleton		n/a	4020-4270 cal BP	Alter *et al.*, 2015
North Sea	Neth.	B2-0492/B2-0494			X	Mandible + Rib	Unknown	n/a	Unknown	Personal communication Arthur Oosterbaan, 11-04-2019
North Sea	Neth.	B2-0430			X	Mandible	Unknown	n/a	Unknown	Personal communication Arthur Oosterbaan, 11-04-2019
Eurogeul	Neth.	B2-0374			X	Jaw (piece)	Unknown	n/a	Unknown	Personal communication Arthur Oosterbaan, 11-04-2019
Central North Sea	Neth.	H1-0766			X	Cranium (fragment)	Unknown	n/a	Pleistocene (Fossil)	Personal communication Arthur Oosterbaan, 11-04-2019
Unknown Polder	Neth.	Z1956/IV (60, 62, 63, 64, 67, 68, 69, 70, 71, and 72)			X	Partial vertebral column (3 vertebrae (of which 1 caudal and the other 2 lumbar/caudal) + 10 vertebral epiphyses	Juv/sub adult	1956	Unknown	This study
Unknown Polder	Neth.	ZW1935/V 344-1			X	Thoracic/Lumbar vertebra	Juv/sub adult	1935	Unknown	This study
Unknown Polder	Neth.	ZW1935/V 244-2			X	Thoracic vertebra (unfused)	Juv/sub adult	1935	Unknown	This study
Unknown Polder	Neth.	8216			X	Mandible (Left)	Juvenile	n/a	Unknown	This study
Kringlevågen	Norway	?	X	X		Cervical+Thoracic vertebra	Adult	2018	1860-1680 cal BP	Hufthammer *et al.*, 2018

Toponym	Country	Specimen	aDNA	ZooMS	Morph.	Element	Estimated age	Find Date	Date	References
Gräsö	Sweden	?			X	Partial skeleton (mandibles and partial postcranial skeleton)	Adult	1859	4395 ± 155 cal BP	Lilljeborg, 1861; Persson, 1986
La Campa de Torres, Gijon	Spain	?	X	X		Scapula		2018	400 BC - 200 BC	Rodrigues *et al.*, 2018
Iulia Traducta, Algeciras	Spain	?	X	X		Vertebral fragment		2018	AD 215-422 cal14c	Rodrigues *et al.*, 2018
Tamuda	Morocco	?	X	X		Bone fragment (cut-/chopmarks)		2018	AD 71-245 cal14c	Rodrigues *et al.*, 2018
Howe	Scotland	Unknown	X	X		Unknown	Unknown	2019	AD 100-400	Personal communication Vicki Szabo, 20-05-2019
Pentuan	England	?	X			Partial skeleton (right mandible, lumbar vertebra, humerus, radius and two metacarpals)	Sub-adult?	1829	1329 ± 195 cal BP	Flower, 1872; Bryant, 1995; Alter *et al.*, 2015
Babbacombe Bay	England	?	X			Vertebra		1861	340 ± 260 cal BP	Gray, 1864
Babbacombe Bay	England	?	X			Vertebra		1865	340 ± 260 cal BP	Gray, 1966; Pengelly, 1865: 1878; Bryant, 1995; Alter *et al.*, 2015
Thames river mouth	England	?			X	Tympanic bulla, os temporale, processus zygomaticus, pars squamosa, os paietale, os alisphenoidale, os pterygoideum, osbasisphenoidale, os occipitale	Juvenile	1978	2024 ± 110 cal BP	Asselberg, 1981; Bryant, 1995
Hjálmarvík	Iceland	?	X	X		Unknown	Unknown	n/a	AD 1300-1477	Personal communication Vicki Szabo, 20-05-2019
GUS	Greenland	?	X	X		Unknown	Unknown	n/a	Unknown	Personal communication Vicki Szabo, 20-05-2019
Tom's River	USA	USNM 187448	X			Mandible (Left)	Adult (estimated length 14 m)	1850s	455 ± 90 cal BP	Mead and Mitchell, 1984
Myrtle Beach	USA	USNM 23260			X	Mandible (Posterior section)	Adult (about 14 months)	1959	865 ± 165 cal BP	Mead and Mitchell, 1984
Chesepeake Bay	USA	USNM 187449			X	Skull (squamosals, exoccipitals, parietals and the basioccipital)	Juvenile	1969	10140 ± 125 cal BP	Mead and Mitchell, 1984
Nags Head	USA	USNM 244465	X			Mandible (Left)	Juvenile	1970s	865 ± 50 cal BP	Mead and Mitchell, 1984
Corolla	USA	USNM 244038			X	Skull (fragment of right squamosal)	Juvenile	1976	2415 ± 90 cal BP	Mead and Mitchell, 1984
Southampton	USA	USNM 244307			X	Mandible (left, fragment)	Juvenile	1977	275 ± 35 cal BP	Mead and Mitchell, 1984
Corolla	USA	USNM 299838			X	Skull	Adult	1977	Unknown	Mead and Mitchell, 1984

Toponym	Country	Specimen	aDNA	ZooMS	Morph.	Element	Estimated age	Find Date	Date	References
Rehobeth	USA	USNM 256749	X			Squamosal	Adult	1978	Unknown	Mead and Mitchell, 1984
Cape Lookout	USA	?			X	Skull	Juvenile	1979	1190 ± 245 cal BP	Mead and Mitchell, 1984
Jupiter Island (Hobe Sound)	USA	UF 99000		X		Skull (including periotic bone)	Juvenile (less than 30 weeks)	1983	1500-2150 cal BP	Garrison *et al.*, 2019
Jacksonville Beach	USA	UF 69000		X		Skull (Braincase)	Juvenile (Between 30 and 52 weeks)	1970s	1570-2220 cal BP	Garrison *et al.*, 2019
JY Reef	USA	?		(X)	X	Mandible		2006	~36000 cal BP	Noakes, Garrison and McFall, 2009; Cherkinsky, 2009; Garrison *et al.*, 2012; Garrison *et al.*, 2019
JY Reef (30 km offshore St. Catherine's Island)	USA	GMNH-27370			?	Vertebra		2006	41490-42070 cal BP	Garrison *et al.*, 2019
JY Reef (30 km offshore St. Catherine's Island)	USA	GMNH 27372		(X)	X	Mandible (left)	Juvenile	2006	40230-41550 cal BP	Garrison *et al.*, 2012; 2019
JY Reef (30 km offshore St. Catherine's Island)	USA	GMNH-27373			?	Vertebra		2006	38350-39140 cal BP	Garrison *et al.*, 2019
JY Reef	USA	GMNH 27375/ Georgia Museum of Natural History (No. 4214)		(X)	X	Mandible (left)	Juvenile	2006	36240-37460 cal BP	Garrison *et al.*, 2019
South Atlantic Bight	USA	?			?	Unknown		2006	48550-50000 cal BP	Garrison *et al.*, 2019

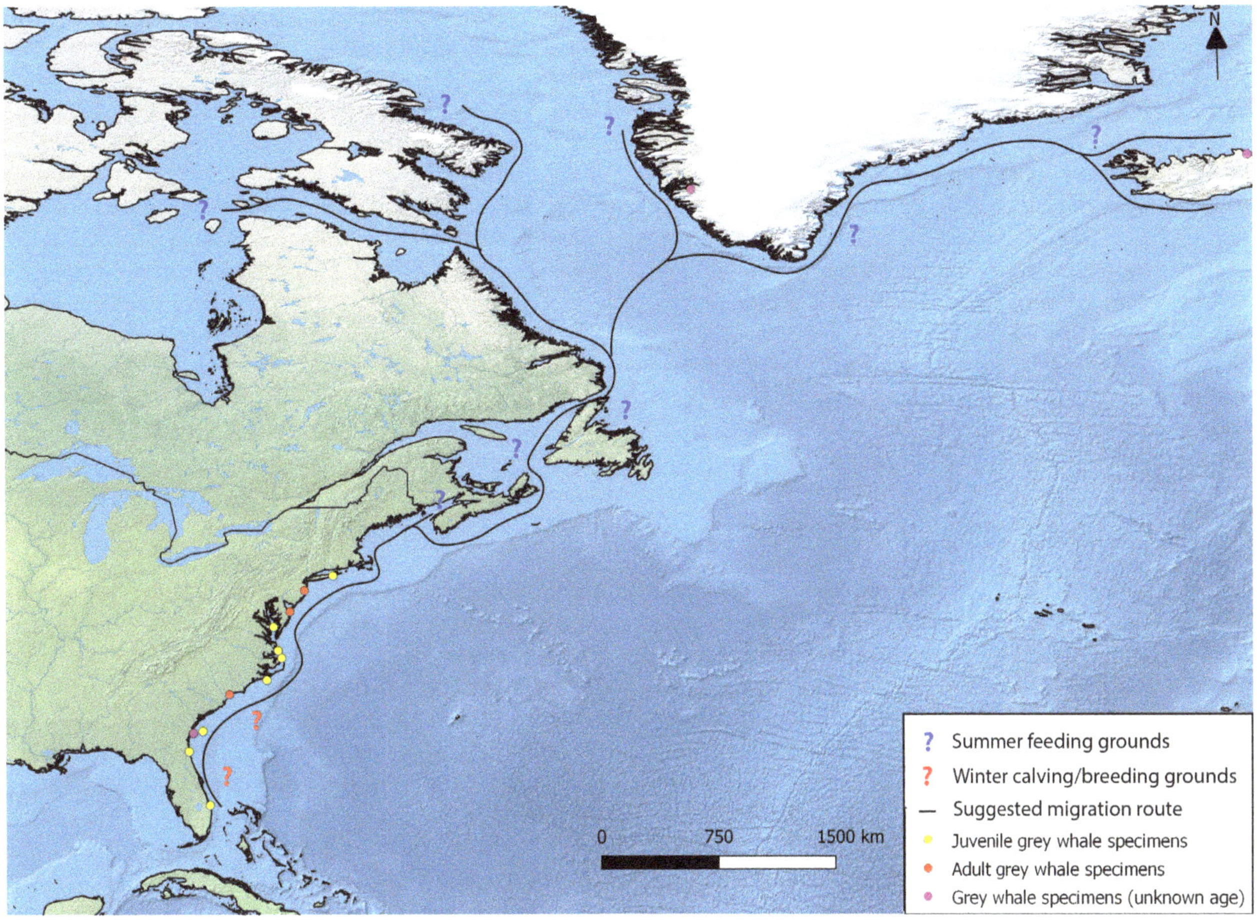

Figure 89. Grey whale findings from the western North Atlantic and suggested migration route. Created by author.

peninsula, Monterey Bay (California), and Glacier Bay (Alaska) (Barrett-Lennard *et al.*, 2011; Melnikov and Zagrebin, 2005; Goley and Straley, 1994).

Pliny the Elder is not the only one who describes killer whale predation on large whales. Olaus Magnus in his twenty-first book, part of the *Description of the Northern People*, published in 1555, also describes a "grampus" attacking a large whale, though he probably heavily relied on Pliny's description. Olaus Magnus states that the "grampus" attacked the genitals and the calf of the whale (Olaus Magnus, 2010, 1091). Calves are indeed frequently attacked by killer whales. Furthermore, it is stated that the "grampus" tried to thrust the whale into the shallows. While the killer whale does not try to thrust its prey into the shallows, it has been observed that large whales often seeks shelter in shallow water to flee from killer whale predation (Barrett-Lennard *et al.*, 2011; Melnikov and Zagrebin, 2005; Goley and Straley, 1994).

Rodrigues *et al.* (2018) argued that the killer whale ecotype, that specialized on whale hunting, have disappeared from the region. Killer whales are still present in the Gibraltar region, but prey on bluefin tuna. A recent inventory of cetacean remains from the collection of Arie and Ineke Vonk, located on Texel, the Netherlands might provide an idea of which species were present in the North Sea in the past. Around 1980 Arie Vonk started to collect subfossil remains of cetacean dredged from the bottom of the southern North Sea by fishermen. Hereby he did not discriminate between species or size of the specimens, and all cetacean remains were added to his collection. Roughly 5500 specimens could be identified to species. Over 4500 specimens were identified as harbour porpoise, which is by far the most common cetacean species in the southern part of the North Sea, but at least 15 other species were identified in his collection. Six grey whale (0.1%) specimens are part of his collection, actually suggesting that the species was not abundant in the southern North Sea area. Furthermore, the killer whale makes up just 0.4% of Vonk's collection (21 specimens; Oosterbaan, 2018). With so few killer and grey whale specimens recovered from the southern North Sea area it appears that killer whale predation on grey whale, did not or at least rarely occurred in the area. It might however have occurred (more frequently) in southwestern European waters as suggested by Rodrigues *et al.* (2018).

It is however, striking that of the 5500 specimens in Vonk's collection merely 6 were identified as grey whale, while as part of this study just 38 specimens from archaeological contexts were analysed and 5 grey whale specimens were identified. There can be several reasons for this discrepancy. The most probable reason might be identification problems. Numerous cetacean remains are extremely weathered allowing no identification based on morphology. The 38 specimens that were analysed as part of this study, were all analysed using molecular identification (ZooMS). As a result, identification was optimized, even allowing identification of specimens that morphologically could not be identified to species. The application of ZooMS analysis on specimens from Vonk's collection might result in the identification of more grey whale specimens. All 6 specimens in Vonk's collection are mandibles or cranial fragments, which are relatively easy to identify. Postcranial elements are harder to identify and among these there might be more grey whale specimens. More detailed analysis of these remains is necessary to confirm this.

- Active whaling or opportunistic scavenging

Whether active whaling was practiced on the grey whales in the Netherlands remains unclear. Historical sources do suggest that some form of whaling was already undertaken in the northern part of the Netherlands during the Medieval Period. Albertus Magnus, a German Dominican friar and Catholic bishop (AD 1193-1280) visited the Frisian area in the northern part of the Netherlands and Germany. There he witnessed the catching of a whale by the Frisian locals (Albert the Great, 1987, 338-342). He states that several species were exploited in the region, though that very large species were not frequently targeted. Even though the work by Albertus Magnus postdates the grey whale samples identified as part of this study, it might be that the Frisians exploited the grey whales in the region, as it only postdates the samples by a hundred years. Indeed, four of the five identified specimens are from the northern part of the Netherlands in the proximity of the Wadden Sea, and the southern North Sea, which is, based on the results of this study, assumed to be an important foraging ground for the grey whale. The Frisians conserved the oil, rendered the whale blubber and exploited the meat, bone, and baleen (Albert the Great, 1987, 338-342).

The Strait of Gibraltar, an area where Rodrigues *et al.* (2018) identified two other grey whale specimens, was a centre of fish processing industry during the Roman period. It could well be that the Romans targeted whales in the region. Rodrigues *et al.* (2018) indicates that the Romans in the area had the means, the motive, and the opportunity to hunt whales in the region considering it was a strategic place for whaling endeavours as they could target the whales entering or exiting the Mediterranean basin.

Another group that potentially could have hunted the grey whale are the Basques. Aguilar (1986) noted that possibility that the Basques in northern Spain and southwestern France also targeted the grey whale, though the North Atlantic right whale was the main target of the whalers. The whaling season in the Basque region lasted from October/November to February/March, with a peak around January. The adults might be targeted during November and December on their way south to the breeding and calving areas, and up north again with their calves during March. Zooarchaeological cetacean material from this region has not been analysed using ZooMS or aDNA research with the exception of a vertebral fragment indeed identified as grey whale coming from La Campa de Torres, Gijon, Spain. Though this specimen dates to 400-200 BC, and thereby predates the Basque whaling period, it clearly indicates that grey whales were present in the region (Rodrigues *et al.*, 2018). The Basques are known to have targeted the calves of the whales they hunted (Aguilar, 1986). This technique might have been undertaken by other whalers as well. While a considerable portion of the grey whale remains from the North Atlantic are from juvenile individuals, most of these do not originate from archaeological contexts, suggesting that this was not the result of whaling activities.

On the American side a different story emerges. Native American people on the Pacific coast are known to have frequently targeted grey whales on several locations along the grey whale's annual migration route, including Queen Charlotte Island, Vancouver Island, Cape Flattery, and off Kamchatka. Though, not all the people rejoiced the exploitation of the whales. The Tlingit had a strict taboo against the consumption of whales (Jones, 2013).

On the Atlantic side of the American continent whaling was not as frequently undertaken. It appears that prior to the arrival of Europeans, large whales were virtually unexploited. This can partially be ascribed to the low indigenous population density and a large population of various large whale species (Bolster, 2008). Native Americans in the New England area believed that the giant Maushop (a culture hero of the Mehegan and Wampanoag tribes) caught whales and deposited them onshore. This indicates that the native Americans in the regions relied on the opportunistic exploitation of stranded individuals. There is additionally no proof that active whaling was performed (Russell, 2001, 575).

This changed however with the arrival of the Europeans. The Basques arrived in North America around 1520 AD and set up several whaling stations on the Strait of Belle Isle. An extensive study by McLeod *et al.* (2008) in which they performed aDNA analysis on 218 zooarchaeological samples, determined that they primarily targeted the bowhead whale. None of the specimens was identified as grey whale, suggesting that the Basques did not encounter and hunt the grey whales in North American waters.

Shortly after the Basques, other European people made their way to American waters. After the permanent English settlement, the Europeans gained the right to appropriate drift and stranded whales on Long Island and Martha's Vineyard. On Nantucket however, the Europeans agreed that "all drift fish belong to the Indian sachems", which was codified into law in 1673. The new settlers quickly saw the potential the newly discovered waters had and during the 1650s and 1660s, 33 shore whaling stations were set up in Massachusetts. Whaling proved to be a prosperous enterprise and the activities intensified quickly. Next to the North Atlantic right whale, the grey whales were exploited as well and from each thirty-six barrels of oil could be extracted. In *The History of Nantucket* by Obed Macy, it is claimed that the first whale killed in Nantucket was in fact a "scragg". This suggest that whalers both targeted the North Atlantic right whale and the grey whale. In 1705 a missionary reported the Delaware Bay and Long Island Sound contained large numbers of "scrags", indicating that the number of whales was still high in the region (Russell, 2001, 575).

This changed soon after, and as early as 1720 clear signs that the whale stock began to get depleted were visible, and by 1740 the whaling shores were "fished out" (Bolster, 2008). The grey whale became extinct in the early eighteenth century, which corresponds with the plummeting whale stocks, reported by the Nantucket whalers.

All these sources make it clear that human exploitation was at least partially the cause of the disappearance of the grey whale from the region. Hunting might have accelerated the extirpation of the species (Bolster, 2008). More research on zooarchaeological samples from the east coast of North American will be necessary to confirm this.

The fact that ZooMS has been able to identify several grey whale specimens deriving from European archaeological contexts suggests that in the future more specimens might turn up. Many cetacean specimens remain unstudied still, but the presence of several specimens identified as part of this study suggests that humans had at least some effect on the grey whale population along the European coastline. Whether they were exploiting the last individuals of a species in decline or were the main reason the species eventually disappeared from European waters remains unclear.

Conclusion

More and more grey whale specimens turn up in both the North Atlantic fossil and archaeological record, increasing interest in the species and allowing the unravelling of the history of the Atlantic grey whale population. ZooMS and aDNA studies especially have the potential to contribute to our understanding of the spread of the grey whale from the Pacific to the Atlantic and the reasons for the eventual demise of the Atlantic population. The specimens from the Groningen Institute of Archaeology need to be analysed using both techniques in order to confirm the grey whale identification as well.

ZooMS and aDNA furthermore should be undertaken on seventeenth and eighteenth century zooarchaeological cetacean remains from the Nantucket region in order to validate the theory that the Nantucket whalers also targeted grey whales. This will allow us to better understand what conditions eventually led to the North Atlantic grey whale's demise, and whether anthropogenic factors played a substantial role in there as well.

With the on-going climate change and the retreating of the Arctic ice the Northwest passage might open up again, allowing grey whales once again to migrate from Pacific to Atlantic waters. Indeed, in 2010 a grey whale was sighted off the coast of Israel and another one off Spain (Scheinin *et al.*, 2011). Furthermore in 2013 a grey whale was confirmed to have been sighted off the coast of Namibia in the South Atlantic. This suggests a climate driven change in distribution (Alter *et al.*, 2015). Predictive habitat model analysis for the year 2100 confirmed that the grey whale habitat might expand into the Atlantic (Alter *et al.*, 2015). The grey whale might eventually return to the Atlantic Ocean again and establish a population there, potentially even reclaiming the southern North Sea where they appear to have been so abundant in the past.

6

Social Implications of Cetacean Exploitation in Medieval Northern and Western Europe

In order to answer the main question this study is concerned with, "What are the social implications of cetacean exploitation in medieval northern and western Europe?", this study has relied on a variety of historical and zooarchaeological sources. Since the geographical as well as temporal span of this study is considerable, identifying uniform social implications of cetacean exploitation is hard if not impossible. The data suggests that the various people and cultures assessed as part of this study, all viewed cetacean exploitation in a different manner. Social emphasis might be put on the practice of whaling itself, the consumption of cetacean meat, or the usage of raw material (bones, teeth, or baleen) for the creation of various artefacts and tools.

The combination of zooarchaeology and the structural historical research allowed the overcoming of the processual versus post-processual schism within the theoretical archaeology (human behaviour ruled by environmental conditions versus cognitive aspects) and suggests that both environmental and ideological conditions played an important factor for the exploitation and consumption of cetaceans. Indeed, edible resources are not automatically classified as food, as food is culturally defined as such, indicating that different cultural groups might develop a different diet that could well have included cetacean meat.

The event that probably led to an increase in cetacean exploitation for medieval Europe, is the spread of Christianity and its accompanied dietary practices. The Rule of Saint Benedict became adopted by monasteries in major parts of Western Europe from the Carolingian period onwards. These dietary restrictions meant that members of clergy were to abstain from eating dairy products, eggs, and meat of four-legged terrestrial mammals, during particular fasting periods such as Lent. Fish and (semi-) aquatic mammals, such as cetaceans, were however allowed (Johnston, 2011, 232-233).

Indeed, following the eighth century, higher numbers of cetacean remains derive from ecclesiastical contexts, suggesting members of the clergy consumed cetacean meat as part of strict dietary practices. From approximately the onset of early tenth century AD, the proportion of high-status and urban sites increased, suggesting that members of the nobility as well as people in urban settings, became interested in the consumption of cetacean meat as well. Cetacean meat was probably an expensive commodity and only available to the rich of urban sites. By consuming cetacean meat, they copied the diet of the social elite and in this way showed their riches and wealth.

That cetacean meat was perceived as a high-status food is clear from historical sources, but this might merely have been attributed to various regions, periods, or cultural groups. In medieval Europe, stranded cetaceans were treated as a royal right. This was the case for France from AD 850 onwards, Norway AD 1100 AD onwards, England AD 1116 onwards, Spain AD 1150 onwards, Denmark AD 1200 onwards, Iceland AD 1254 onwards, Ireland AD 1295 onwards, Scotland AD 1324 onwards, and Flanders AD 1384 onwards. Any cetaceans stranded along the shore of the nations mentioned, were by law the property of the ruler of that nation (often King of Queen). For other parts of Europe, including Portugal, Flanders, the Netherlands, and Germany, these rights appear to have not been present (or in any case cetaceans were not explicitly mentioned in any laws).

Feudalism flourished between the ninth and fifteenth century AD in major parts of Europe. The three estates of the realm (nobility, clergy, and peasantry) were bound by manorialism. European nations were subdivided into fiefs, ruled by vassals of the ruler of that nation. The right to stranded cetaceans was frequently treated as a seigneurial right, meaning that the King often granted rights to stranded cetaceans to his vassals. However, in other instances the King kept part of the stranded cetaceans, frequently the head (which contained the tongue, which was perceived a delicacy). The vassals were members of either the nobility or clergy and historical sources indicate that both had an interest in cetacean meat.

From historical sources it is clear that the peasantry also (illegally) attempted to get access to stranded cetaceans. The case study concerned with England indicated that based on records from the *Calendar of Patent Rolls*, the elite attempted to fine or punish the peasants that exploited stranded cetaceans in their fiefs. The zooarchaeological data indicates that from the High Medieval Period onwards, cetaceans are indeed less frequently recovered from "rural" or "small settlement" sites, and more frequently from "high-status" sites. This indeed suggests that stranded cetaceans from the High Medieval Period onwards were less frequently available to peasantry and that the social elite tried to monopolize the exploitation of the stranded individuals.

That cetacean remains are still frequently recovered from "rural" or "small settlements" (in fact, this site type is the most strongly represented for the entire period of AD 400-1600), can be ascribed to the fact that a much larger portion of society belonged to peasantry than to the nobility and the clergy, resulting in a higher proportion of rural sites. However, the recovery of cetacean remains from rural sites is also a clear sign of poaching activities. Many parts of the various European countries were not densely populated, making it relatively easy for peasants to exploit stranded cetaceans and hid them from the elite. If caught, peasants could be fined and punished, making it a risky undertaking. It is obvious that peasants were in this way undermining elite control.

Van Neer and Ervynck's (2004) research on the social significance of herring consumption indicated that context is vital to understanding the associated status of a particular species. In the case of the herring, Van Neer and Ervynck (2004) determined that preservation (e.g. gutting, salting or fresh herring), distance from the sea, and size were all aspects that played a role in the determination of whether herring was perceived as a high-status food source. This might have been the case for cetaceans as well. From the High Medieval Period, cetacean remains turn up more frequently further inland. An example is the site of Sveigakot in north-eastern Iceland, from which several cetacean remains have been retrieved. Whale bones are frequently recovered from Icelandic sites, but the site of Sveigakot is located 70 km away from the ocean. Most of these whale remains have been crafted into various artefacts, but a porpoise bone (*Phocoena* sp.) dating to the late ninth to the early tenth century, which is most likely not used for the creation of tools or artefacts as this species is small, might indicate that cetaceans were also consumed at the site. The location of the site demonstrates the length people were willing to go through to get access to cetacean meat through the interconnected web of economic relationships present in Iceland shortly after *Landnám* (Dugmore *et al.*, 2005).

Other sites located further inland, include several sites within London, England (including Westminster Abbey, Bermondsey Abbey, Westminster Sub-Vault of the Misericorde, Westminster Songschool, and Adelphi Building); Erfurt, Germany; Lödöse, Sweden; Saint-Martin-de-Boscherville and La Cour Napoléon du Louvre, France; Sveigakot, Iceland and Oxford, England (Gardiner, 1997; Pipe, Rielly and Ainsley, 2011, Pipe, 1995; personal communication Kevin Rielly, 2018; Prilloff, 2002; Lepiksaar, 1975; Clavel, 2001; Meniel and Arbogast, 1989; Dugmore *et al.,* 2005; Merples, 1976). At all these sites harbour porpoise remains have been identified and are all dated to the High Medieval Period (with the exception of Adelphi Building, London which is dated to the Early Medieval Period).

In France, several whale bones have also been recovered from inland sites. Just as is the case for the other inland sites, these all date to the High or Late Medieval Periods. This includes several remains from Paris (including the already mentioned harbour porpoise remains from La Cour Napoléon du Louvre). Additionally, excavations at two sites with an ecclesiastical context (Saint Gilles (Chazottes, 2017) and Hôtel de Beauvais (Du Bouëtiez, Clavel and Ravoire, 2013) and one with a high-status context (La Tour des Salves (Clavel, 2001), have resulted in the recovery of several cetacean remains, suggesting that the clergy and nobility in France had developed an interest in cetacean meat during the High and Late Medieval Period, resulting in cetacean remains being transported further inland. Furthermore, a trictrac gaming pieces has been recovered from the more inland situated site of Tours (Motteau, 1991, 138). Though this last case might not be indicative of cetacean consumption but might suggest that whale bone was a precious raw material used for the creation of valuable artefacts.

This rise in cetacean remains deriving from more inland oriented sites can presumably be ascribed to the already discussed spread of Christian dietary restrictions, but also to the advancements made in preservation techniques (Barrett, Locker and Roberts, 2004a, 2417-2421; Fagan, 2006, 15-57). Salting, as a form of preservation, was already used prior to the Medieval Period. However, it was used more frequently from the High Medieval Period onwards. This allowed cetacean meat to be transported further in land, which prior to this development was restricted to coastal regions. It might well be that development underlay the interest by the social elite in cetacean meat, as they too were hindered by the travel time from the coast to their manors.

This eventually led to longer preservation, making cetacean meat also available to urban populations. Indeed, from the High Medieval Period onwards a rise in archaeological sites with an urban character with cetacean remains can be noted as well. Historical sources indicate that cetacean meat was sold at several European markets, suggesting a certain commercialization of cetacean meat. However, the species being exploited remain unclear based on historical sources.

Zooarchaeological analysis of cetacean remains, and the identification of those remains to the species level, is seriously hampered by various taphonomic factors, including the often-fragmented state of the osteological remains, the lack of reference collections and manuals, and the osteological morphological similarities between several of the species. These factors and the problems identifying those remains acquired through active whaling and those of the individuals exploited through opportunistic scavenging, seriously hinders our understanding of past cetacean exploitation.

As part of this study, several zooarchaeological remains from the Netherlands and Belgium were analysed using ZooMS. This analysis revealed a wide range of species being exploited, though some species predominate. High numbers of North Atlantic right whale remains were identified. This species is most frequently linked to medieval cetacean exploitation enterprises. This might suggest that the species was frequently exploited in medieval Europe or that individuals stranded regularly along the European shore.

As several medieval cultures and groups are associated with cetacean exploitation through the analysis of historical sources, and zooarchaeological material has been recovered from all those regions, a case can be made that some form of whaling was practiced for all those regions. However, there is quite a bit of variation between the species targeted, the time periods the exploitation was undertaken, what methods were used to perform whaling, and on what scale the exploitation was practiced. Therefore, all the cultures considered should be analysed on a case by case basis.

- Normans

Zooarchaeological cetacean material is surprisingly rare for Normandy, even though the Normans are one of the cultures most frequently associated with medieval whaling practices. Historical sources suggest that whaling was undertaken between at least the onset of the tenth century until the twelfth century. The low number of zooarchaeological cetacean material from Northern France is therefore surprising. This low number can potentially be ascribed to a language-barrier or to fewer zooarchaeological studies being conducted in the region in comparison to other regions. Dr Tarek Oueslati of the Université de Lille, is currently performing analysis of cetacean remains deriving from the Normandy region, dating to the Roman and Medieval Periods. These remains are currently being analysed using ZooMS and will potentially reveal more details regarding the history of Norman whaling practices (personal communications Dr Tarek Oueslati of the Université de Lille, April 2019).

- Basques

That few cetacean remains derive from the Basque region, is even more surprising. The Basques are the cultural group most frequently associated with medieval whaling and are often described as the first to perform commercial whaling, as opposed to aboriginal whaling. Active whaling initiated from at least the mid-eleventh century and was undertaken until at least the end of the Medieval Period. Grau-Sologestoa and García-García (2018) have noted that recently medieval zooarchaeological studies are more frequently being conducted in the Basque region and Spain itself, but cetacean remains still do not frequently turn up. Historical sources clearly indicate that whaling was frequently undertaken by the Basques and that whaling was an integral part of Basque culture. The lack of zooarchaeological cetacean material is therefore intriguing but might be the result of still relatively few excavations being undertaken in the region or not being undertaken in the right areas. It might well be that the Basques left osteological remains at the foreshore, where it is unlikely that remains are going to be found during archaeological excavations.

Even though zooarchaeological remains are rarely recovered from Normandy and the Basque region, limiting out understanding of which species were exploited, it is clear from historical sources that the activity itself was frequently undertaken and a vital part of local economies. The social elite attempted to get regular access to cetacean meat, by claiming a part of the caught specimens. This was again both the case for the Basques and the Normans.

- Anglo-Saxons and the English

One of the oldest sources specifying whaling activities for medieval Europe is Bede's *Historia Ecclesiastica*, dating to AD 731. Other sources also seem to indicate that whaling activities were already undertaken in Great Britain from the Early Medieval Period onwards. Furthermore, numerous historical sources indicate that stranded cetaceans were a seigneurial right and the exploitation was strictly forbidden to be carried out by peasants. Additionally, whale meat was also frequently sold in London, suggesting some form of commercialisation as well.

Zooarchaeological sources confirm that whaling was already undertaken during the Anglo-Saxon period. Cetacean remains are frequently recovered from contexts in London, and the sites of Hamwic and Flixborough have also provided numerous remains. However, these three sites all show very different patterns. For London, a large variety of species are identified, suggesting that the species were exploited elsewhere and were transported to the London markets. From Hamwic large quantities of large whale remains have been identified, and several of these have been worked. This might indicate an active whaling community being present in Hamwic, potentially targeting the North Atlantic right whale, after which the bone remains were used for the production of several tools and artefacts. For Flixborough, numerous common bottlenose dolphin remains are identified, suggesting that the community there practiced a very different form of whaling and primarily targeted the common bottlenose dolphin population present in the area.

These zooarchaeological and historical sources clearly indicate that whaling was practiced in several part of Great Britain and not one uniform technique was practiced in order to hunt the cetaceans. Peasants still frequently tried to get access to stranded cetaceans as confirmed by zooarchaeological remains turning up in "rural" site contexts. Historical sources indicate that whenever caught, they risked a penalty for poaching the carcasses.

- Danes

In other regions organized groups performed hunting on cetaceans as well. This was for example the case for Denmark. Besides stranded cetaceans being a seigneurial right in Denmark since at least AD 1241, harbour porpoise hunting is known to have been performed in Denmark, more precisely in Middelfart. In the winter season the harbour porpoise was hunted in the Gamborg Fjord, from at least AD 1593, though it has been suggested that hunting might have predated this (Middelfart Museum, 2014). The hunters organized themselves in a guild. Zooarchaeological remains of the harbour porpoise occasionally turn up in

Denmark, but not in large quantities. The site of Ørkild near Svenborg, just like Middelfart located on the island of Funen, produced harbour porpoise remains which might possibly have been caught at Middelfart. Although, harbour porpoises are abundant in Danish waters and could have been caught in the direct vicinity of Ørkild as well. Other remains might potentially derive from opportunistically exploited stranded cetaceans.

- Polish

A similar harbour porpoise hunting guild was present in Poland. Both historical as well as archaeological sources suggest that the harbour porpoise was frequently exploited along the Polish coastline from at least the fourteenth century to the mid sixteenth century (Makowiecki and Makowiecka, 2014). This seems to have been a well-organized enterprise and the fishermen involved in harbour porpoise hunting had to pay a yearly fee (Ropelewski, 1957). Zooarchaeological research on material from Gdansk confirmed that the harbour porpoise was frequently exploited. The harbour porpoise is probably the only species regularly exploited, though a killer whale fragment was recovered from Kołobrzeg, but this specimen might have been brought to the site from another region (Makowiecka and Makowiecki, 2007).

This case, and the Danish porpoise hunting guild, demonstrate that a certain commercialization of harbour porpoise hunting developed during the Late Medieval Period, not being restricted to the social elite. It might still be that harbour porpoise was still an expensive product, in this way still limiting its consumption to the upper strata of Medieval society. Additionally, during the Late Medieval Period, the North Atlantic right whale population was already severely depleted, resulting in whalers switching to this smaller species.

- Portuguese

Historical sources from Portugal make it clear that whaling was also practiced in Portugal from at least the beginning of the twelfth century AD until at least the late fourteenth century AD. Few zooarchaeological remains of cetaceans are recovered from Portugal, but 80 remains from Peniche suggest that the North Atlantic right whale was probably actively caught (Teixeira, Venâncio and Brito, 2014). The was the species also caught by the Basques and the Portuguese might have learned how to hunt them by copying the techniques practiced by the Basques. Historical sources do not indicate that stranded cetaceans were seigneurial rights of the social elite, but tithes were levied as part of the feudal system that was set in place in thirteenth-century Portugal, indicating that whales were a prized resource and were associated with the social elite (Brito, 2011).

- Sami

In the most northern part of Norway, in Finnmark, over 700 slab-lined pits ("*Hellegroper*") have been identified that were used for the processing of blubber from marine mammals. These pits have frequently been associated with large numbers of seal and whale bones, suggesting the exploitation of these animals and the extraction of blubber for the production of oil (Amundsen *et al.*, 2003; Nilsen, 2017). These whale bones are presumably from North Atlantic right whales or bowhead whales (Henriksen and Roll Valen, 2013, 384). Many of the pits have been dated to 600-900 AD (Nilsen, 2017). Henriksen and Roll Valen (2013, 385) have suggested that the oil economy relied on drift whale carcasses and that active whaling was not undertaken. However, in order to facilitate over 700 slab-lined pits, of which the majority was in use between AD 600-900, a large supply of whale and seal carcasses was necessary. Whale populations were of course higher prior to post-medieval whaling activities undertaken by primarily the Dutch and the English in Svalbard region, and indeed a lot of oil can be extracted from just one carcass, potentially suggesting indeed that drift whale might have been the primary source exploited.

However, the possibility that active whaling was undertaken in the region cannot be completely ruled out. The large number of bone material from the area indeed suggests that active whaling might have been undertaken in the area. The whaling techniques practiced however remain unclear. If whaling was actively undertaken in the area, then it pre-dated the Norse whaling endeavours further to the south that mainly date to the ninth to eleventh centuries AD. If this is the case, Norse whaling might have been influenced by the whaling undertaken from the seventh to the ninth centuries AD in the most northern part of Norway by the Sami.

- Norse

The Norse are also thought to have performed whaling, and Lindquist (1997) has noted that the techniques practiced by the Norse were varied. In Iceland this appears to have included spearing a large whale (most probably a large rorqual) and waiting patiently for the animal to succumb from the damage inflicted by the spear and wash up along the shore. Furthermore, primarily in Norway, whales were trapped in inlets along the shore and were killed by spears dipped in rotting whale meat causing blood poisoning in the targeted whales.

These techniques are different to the techniques practiced by the Basques or Normans and it has been suggested that the Norse targeted rorquals more frequently. Recent ZooMS and aDNA analysis by Szabo and her team indeed seem to confirm this and they have identified numerous blue whale remains coming from Iceland (pers. comm. V. Szabo, 2019). For Norway, Hufthammer *et al.* (2018) have performed research, but the majority of the cetacean remains identified have not been published yet. It will be valuable to compare the cetacean remains from the Norse region with those of more southern whaling cultures, including the Basques and Normans, to see whether species composition for the regions is different.

Hennius *et al.* (2018) performed ZooMS analysis on gaming pieces from Sweden and identified five of these gaming pieces as North Atlantic right whale. Hennius *et al.* have suggested that the bones derived from Sami hunters in Finnmark and were traded to the Norse in more southern regions, who used it for the creation of artefacts including gaming pieces. The Norse however appear to have performed active whaling themselves as well, especially along the coast of Norway. The acquired whales were probably stripped of their valuable resources and their bones were used for the creation of artefacts in the Norse regions, including weaving swords, cleavers, and whale bone plaques. It might also be possible that the bones of individuals caught in Norway were transported to Sweden, and that those bones were used for the production of gaming pieces in that area, and not the bones from the by the Sami caught individuals.

The artefacts created from whale bone by the Norse, have frequently been recovered from grave contexts, predominantly located in Norway and Sweden. Whale bone plaques, weaving swords, gaming pieces, and cleavers are frequently recovered from graves with a high-status signature, indicating that whale bone was a valuable resource. This tradition ended at the end of the eleventh century, with the spread of the Christianity and the accompanied burial practices. The decline in numbers of sites as well as the estimated frequency density data (figures 57 and 58) for the late Early Medieval Period can probably be ascribed to the cessation of burial practices with whale bone artefacts, following their conversion to Christendom.

- Norse-Gaels

Based on historical sources, the Norse-Gaels are also known to have performed whaling from at least the mid-eleventh century onwards. Cetacean remains are frequently identified from Irish sites dating to the ninth to the twelfth century and are especially frequently recovered from Dublin and Cork. This clearly suggests that active whaling was practiced in the region as well, though it remains unclear which species were the primary target. It is however likely that the North Atlantic right whale was also the main target for the Norse-Gaels, but ZooMS or aDNA analysis needs to be carried out in order to confirm this.

- Frisians

Two whale bone weaving swords have been recovered from the terps of Rottum and Leens in the Netherlands. These two specimens were analysed using ZooMS and were identified as grey whale and North Atlantic right whale respectively. Weaving swords are frequently found in female graves in various different sizes in the Norse region and have been interpreted to have been status symbol (Petersen, 1951; Sjövold, 1974). It is possible that the weaving swords found at the two terp sites were constructed in Norway and traded with the Frisians. Interaction between the two groups is known to have occurred regularly from at least the seventh century onwards (Lebecq, 1989, 45). Although the two Frisian swords appear to be typologically different, indicating that the Frisians might have constructed those themselves, or attempted to copy the Norse weaving swords.

Although it has often been argued that cetacean remains recovered from terp sites most likely derived from stranded cetaceans in the Wadden Sea and southern North Sea area. Albertus Magnus however seems to suggest that in the thirteenth century, just after the endykement of large areas of the northern parts of the Netherlands and Germany, active whaling was actually undertaken (Albert the Great, 1987; 338-342).

Zooarchaeological analysis however indicated that a wide variety of species are represented in the material deriving from the terps, including sperm whale, northern bottlenose whale, killer whale, and fin whale. These species are unlikely to have been caught in the region, although remains of North Atlantic right whale and grey whale were also identified. These species might have been targeted by the Frisians, and the hunting methods described by Albertus Magnus closely relate to the hunting techniques practiced by the Basques, who frequently exploited the North Atlantic right whale. This indeed clearly indicates that whaling was occasionally undertaken by Frisians, though they also eagerly exploited stranded cetaceans as well.

- Dutch

From other parts of the Netherlands, cetacean remains have also occasionally been identified at medieval sites. Harbour porpoise, North Atlantic right whale, and grey whale remains are relatively frequently identified, suggesting that occasional hunting was undertaken. For the Netherlands it is not recorded that stranded cetaceans were a royal or seigneurial right, but a letter dating to mid thirteenth century specifies rights to beach finds between William II, Count of Holland and Zeeland, and Margaret of Constantinople, Countess of Flanders (De Grootte, 1999). Although cetaceans are not mentioned in this letter it is clear that the elite had an interest in beach finds, which might well have included cetaceans. The presence of zooarchaeological cetacean remains at "high-status" and "ecclesiastical" sites, confirms that the social elite in the Netherlands had an interest in cetacean meat, just like in other regions of north-western Europe. Active whaling might have occasionally been undertaken, targeting the harbour porpoise and the North Atlantic right whale. Though any hunting was probably not as frequently undertaken as for example the Basques or the Normans.

- Flemish

For Flanders, medieval sources seem to indicate that active whaling was undertaken from the late ninth century onwards, though from the twelfth century onwards whaling endeavours are more frequently mentioned and could only be undertaken if the whalers were granted the rights to do so. Historical sources furthermore indicate the social elite's interest in cetaceans from at least the early twelfth century onwards and the Count of Flanders is known to have claimed stranded cetaceans from at least AD 1384 onwards, though this might well have been earlier (De Groote, 1999; Steevens, 2014). Zooarchaeological cetacean remains from Flanders are however rare. Several remains have however been recovered from "high-status" or "ecclesiastical" sites, indicating the interest in cetacean meat by the Flemish social elite. This might suggest that stranded cetaceans were a seigneurial right, though historical sources also indicate that cetacean meat was sold at several Flemish markets, including Damme, Boulogne, Calais, and Bruges (De Smet, 1981; Espeel, 2016). The meat was probably still highly prized and therefore only available to the richest.

Apart from being recovered from archaeological contexts, whale bone material is also still present at various castles and ecclesiastical institutions all over Europe nowadays. A list by Svadberg (2008), based on the works of Redman (2004, 2009, 2010a, 2010b, 2013, 2014), indicates that several churches in Europe still hold cetacean remains. This is recorded for churches in Germany, Sweden, Great Britain, Poland, Italy, France, the Netherlands, Finland, Denmark, Iceland, Portugal, Slovenia, and Spain, but also for churches in Switzerland and the Czech Republic. Cetacean remains are also kept at various castles in especially Great Britain, the Netherlands, and Germany, but also in Italy, Sweden, Italy and France.

A record from the Dominican Abbey of Stockholm dating to AD 1489, indicates that a large whale stranded in Roslagen, Sweden and that peasants stripped the carcass of its blubber to produce oil, while parts of the skeleton were distributed to churches in the vicinity. One of these churches was the church in Edebo parish. The bones were identified as North Atlantic right whale. The bones might have been transported to churches with the meat still attached to it and might have been used to extract oil from. However, the bones have also been used for various other purposes, including chairs, seats, chopping blocks, and various other artefacts and tools. Furthermore, large whale bones were placed in churches as proof that giants once roamed the lands and were killed by the Giant Flood mentioned in Genesis (Svanberg, 2018). It might be that the whale bones deriving from ecclesiastical contexts from the Medieval Period were merely there as proof of giants, although historical sources clearly indicate that cetacean meat was also valued by the clergy.

It is clear from the analysis of both historical sources as well as zooarchaeological sources that social aspirations and tensions were expressed by the elite in major parts of northern and western Europe, through the exploitation and consumption of cetaceans. The social emphasis placed on the consumption of cetaceans exceeded their utilitarian value. While indeed in some cases it can be argued that active whaling was undertaken, opportunistic scavenging might still have been the main source of acquirement of cetacean meat.

These strandings are not the direct result of anthropogenic factors, but climate change might have played a role in regard to numbers of cetaceans turning up in the archaeological record. Climate change has an effect on animal populations and their natural range. This is also the case for cetaceans. Variations in water temperature, water circulations, water salinity and current, all affect cetaceans. Mannino *et al.* (2015) have indicated that during periods of climate variability, cetaceans are suspect to (mass-) strandings. In regard to the Medieval Period in Europe, there are two periods with climate variability. These two periods are the Medieval Warm Period and the Little Ice Age. Though these periods should not be taken as absolute climate change, overall the climate was warmer and drier during the Medieval Warm Period and colder and wetter during the Little Ice Age.

The Medieval Warm Period was a period of milder winters and warmer summers. Additionally, the sea-surface temperature increased by 1 to 4 degrees Kelvin. This might have resulted in more southern subtropical species coming further north. Moreover, the Warm Period corresponds with the onset of the Norse expansion to the Scottish Isles, Iceland, Greenland and even modern-day Canada. The peak in number of sites with cetacean remains at AD 950 might be explained by an increase in strandings as a result of the climatic variability caused by the Medieval Warm Period. Coastal communities might have exploited the stranded cetaceans more frequently, leading to more osteological remains ending up in the archaeological record. Alternatively, the climate variability, might have led to food scarcity, resulting in people attempting to get access to alternative resources, in this case stranded cetaceans.

Right after AD 950, numbers of sites decrease again, suggesting that a possible increase in interest in cetacean exploitation was short-lived. Numbers are still relatively high, but the onset of the Medieval Warm Period, is by far the best represented period based on the number of sites as well as the estimated frequency density.

The second period with climatic variability is the Little Ice Age, commencing around AD 1250/1300. Interestingly, this is another, smaller peak in the number of sites, as well as the estimated frequency density. The Little Ice Age was a period with stronger winters and colder summers. The sea surface temperature appears to have been 5 degrees Kelvin below today's average around Iceland and 2 to 3 degrees Kelvin lower than in the North Atlantic (Szabo, 2008, 82). During this period of climate shift, whales might have been much more valuable than before. Again, this might have been a period for which strandings occurred more regularly than before, leading to an increase in cetacean remains deriving from archaeological contexts.

Figure 90. A whale-oil factory of the Nordic or Greenland Company on Amsterdam Island at the north end of Spitzbergen (Svalbard) by Cornelis de Man dating to the AD 1625-1650. Object number: SK-A-2355. Digital image courtesy of the Rijksmuseum Open Data Policy. Available at: <http://hdl.handle.net/10934/RM0001.COLLECT.8964>.

Shortly following the onset of the Little Ice Age, whaling endeavours declined. Whether this was a direct result of this remains unclear. Another possibility is that whale stocks had already severely declined after this point, as was suggested by the ceasing of Norman whaling enterprises and a shift in whaling regions by the Basques. Furthermore, the European Great Famine (AD 1315-1317) and the Black Death (AD 1347-1351) decimated Europe's population, leading to fewer whaling endeavours being practiced.

Consumption of whales, dolphins, and porpoises appears to have declined in the early modern period of England even further (Gordon, 2015, 220) and the data that that this study has produced, confirms this for other regions as well. It is likely that this is the direct result of fewer whales (most likely North Atlantic right whales). By this point whaling might already have seriously depleted the European stock of this this species. Furthermore, the grey whale, which disappeared from European water around the High Medieval Period, might have left whalers with no other option than to cease their whaling activities.

This however changed during the late sixteenth, as Willem Barentsz discovered Spitsbergen/Svalbard and a new whale stock present there, leading to whaling being performed at a previously unseen industrialised level (figure 90). Over the next few centuries new stocks were discovered, and new whaling methods were developed, allowing more cetacean species to being targeted. Eventually this led to the overexploitation of numerous populations and species, up to the point that several populations vanished, and several species are now even close to extinction.

Conclusion

This study displays the potential an interdisciplinary approach on the history of medieval whaling can have by relying on both historical and zooarchaeological sources. Research on medieval historical sources has frequently been undertaken, however in order to fully understand the complexities of medieval whaling, research can only progress though the analysis of zooarchaeological samples. Based on this an answer was sought to the main research question of this study "What are the social implications of cetacean exploitation in medieval Northern and Western Europe?".

Regarding the first sub-question of this study "Which cetacean species were present in the eastern North Atlantic Ocean during the Medieval Period?", a thorough analysis of primarily biological and ecological aspects of cetaceans was undertaken. At least 35 different cetacean species were present in the North Atlantic Ocean during the Medieval Period, of which one species (the grey whale) is no longer present in the North Atlantic. Prior to the period of industrialised whaling, populations are assumed to have been much higher than they are now. Szabo (2008) noted that over a millennium, the size of the populations of the various cetacean species, their distribution, and even their behaviour might have changed. Humans have had interest in cetaceans for millennia already as numerous products can be extracted from their carcasses, including meat, baleen, ivory, bone, oil, and ambergris. For these products, cetaceans were already targeted in the Medieval Period.

In regard to the second sub-question "What do medieval sources mention about human-cetacean interaction and how should these be interpreted?", it is clear that several cultural groups in medieval Europe performed whaling from at least the Early Medieval Period onwards. The combination of Christian dietary restrictions, the spread of preservation techniques, and potentially climatic variability led to an increase of interest in cetacean exploitation. Especially the Basques and the Normans have frequently been associated with cetacean exploitation through the study of historical documents. The Normans practiced whaling from at least the mid/late ninth century up until at least the twelfth century, while the Basques started their shore whaling endeavours around the early eleventh century AD and lasted for several centuries in which the Basques continuously set out in search for new whaling grounds (Musset, 1964; Aguilar, 1986; Proulx, 1986, 15). Several other cultures have however also been associated with whaling including the Flemish, Anglo-Saxons, Portuguese, Frisians, Norse, Dutch, and Polish.

For many of these regions, the North Atlantic right whale was probably the main target. In addition, based on historical sources, it is apparent that for many parts of Europe, the social elite attempted to claim part of any caught or stranded whale ("wreck of sea right") from at least the High Medieval Period onwards. Both the nobility and clergy of medieval Europe are frequently associated with the consumption of cetacean meat, presumably the result of Christian dietary restrictions, as well as the development of salting as a method of preservation, allowing cetacean meat to be transported further inland.

As part of this study, data regarding zooarchaeological remains of cetaceans dating to the Medieval Period (AD 400-1600) were assessed to answer the sub-question "At which medieval sites were cetacean remains found?", to eventually compare the zooarchaeological data with the historical data. A total of 406 medieval sites with cetacean remains have been identified, located all over northern and western Europe. At least 5528 cetacean specimens (NISP) have been identified as a total, belonging to at least 18 different species, indicating that just over half of the 35 species present in the North Atlantic Ocean were exploited in medieval Europe, although this number may potentially be even higher.

Even though this is an extensive study, it is highly likely that not all whale bone from medieval contexts are included in this study. Especially whale bone artefacts are likely to have been excluded, as these are often treated as another category than other zooarchaeological remains. This was probably especially the case for Scandinavia, as well as the Basque region, where several artefacts have been uncovered, but the language barrier and a lack of visibility through publications, prevents further analysis.

Although whale bone is frequently recovered from regions all over Europe, it is often not clear which species they represent and whether these remains derive from actively caught or opportunistically scavenged stranded cetaceans. As part of chapter three the sub-question "How can zooarchaeological cetacean remains be studied?" was dealt with. The fragmented state of the specimens, as well as the fact that many specimens were worked into artefacts or tools, hinders identification. Furthermore, as there are 35 cetacean species in European waters, identification is hard as several of these species display a comparable osteological morphology.

aDNA research is more frequently undertaken on cetaceans, though it is still relatively expensive to perform. It has especially the potential to reconstruct the history of whale populations, as was for example done by Alter *et al.* (2015) on the Atlantic grey whale. A similar study is currently being undertaken by Szabo and McLeod on the North Atlantic right whale and will potentially reveal what happened to the species over the past millennium and how it has been affected by medieval whaling practices (personal communication Szabo and McLeod, 2018).

ZooMS has proven itself to be a ground-breaking technique with a lot of potential. Though it is less precise than aDNA, it is possible to reconstruct the past range of species, and it can potentially be useful to determine whether active whaling was undertaken in a particular region or period. ZooMS has already been undertaken on a select amount of cetacean material and valuable information has been published. As part of this study, 38 samples from the medieval Netherlands and Flanders were analysed using ZooMS, revealing which species were exploited in those regions.

For the Netherlands and Flanders, 12 North Atlantic right whale specimens were identified, indicating that the species was once abundant in the southern North Sea area. This species is assumed to have been the primary target of medieval whalers and this appears to have also been the case for the Netherlands. For the Dutch region, the Frisians are thought to have undertaken whaling, though on a smaller scale than the Basques or Normans. However historical sources indicate that large species were also targeted by the Frisians for at least the thirteenth century AD. It is likely that the centuries prior to this, whaling was also undertaken, probably primarily targeting the North Atlantic right whale. The Flemish are also known to have undertaken whaling, probably on a more regular basis than the Frisians. They also likely targeted the North Atlantic right whale, though zooarchaeological sources from the Flemish region are rare. Furthermore, several other species have been identified for the Dutch and Flemish regions as well, indicating that a wide variety of species were exploited, suggesting that opportunistic scavenging was also an important method of procurement of cetacean meat.

Moreover, the Osteological Reference for Cetaceans in Archaeology (ORCA) – Manual (van den Hurk, forthcoming), has proven itself to be a valuable source for the identification of cetacean remains, especially when combined with ZooMS analysis. The manual contains data regarding the mandibles, vertebral columns, scapulae, humeri, radii, and ulnae of the 35 most commonly found cetacean species in the eastern North Atlantic Ocean. The manual primarily focuses on osteometric data, and is based on 51,000 measurements, performed on approximately 1000 individual cetacean specimens, stored at ten different museums and institution, including the Smithsonian Institution in Washington DC, USA, and the Statens Naturhistoriske Museum, Copenhagen, Denmark. The identifications made using the ORCA-Manual still need to be tested using aDNA analysis, in order to verify them, and validate the use of the manual.

Identification to species level is necessary in order to understand the complexities of cetacean exploitation, not only in medieval Europe, but also for other periods and regions. In the case of medieval Europe, of the 5528 cetacean specimens assessed as part of this study, 4627 (83.7%) (coming from 257 of the 406 sites (63.3%)) were merely identified as "unknown cetacean". This clearly demonstrates that identification of cetacean material is extremely hard to be undertaken and is hampered by a large variety of taphonomic factors, as was discussed and outlined in table 3. This greatly delimits our understanding of past cetacean exploitation.

By far the hardest part of dealing with cetaceans in zooarchaeology, even harder than the identification of their remains to the species level, is determining which remains derived from actively caught individuals and which from opportunistically exploited individuals. The species composition of a zooarchaeological assemblage, ethnographic accounts, the presence of suitable hunting tools at a site, and the access to suitable boats and ships, are all potential aspects that might prove that active whaling was undertaken. Future research should especially look into possibilities to identify spear and harpoon remains from archaeological contexts and connect those to zooarchaeological cetacean remains, in order to suggest whether or not active whaling was practiced at a site. However, the most reliable indication of active whaling is still the presence of a spear or harpoon imbedded in a bone but known cases of this are extremely rare.

Thecombinationofhistoricalsourcesandzooarchaeological sources can also serve to argue that active whaling was undertaken, as was the case for the Frisians and the Flemish discussed earlier on. Several historical sources indicate that whaling endeavours were varied for several medieval cultures, with the Basques performing active whaling in boats, while the Norse performed whaling by trapping them within a fjord, sound, or enclosed area. The two might have targeted different species, though until now analysis of this material has not been undertaken, leaving it unclear whether there was a difference between the species targeted, but ZooMS has the potential to reveal the complexities of medieval cetacean exploitation.

The lack of cetacean bone material from the Basque Country, both the French as well as the Spanish parts, as well as Normandy, is interesting. The lack of osteological remains can potentially be ascribed to a language barrier, few excavations being undertaken on the foreshore where the osteological remains are probably located, or the Basques and the Normans simply did discard cetacean bone in the water, like was the case for the Basques in Red Bay, Canada (Grenier, Stevens and Bernier, 2007). Though even there, whale bone has been identified at the site itself as well. More archaeological research is needed in the Basque region and Normandy to fully understand the complexities of cetacean exploitation and its social implications.

From other regions, whale bone is frequently recovered, especially from the southern Norse Sea area. For medieval

England, historical sources imply that whaling was already undertaken during the Anglo-Saxon period, and zooarchaeological sources from Hamwic and Flixborough indeed suggest that in some areas cetacean exploitation was regularly undertaken. However, different methods were undertaken at Hamwic and Flixborough, as the majority of the remains found at Hamwic belong to large species, while the remains from Flixborough have primarily been identified as common bottlenose dolphin.

The combination of zooarchaeological and historical sources suggests that the clergy of large parts of northern and western Europe, already developed a taste for cetacean meat during the seventh century AD, while the nobility appears to have followed during the early tenth century. This signifies a cultural shift that eventually led to economic and legal shifts. From that point onwards, stranded cetaceans were treated as a "wreck of sea right", restricted to the King, Queen, or their vassals (either members of the nobility of clergy with jurisdiction over a fief). Historical sources from England, suggest that considerable effort was put into claiming of stranded cetaceans by the social elite, but peasants attempted to illegally exploit stranded cetaceans as well. When caught, these peasants were punished or had to pay a fine. Based on this, it is clear that for medieval England, cetacean meat was a high-status food source, normally restricted to the social elite, though peasants were clearly undermining the social elite's power by poaching activities.

For other European regions, similar "wreck of sea rights" were set in place, from as early as the late ninth century AD for France. For Flanders, this was also the case and zooarchaeological sources imply that cetacean meat was often consumed by the social elite, but it was also frequently sold at markets in urban centres, including Bruges and Calais, but also other large cities in Europe, including London and potentially Paris, which might be the direct result of the spread of salting as a preservation method.

The presence of cetacean meat at these urban centres from the tenth century onwards, implies that cetacean meat was not solely restricted to the social elite, though it might still have been perceived as a precious food source only available to the wealthy (again most likely the social elite). This is confirmed by zooarchaeological analysis, as the number of urban sites with cetacean remains rises during the High Medieval Period. The presence of whale bone on urban sites represent accidental losses, symbolic depositions, production debris, butchery waste, or various other processes.

Besides social aspects, climatic aspects might have also influenced cetacean exploitation. In regard to the Medieval Period in Europe, there are two periods with climate variability. These two periods are the Medieval Warm Period and the Little Ice Age. Though these periods should not be taken as periods with absolute climate changes, overall the climate was warmer and drier during the Medieval Warm Period and colder and wetter during the Little Ice Age. Cetacean strandings are known to occur more frequently during periods of climatic variability. This might have been the case for the Medieval Warm Period and the Little Ice Age as well. Especially high numbers of cetacean remains have been identified dating to the period around the onset of Medieval Warm Period. This might suggest that cetaceans stranded more frequently during this period and medieval people eagerly took advantage of this. For the Little Ice Age another smaller peak in the number of remains can be observed, suggesting a similar situation for this period as for the Medieval Warm Period. This might indicate that both social as well as climatic changes affected cetacean exploitation.

Accessibility is another important aspect and the most common species in northern European waters, the harbour porpoise, was also frequently exploited. In fact, the species appears to be the most frequently exploited species in medieval Europe and has been recovered from a large variety of contexts. The meat of this species appears to have been appreciated by the social elite (both the clergy and the nobility), though its meat was also sold at several medieval European markets, indicating some form of commercialisation took place as well, and in northern Poland and in Denmark special harbour porpoise hunting guilds were created.

This study indicates that social implication regarding cetacean exploitation varied greatly between medieval societies. Social value was put on the activity of whaling itself, the consumption of cetacean meat, the use of whale bone as a raw resource, or a combination of these. For most parts of Europe this was evident from the start of the High Medieval Period until the thirteenth or fourteenth century.

Therefore, this study suggests that the periodisation by Gardiner (1997), that was assessed as part of this study, in case it is going to be applied to northern and western Europe, should be adapted. During the Early Medieval Period (AD 400-1000) cetacean exploitation seems to have been undertaken infrequently by most medieval cultures. The Norse appear to be an exception to this, but the technique they practiced, trapping and poisoning animals in an enclosed area, is much different to those practiced by other cultures. The Norse additionally used the bone extracted from large whales for the creation of several artefacts including weaving swords, gaming pieces, cleavers, and plaques. These artefacts have frequently been retrieved from graves with a high-status signature. Grave goods are reflections of conscious acts included in the funerary rituals, in this case indicating that social value was placed on whale bone by the Norse.

The second phase (AD 1000-1350) starts with the onset of active whaling by more cultures, most predominantly the Basques and the Normans. This period can potentially be viewed as the period with the first large scale whaling endeavours. The main target was the North Atlantic right whale, and the species was targeted relentlessly by not only the Basques and the Normans, but also by the Portuguese, the Flemish, the Anglo-Saxons, and the Norse, along its migration route. These whaling practices eventually led to a collapse of the North Atlantic right whale population, resulting in the Normans abandoning their whaling endeavours in the thirteenth century, the Flemish in the thirteenth/fourteenth century, the Anglo-Saxons in the late fourteenth century (though they probably relied more on stranded cetaceans), the Portuguese in the late fourteenth century, and the Basques venturing to new whaling areas in the mid-fourteenth century. It has been suggested by Gardiner (1997) that the North Atlantic right whale population went down during the twelfth century, however based on Basque whaling records it appears that the right whale population only declined from the early fourteenth century onwards. The zooarchaeological data analysed as part of this study indicates that sites with cetacean remains declined right after the early fourteenth century. This might have been a direct result of a decline in the right whale population, resulting in whalers in some regions surrendering their whaling endeavours.

In the subsequent third phase (AD 1350-1600), the Basques appear to be the only culture to frequently exploit cetaceans. They were only able to do so by continuously searching for new whaling grounds, targeting new stocks of North Atlantic right whale or switching to the more northern species of the bowhead whale. For most parts of Europe, cetaceans seem to have been occasionally exploited, but as the North Atlantic right whale appears to have been already rare during this period, exploitation happened far less frequently than in the previous phase.

Following this phase, a new period of whaling can be identified, for which the discovery of Svalbard/Spitsbergen by Willem Barentsz in AD 1596 can be viewed as the starting point. This period saw a period of intensified whaling, leading to the disappearing of several populations of North Atlantic right whale and bowhead whale. This phase will be called phase 4 (AD 1600-1750), before a new phase commenced during which even more stocks and species were exploited, resulting in a severe drop in worldwide cetacean populations.

One of the most interesting findings done as part of this study, is the finding of four, or potentially five, grey whale specimens, using ZooMS, morphological, and osteometric identification based on the ORCA-Manual (van den Hurk, forthcoming). Grey whale remains are frequently found in the Netherlands, although previously only sub-fossil remains were identified. This study has produced the first cases of archaeological grey whale remains. Extensive zooarchaeological analysis of medieval remains from Iceland, Greenland, northern Scotland (personal communication Vicki Szabo, 2018) and Norway (Hufthammer *et al.*, 2018) have produced both only one grey whale specimen for each country, suggesting that the coastal region of the Netherlands might have been an important foraging ground for the species. The presence of grey whale remains in archaeological contexts suggests that hunting by humans contributed to the demise of the species, but more findings of remains of the species might reveal whether anthropogenic factors, low genetic diversity, climate change, or a combination of the three, are the cause of the extirpation of the species from the North Atlantic. More research, that should focus on the application of both ZooMS and aDNA analysis, is necessary in order to fully understand what triggered the vanishing of the grey whale from the North Atlantic. This topic is very current, as recent sightings of several grey whales in the North Atlantic have been made, possibly signalling the return of the species.

The North Atlantic right whale however, somehow managed to survive at least a millennium of excessive whaling practices. The population on the European side is probably completely decimated, but on the American side a small population still survives. The species is now endangered and protected by law but still suffers from anthropogenic factors. Though whaling is no longer undertaken, the species is now threatened by ship strikes or entanglement in fishing gear. Whether the North Atlantic right whale will face a similar fate as the North Atlantic grey whale, the future will tell, though active protection of this endangered species is necessary in order for it to survive. If not, the North Atlantic right whale, will be the first baleen whale species to go extinct as a direct result of anthropogenic activities.

References

Aaris-Sorensen, K., 2009. Diversity and dynamics of the mammalian fauna in Denmark throughout the last glacial-interglacial cycle, 115-0 kyr BP. *Fossils Strata,* 57, pp.1-59.

Aaris-Sorensen, K., Lund Rasmussen, K., Kinze, C. and Strand Petersen, K., 2010. Late Pleistocene and Holocene whale (Cetacea) from Denmark and adjacent countries: species, distribution, chronology, and trace element concentrations. *Marine Mammal Science,* 26(2), pp.253-281.

Aguilar, A., 1981. The Black Right Whale, *Eubalaena glacialis*, in the Cantabrian Sea. *Rep. Int. Whal. Commn,* 31, pp.457-459.

Aguilar, A., 1986. A review of old Basque whaling and its effect on the right whale (*Eubalaena glacialis*) in the North Atlantic. *Rep. Int. Whal. Commn*, special issue 10, pp.191-199.

Albarella, U. and Davis, S.J.M., 1996. Mammals and birds from Launceston Castle, Cornwall: decline in status and the rise of agriculture. *Circaea, The Journal of the Association for Environmental Archaeology*, 12(1), pp.1-156.

Albert the Great, 1987. *Man and the Beasts. De animalibus (book 22-26)*. Translated by: J.J. Scanlan. Binghamton, New York: Medieval & Renaissance texts & studies.

Alter, E., Meyer, M., Post, K., Czechowski, P., Gravlund, P., Gaines, C., Rosenbaum, H.C., Kaschner, K., Turvey, S.T., van der Plicht, J., Shapiro, B., and Hofreiter, M., 2015. Climate impacts on trans-ocean dispersal and habitat in grey whales from the Pleistocene to 2100. *Molecular Ecology*, 24(7), pp.1510-1522. DOI 10.1111/mec.13121.

Alter, S.E., Newsome, S.D. and Palumbi, S.R., 2012. Pre-Whaling Genetic Diversity and Population Ecology in Eastern Pacific Gray Whales: Insights from Ancient DNA and Stable Isotopes. *PLoS ONE*, 7(5), e35039.

Alter, S.E., Rynes, E. and Palumbi, S.R., 2007. DNA evidence for historic population size and past ecosystem impacts of gray whales. *Proceedings of the National Academy of Sciences,* 104(38), pp.15162-15167.

Amaral, A.R., Coelho, M.M., Marugán-Lobón, J. and Rohlf, F.J., 2009. Cranial shape differentiation in three closely related delphinid cetacean species: Insights into evolutionary history. *Zoology*, 112(1), pp.38-47.

Amundsen, C., Hendriksen, J., Myrvoll, E., Olsen, B. and Urbanczyk, P., 2003. Crossing borders: Multi-room houses and inter-ethnic contacts in Europe's extreme north. *Fennoscandia archaeologica*, XX, pp.79-100.

Ardouin, S., Hadjouis, D. and Arroyo, M., 2009. *Etude archéologique sur la Seine dans le Val-de-Marne du Paléolithique au haut Moyen Age, Service Archéologie du 94.*

Armit, I., 1996. *The archaeology of Skye and the Western Isles*. Edinburgh: Edinburgh University Press.

Arneborg, J., 2000. Greenland and Europe. In Fitzhugh, W.W. and E.I. Ward, eds. *Vikings: North Atlantic Saga.* London: Smithsonian Institution Press. pp.312-317.

Arneborg, J., Heinemeier, J., Lynnerup, N., Nielsen, H.L., Rud, N. and Sveinbjörnsdóttir, Á.E., 1999. Change of diet of the Greenland Vikings determined from stable carbon isotope analysis and 14C dating of their bones. *Radiocarbon*, 41(2), pp.157-168.

Arnold, E.F., 2016. Washed ashore: marine mammals from medieval times to today. *Origins; Current Events in Historical Perspective*. Available at: <http://origins.osu.edu/connecting-history/washed-ashore-marine-mammals-medieval-times-today> [Accessed 18-04-2019].

Ashby, S.P., 2002. The role of zooarchaeology in the interpretation of socioeconomic status: a discussion with reference to Medieval Europe. *Archaeological Review from Cambridge*, pp.37-59.

Asselberg, R., 1981. Een subfossiele schedel van een grijze walvis, in het bezit van het Visserijmuseum van Oostduinkerke. *Marswin,* 2, pp.5-11.

Azpiazu, J.A., 2000. Los balleneros vascos en Cantabria, Asturias y Galicia, Itsas Memoria. *Revista de Estudios Marítimos del País Vasco,* 3, pp.77-97.

Baldanza, A., Bizzarri, R., Famiani, F., Monaco, P., Pellegrino and Sassi, P., 2013. Enigmatic, biogenically induced structures in Pleistocene marine deposits: a first record of fossil ambergris. *Geology*, 42(10), pp.1075-1078.

Banham, D., 2004. *Food and Drink in Anglo-Saxon England.* Tempus Publishing Ltd.

Barrett, J.H., 2016. Medieval Sea Fishing, AD 500-1550: Chronology, Causes and Consequences. In: Barrett, J.H. and Orton, D.C., eds. *Cod and Herring: The Archaeology and History of Medieval Sea Fishing.* Oxford: Oxbow Books.

Barrett, J.H., Locker, A.M. and Roberts, C.M., 2004a. The origins of intensive marine fishing in Medieval Europe: the English evidence. *Proceedings of the Royal Society of London B,* 271, pp.2417-2421.

Barrett, J.H., Locker, A.M. and Roberts, C.M., 2004b. 'Dark Age Economics' revisited: the English fish bone evidence AD 600-1600. *Antiquity*, 79(301), pp.618-635.

Barrett, J.H., Orton, D., Johnstone, C., Harland, J., Van Neer, W., Ervynck, A., Roberts, C., Locker, A., Amundsen, C., Enghoff, I.B., Hamilton-Dyer, S., Heinrich, D., Hufthammer, A.K., Jones, A.K.G., Jonsson, L., Makowiecki, D., Pope, P., O'Connell, T.C., De Roo, T. and Richards, M., 2011. Interpreting the expansion of sea fishing in medieval Europe using stable isotope analysis of archaeological cod bones. *Journal of Archaeological Science*, 38(7), pp.1516-1524.

Barrett-Lennard, L.G., Matkin, C.O., Durban, J.W., Saulitis, E.L. and Ellifrit, D., 2011. Predation on grey whales and prolonged feeding on submerged carcasses by transient killer whales at Unimak Island, Alaska. *Mar Ecol Prog Ser*, 412, pp.229-241. https://doi.org/10.3354/meps08906.

Barthelmess K. and Svanberg I., 2006. Linnaeus' whale. A wash drawing of bottlenose whales (Hyperoodon ampullatus) at Hammarby, with remarks on other early depictions of the species. *Lychnos, Annual of the Swedish History of Science Society,* 2006, pp.303–317.

Barthelmess, K., 1997. An unrecorded broadside commemorating cetacean strandings on the Dutch and Flemish coast 1519 to 1617. *Lutra*, 40, pp.9-19.

Barton, D., 1943. A Whale Bone Built into a Wall. *Journal of the Royal Society of Antiquities of Ireland*, 73, p.298.

Bartosiewicz, L., Gyetvai, A. and Küchelmann, H.C., 2010. The Beast in the Feast. In: Pluskowski, A.G., Kunst, G.K., Kucera, M., Bietak., M. and Hein, I., eds. *Bestial Mirrors – Using Animals to construct human identities in medieval Europe, Animals as Material Culture in the Middle Ages 3*. Wien. pp.85-99.

Bately, J., 2007. Text and Translation: The Three Parts of the Known World and the Geography of Europe North of the Danube According to Orosius' Historiae and its Old English Version. In: Bately J. and Englert A., eds. Ohthere's Voyages*: A Late 9th-Century Account of Voyages along the Coasts of Norway and Denmark and its Cultural Context (Maritime Culture of the North 1).* Roskilde: Viking Ship Museum, pp.40–50.

Batey, C.E., 1987. *Freswick Links, Caithness. A re-appraisal of the Late Norse site in its context*. Oxford: B.A.R.

Beker, C., 2012. Aus dem Dunkel eines Magazins ans Licht gebracht: Archäozoologische Untersuchungen zu Hatzum-Boomborg, einer Siedlung der Vorrömischen Eisenzeit in Ostfriesland. *Siedlungs- und Küstenforschung im südlichen Nordseegebiet,* 35, pp.201-294.

Béland, S., 2012. Which Whales Were Hunted? Genetic Species Identification of Bones from Traditional Whaling Sites on Vancouver Island. Unpublished thesis.

Bendrey, R., 2005. The Animal Bone. In: J. Butler, ed. 2005. *Saxons, Templars & Lawyers in the Inner Temple; Archaeological excavations in Church Court & Hare Court*. London: Pre-Construct Archaeology Limited. Pp.64-67.

Benke, H., 1993. Investigations on the osteology and functional morphology of the flipper of whales and dolphins (cetacean). In: Pilleri, G., ed. 1993. *Investigations on Cetacean*, XXIV, pp.9-252.

Bernal-Casasola, D., Gardeisen, A., Morgenstern, P., Kolska Horwitz, L., Piqués, G., Theodoropoulou, T. and Wilkens, B., 2016. Ancient whale exploitation in the Mediterranean: the archaeological record. *Antiquity*, 90, pp.914-927. doi:10.15184/aqy.2016.116

Berta, A. and Deméré, 2009. Mysticetes, Evolution. In: Perrin, W.F., Würsig, B. and Thewissen, J.G.M., eds. 2009. *Encyclopaedia of Marine Mammals*. 2nd ed. London: Academic Press. pp.749-753.

Berta, A., Ekdale, E.G. and Cranford, T.W., 2014. Review of the Cetacean Nose: Form, Function, and Evolution. *The Anatomical Record,* 297, pp.2205-2215.

Biard, V., Gol'din, P., Gladilina, E., Vishnyakova, K., McGrath, K., Vieira, F.G., Wales, N., Fontaine, M.C., Speller, C. and Olsen, M.T., 2017. Genomic and proteomic identification of Late Holocene remains: setting baselines for Black Sea odontocetes. *Journal of Archaeological Science: Reports*, 15, pp.262-271.

Bjelland, O., Bergstad, O.A., Skjaeraassen, J.A. and Meland, K., 2011. Trophic ecology of deep-water fishes associated with the continental slope of the Eastern Norwegian Sea. *Sarsia*, 85(2), pp.101-117.

Blackmore, L. and Cowie, R., 2001. Saxon and medieval Battersea excavations at Althorpe Grove, 1975-8. *Surrey Archaeological Collections*, 88. pp.67-92.

Blanco-Zubiaguirre, L., Ribechini, E., Degano, I., La Nasa, J., Carrero, J.C., Iñañez, Olivares, M. and Castro, K., 2018. GC–MS and HPLC-ESI-QToF characterization of organic lipid residues from ceramic vessels used by Basque whalers from 16th to 17th centuries. *Microchemical Journal*, 137, pp.190–203.

Bloch, D., Dam, M. and Hanusardóttir, M., 2003. Marine foods. Marine mammals as meat sources. In: Caballero, B., ed. 2003. *Encyclopaedia of food sciences and nutrition.* 2nd ed. Academic Press. pp.3733-3739.

Bolster, W., 2008. Putting the Ocean in Atlantic History: Maritime Communities and Marine Ecology in the Northwest Atlantic, 1500-1800. *The American Historical Review,* 113(1), pp.17-47.

Bond, J.M. and O'Connor, T.P., 1999. *The Archaeology of York; the Animal Bones. Bones from medieval deposits at 16-22 Coppergate and other sites in York.* York: York Archaeological Trust.

Born, W., 1989. *Eten door de eeuwen. De geschiedenis van de culinaire cultuur*. Baarn: Uitgeverij Bosch & Keuning.

Bosselaers, M. and Collareta, A., 2016. The whale barnacle *Cryptolepas rhachianecti* (Cirripedia: Coronulidae), a phoront of the grey whale *Eschrichtius robustus* (Cetacea: Eschrichtiidae), from a sandy beach in the Netherlands. *Zootaxa*, 4154-3, pp.331-338.

Boynton, G.R., 2016. *Calendar of Patent Rolls*. Available at: <http://sdrc.lib.uiowa.edu/patentrolls/search.html> [Accessed 13-01-2019]. The University of Iowa Libraries.

Bradley, R.S., Hughes, M.K. and Diaz, H.F., 2003. Climate in Medieval Time. *Science*, 302, pp.404-405.

Bradshaw, C.J.A., Evens, K. and Hindell, M.A., 2006. Mass Cetacean Strandings: A Plea for Empiricism. *Conservation Biology*, 20, pp.584-586.

Brito, C. and Sousa, A., 2011. The environmental history of cetaceans in Portugal: Ten centuries of whale and dolphin records. *PLos ONE*, 6(9), e23951.

Brito, C., 2011. Medieval and Early Modern Whaling in Portugal. *Anthrozoös,* 24(3), pp.287-300.

Brito, C., Jordão, V.L. and Pierce, G.J., 2015. Ambergris as an overlooked historical marine resource: its biology and role as a global economic commodity. *Journal of the Marine Biological Association of the United Kingdom*, pp.1-12.

Brothwell, D.R., 1979. Notes on the Mammal remains in Medieval pits and well at Seaford Church Street. *Sussex Archaeological Collections*, 117, pp.231-233.

Brown, N.M., 2015. *Ivory Vikings; The mystery of the most famous chessmen in the world and the woman who made them*. New York: St. Martin's Press.

Browsher, J., 1999. Saxon foreshore at Temple Place. *London Archaeologist*, 9(3), pp.82-89.

Bryant, P.J., 1995. Dating remains of Gray Whales from the Eastern North Atlantic. *Journal of Mammalogy*, 76(3), pp.857-861.

Bryant, P.J., 1995. Dating remains of grey whales from the Eastern North Atlantic. *Journal of Mammalogy*, 76(3), pp.857-861. DOI 10.2307/1382754.

Buckley, M., Fraser, M., Herman, J., Melton, N.D., Mulville, J. and Palsdottir, A.H., 2014. Species identification of archaeological marine mammals using collagen fingerprinting. *Journal of Archaeological Science*, 41, pp.631-641.

Buitenhuis, H., 2008. Walvisbotten. In: P.J.A. Stokkel, *Een archeologische begeleiding langs de Oudelandertocht, gemeente Wieringermeer (N.-H.). (ARC-Rapporten 2008-128).* Geldermalsen: ARC. pp.4-6.

Buitenhuis, H., 2011. Faunaresten. In: Ufkes, A., ed. *Een archeologische opgraving in de vroegmiddeleeuwse ringwalburg van Domburg, gem. Veere (Z)*. Groningen: ARC bv.

Camphuijsen, K. and Peet, G., 2006. *Walvissen en dolfijnen in de Noordzee*. Leeuwarden/Utrecht: Fontaine Uitgevers BV/Stichting de Noordzee.

Canto, J., Salas-Gismondi, R., Cozzuol, M., Yáñez, J., 2008-09. The Aquatic Sloth *Thalassocnus* (Mammalia, Xenarthra) from the late Miocene of North-Central Chile: biographic and ecological implications. *Journal of Vertebrate Palaeontology*, 28(3), pp.918-922.

Carmiggelt, A., Laarman, F.J. and Waasdorp, J.A., 1998. Het archeozoologisch onderzoek. In: Carmiggelt, A., ed. *Romeinse vondsten van de Scheveningseweg te Den Haag; De dieren- en plantenresten (Haagse Oudheidkundige Publicaties, 4)*, pp.11-37.

Carnandet, J., 1876, ed. *Acta Sanctorum, August 3* (no.37). Paris Rome: Victor Palmé.

Carwardine, M., 1995. *Whales Dolphins and Porpoises*. London: Dorling Kindersley Books.

Castello y Tickell, S. and Barker, J., 2015. *Thames Marine Mammal Sightings Survey Ten Year Report (2004-2014)*. London: Zoological Society of London.

Centre de Recherche sur les Mammifères marins & LIENSs Labaratory, 2014. *Centre de Recherche sur les Mammifères marins & LIENSs Labaratory*. [online] Available at: <http://crmm.univ-lr.fr/index.php/en/home> [Accessed 03 July 2014].

Charlier, R.H., 2004. Cetaceans and Belgian whalers; a brief historical view. In: Marcos, S., Zhu, M., Charlier, R., Gerges, M., Kullenberg, G., Lenz, W., Lu, M. and Zou, E., eds. *Ocean sciences bridging the millennia - A Spectrum of historical accounts*. Paris: UNESCO Publishing. pp.385-397.

Chazottes, M.-A., 2017. Matières du quatoidien, matières de luxe: os, bois de cerf, ivoire, corail, nacre, corne, fanon de baleine et écaille de tortue dans l'artisanat médiéval et postmédiéval en Provence à partir de l'étude conjointe des sources archéologiques, écrites et iconographiques. Unpublished PhD thesis.

Cherkinsky, A., 2009. Can we get a good radiocarbon age from "Bad Bone"? determining the reliability of radiocarbon age from bioapatite. *Radiocarbon*, 51(2), pp.647-655.

Chevallier, J., 2014. Whaling in Medieval France. *Les Leftovers*, January 2014. Available at: https://leslefts.blogspot.com/2014/01/whaling-in-medieval-france.html [Accessed 17-05-2019]

Childe, V.G., 1931. *Skara Brae, a Pictish Village in Orkney*. London.

Christen, A.G. and Christen, J.A., 2011. The unicorn and the narwhal: a tale of the tooth. *Journal of the history of dentistry*, 59(3), 135-142.

CITES, 2019. *Convention on International Trade in Endangered Species of Wild Fauna and Flora*. Available at: <https://www.cites.org/> [Accessed 31-10-2019].

Clark, G., 1947. Whales as an economic factor in prehistoric Europe. *Antiquity*, 21, pp.84-104.

Clarke, M.R., 1978. Structure and proportions of the spermaceti organ in the sperm whale. *Journal of the Marine Biology Association*, 58, pp.1-17.

Clarke, R., 2005. A southern bottlenose whale examined in the Antarctic. *LAJAM*, 4(2), pp.83-96.

Clarke, R., 2006. The origin of ambergris. *Latin American Journal of Aquatic Mammals*, 5, p.7.

Clason, A.T., 1967. The animal bones. In: van Es, W.A., ed. *Wijster. A native village beyond the imperial frontier 150-425 AD. Palaeohistoria*, 11. Groningen: J.B. Wolters.

Clavel, B., 2001. L'Animal dans l'alimentation médiévale et moderne en France du Nord (XIIIe – XVIIe siècles). *Revue Archaeologique de Picardie*, 19, pp.9-204.

Coenen, A., 1585. *Het Walvisboek; Walvissen en andere zeewezens beschreven door Adriaen Coenen in 1585*. ed. Egmond, F. Mason, P. and Lankester, K., 2003. Zutphen: Walburg Pers.

Collett, R., 1912. *Norges Pattedyr (Norges Hvirveldyr I)*. Kristiana, Oslo: H. Aschehoug & Co (W. Nygaard).

Coltrain, J.B., Hayes, M.G. and O'Rourke, D.H., 2004. Sealing, whaling and caribou: the skeletal isotope chemistry of Eastern Arctic foragers. *Journal of Archaeological Science*, 31(1), pp.39-47.

Crabtree, P.J., 1990. Zooarchaeology and complex societies: some uses of faunal analysis for the study of trade, social status, and ethnicity. In. Schiffer, M.B., ed. *Archaeological Method and Theory, 2*. Tuscon: University of Arizona Press. pp.155-205.

Craig, O.E., Steele, V.J., Fischer, A., Hartz, S., Andersen, S.H., Donohoe, P., Glykou, A., Saul, H., Jones, D.M., Koch, E. and Herron, C.P., 2011. Ancient lipids reveal continuity in culinary practices across the transition to agriculture in Northern Europe. *Proceedings of the National Academy of Sciences of the United States of America*, 108(44), pp.17910-17915.

Crovetto, A., 1991. Etude osteometrique et anatomo-functionelle de la colonne vertebrale chez les grands cetacés. *Invest. Cetacea*, 23, pp.7-189.

Crowell, A.L., 1997. *Archaeology and the Capitalist World System. A study from Russian America*. New York: Plenum Press.

Cumbaa, S.L., 1986 Archaeological evidence of the 16th century Basque right whale fishery in Labrador. *Report of the International Whaling Commission*, 10, pp.187-190.

Cumbaa, S.L., Brown, M.W. and White, B.N., 2002. Zooarchaeological and molecular perspectives on Basque whaling in 16th century Labrador. *Canadian Zooarchaeology*, 20, pp.6-15.

Curl, J., 2006. Animal, bird and fish bone. In: Wallis, H., ed. *Excavations on the Site of Norwich Cathedral Refectory 2001-3*. Norfolk: Archaeology and environment Division, Norfolk Museums and Archaeology Service.

Curle, C.L., 1982. Pictish and Norse Finds from the Brough of Birsay 1934-74. *Society of Antiquaries of Scotland. Monograph Series* 1. pp.1-141.

Dahl, S., 1971. the Norse settlement of the Faroe Islands. *Medieval Archaeology*, XIV, pp.60-73.

Dalton, O.M., 1928. Early chessmen of whale's bone excavated in Dorset. *Archaeologia*, 77, pp.77-86.

Davis, S., 1987. *The archaeology of animals*. New Haven: Yale University Press.

De Baets, P., 2013. Walvissen op de Vlaamse kust en in het Scheldebekken. *Biekorf*, 113(4), pp.385-413.

De Groote, M., 1999. De baljuwsrekening: venster op de praktijk van het middeleeuws strandrecht in Vlaanderen. *Madoc*, jaargang 1999, pp.258-267.

De Grossi Mazzorin, J. and Minniti, C., 1999. Diet and religious practices: the example of two monastic orders in Rome between the XVIth and XVIIIth centuries. *Anthropozoologica*, 30, pp.33-50.

De Gryse, R., 1940-1945. De Vlaamsche walvischvangst in de middeleeuwen. *Biekorf*, jaargang 46, pp.20-23.

De Haan, C. and Oosterman, J., 1996. *Is Brugge groot?* Amsterdam: Em. Querido's Uitgeverij.

De Smet, W.M.A., 1979. Geschiedenis en huidige toestand van het onderzoek over cetaceeen in België. *Lutra*, 20, pp.25-36.

De Smet, W.M.A., 1981. Evidence of whaling in the North Sea and English Channel during the middle ages. *Advisory Committee on Marine Resources Research Working Party on Marine Mammals, General Papers and Large Cetaceans, Mammals of the Seas*, 3. Rome: United Nations Food Agricultural Organization. pp.301-309.

Degerbøl, M., 1933. Danmarks Pattedyr i Fortiden i sammenligning med recente Former. *Vidensk. Medd. Nat For.*, 95. Copenhagen.

Delaunay, P., 1997. *La Zoologie au seizième siècle*. 2nd ed. Paris: Hermann.

Delhaye, M., 1998. L'épave médiévale de Cavalaire: un exemple de l'évolution navale architecturale avant la Renaissance. *Itsas memoria. Revista de Estudios Marítimos del País Vasco*, 2. Donostia-San Sebastián: Untzi Museoa-Museo Naval. Pp.43-48.

Deméré, T.A., McGowen, M.R., Berta, A. and Gatesy, J., 2008. Morphological and molecular evidence for a stepwise evolutionary transition from teeth to baleen in mysticete cetaceans. *Syst. Biol*, 57, pp.15-37.

Department VI of the Royal Belgian institute of Natural Sciences, 2014. *Management Unit of the North Sea Mathematical Models*. [online] Available at: <http://www.mumm.ac.be/EN/Management/Nature/search_strandings.php> [24 October 2017].

Dobney, K., Jaques, D., Barrett, J. and Johnstone, C., 2007. *Farmers, Monks and Aristocrats; The Environmental Archaeology of Anglo-Saxon Flixborough*. Oxford: Oxbow Books.

Douglas, A., Jarrett, C., Rielly, K., Riddler, I. and Trzaska-Nartowski, N., 2015. *Medieval Craft-Working in Oxford: Excavations at 15-17 Clarendon Centre*. Wallingford: Pre-Construct Archaeology. pp. 64-83.

Drinkwater, K.F., Belgrano, A., Borja, A., Conversi, A., Edwards, M., Greene, C.H., Ottersen, G., Pershing, A.J. and Walker, H., 2003. The response of marine ecosystems to climate variability associated with the North Atlantic Oscillation. In: Hurrell, J.W., Kushnir, Y., Ottersen, G. and Visbeck, M., eds. *The North Atlantic Oscillation: climatic significance and environmental impact*. Washington DC: American Geophysical Union. Pp. 211-234.

Driver, J.C., 2004. Food, status and formation processes: a case study from medieval England. In: O'Day, S.J., 2004. *Behaviour Behind Bones. The Zooarchaeology of ritual, religion, status and identity. Proceedings of the 9th ICAZ conference, Durham 2002*. Oxford: Oxbow Books. pp.244-251.

Du Bouëtiez, E., Clavel, B. and Ravoire, F., 2013. Vestiges médiévaus et modernes découverts sur le site de l'hôtel de Beauvais à Paris (4e arrondissement). *Revue Archéologique d'Île de France,* 5, pp.193-254.

Dudley, P., 1725. An essay upon the natural history of whales. *Philosophical Transactions of the Royal Society London*, 33, pp.256-269.

Dugmore, A.J., Church, M.J., Buckland, P.C., Edwards, K.J., Lawson, I., McGovern, T.H., Panagiotakopulu, E., Simpson, I.A., Skidmore P. and Sveinbjarnardóttir G., 2005. The Norse landnám on the North Atlantic islands: an environmental impact assessment. *Polar Record* 41(216), pp.21-37.

Edvardsson, R. and McGovern, T.H., 2005. Archaeological Excavations at Vatnsfjördur 2003-04. *Archaeologica Islandica*, 4, pp.16-30.

Ekdale, E.G., Berta, A. and Deméré, T.A., 2011. The Comparative Osteology of the Petrotympanic Complex (Ear Region) of Extant Baleen Whales (Cetacea: Mysticeti). *PLoS ONE*, 6(6), e21311. doi:10.1371/journal.pone.0021311

Ervynck, A., 2004. Orant, pugnant, laborant. The diet of the three orders in the feudal society of medieval north-western Europe. In: O'Day, S.J., 2004. *Behaviour Behind Bones. The Zooarchaeology of ritual, religion, status and identity. Proceedings of the 9th ICAZ conference, Durham 2002*. Oxford: Oxbow Books. pp.215-223.

Ervynck, A., van Neer, W. and Pieters, M., 2004. How the North Was Won (and Lost Again): historical and archaeological data on the exploitation of the North Atlantic by the Flemish Fishery. In: Housley, R.A. and Coles, G., ed. *Symposia of the Association for Environmental Archaeology*, 21 (Oxford 2004). pp.230-239.

Espeel, S., 2016. De Brugse stadsmagistraat in de late 14e eeuw. Een prosopografische studie voor de periode 1359-1375. Unpublished Masters Thesis. University of Ghent.

Espinoza, E.O. and Mann, M.J., 1991. *Identification guide for ivory and ivory substitutes*: Baltimore: World Wildlife Fund and Conservation Foundation.

Esser, E., Beerenhout, B. and Kootker, L.M., 2010. Dierlijk botmateriaal. In: Dijkstra, M.F.P. and Brandenburgh, C.R., eds. *Bodemonderzoek en Bouwhistorie in Leiden 1; Leiden-Aalmarktschool Archeologisch en bouwhistorisch onderzoek naar 800 jaar bewoning langs de Oude Rijn, ter plaatse van het voormalige St. Catharinagasthuis*. pp.172-231.

Evans, K., Thresher, R., Werneke, R.M., Bradshaw, C.J.A., Pook, M., Thiele, D. and Hindel, M.A., 2005. Periodic variability in cetacean strandings: links to largescale climate events. *Biology Letters*, 1(2), pp.147-150.

Evans, P.G.H. and Bjørge, A., 2013. Impacts of climate change on marine mammals. *Marine Climate Change Impacts Partnership: Science Review*, 2013.

Evans, S., Briz I Godino, I., Álvarez, M., Roswell, K., Collier, P., Prosser de Goodall, R.N., Mulville, J., Lacrouts, A., Collins, M.J. and Speller, C., 2016. Using combined biomolecular methods to explore whale exploitation and social aggregation in hunter–gatherer–fisher society in Tierra del Fuego. *Journal of Archaeological Science: Reports,* 6, pp.757-767.

Fagan, B., 2000. *The Little Ice Age: How the climate Made History, 1300-1850*. New York: Basic Books.

Fagan, B., 2006. *Fish on Friday; Feasting, Fasting and the Discovery of the New World.* New York: Basic Books.

Fielding, R., 2018. *The wake of the whale: Hunter societies in the Caribbean and North Atlantic*. Cambridge: Harvard University Press.

Fischer, P., 1881. Cétacés de Sud-Ouest de la France. *Actes de la Société Linnéenne de Bordeaux, 4th ser.*, 35, pp.5-219.

Fiskeripublikasjoner, n.d. *Fangst av hval i Vikingtid og Middelalder*. Available at: <https://www.kyst-norge.no/?k=2909&id=16004&aid=8396&daid=2603> [Accessed at 16-05-2019].

Fitzhugh, W., 2000. Puffins, Ringed pins, and runestones: Viking passage to America. In. Fitzhugh, W. and Ward, E. eds. *Vikings: The North Atlantic Saga*. Washington DC: Smithsonian Institution Press. pp.11-25.

Flower, W.H., 1872. On a subfossil whale (*Eschrichtius robustus*) discovered in Cornwall. *Annals and Magazine of Natural History,* 9(54), pp.440-442. DOI 10.1080/00222937208696615.

Fontaine, M.C., Tolley, K.A., Michaux, J.R., Birkun, A., Ferreira, M., Jauniaux, T., Llavona, Á., Öztürk, B., Özturk, A.A., Ridoux, V., Rogan, E., Sequeira, M., Bouquegneau, J-M. and Baird, S.J.E., 2010. Genetic and historic evidence for climate-driven population fragmentation in a top cetacean predator: the harbour porpoises in European water. *Proceedings of the Royal Society B.,* 277.

Foote, A.D., Kaschner, K., Schultze, S.E., Garilao, C., Ho, S.Y.W., Higham, T.F.G., Stokowska, C., Es, H. van der, Embling, C.B., Gregersen, K., Johansson, F., Willerslev, E. and Gilbert, T.P., 2013. Ancient DNA reveals that bowhead whale lineages survived Late Pleistocene climate change and habitat shifts. *Nature Communications,* 4, 1677.

Foote, A.D., Liu, Y., Thomas, G.W.C., Vinar, T., Alföldi J., Deng, J., Dugan, S., van Elk, C.E., Hunter, M.E., Joshi, V., Khan, Z., Kovar, C., Lee, S.L., Lindblad-Toh, K., Mancia, A., Nielsen, R., Qin, X., Qu, J., Raney, B.J., Vijay, N., Wolf, J.B.W., Hahn, M.W., Muzny, D.M., Worley, K.C., Gilbert, M.T.P. and Gibbs, R.A., 2015. Convergent evolution of the genomes of marine mammals. *Nature Genetics*, 34(3), pp.272-276.

Foote, P., 1996. *Olaus Magnus Historia de Gentibus Septentionalibus Romae 1555.* London: The Hakluyt Society.

Foote, P.G., 2017. *Olaus magnus. A Description of the Northern Peoples. 1555. Volume III.* Oxon: Routledge.

Fordyce, R.E. and Barnes, L.G., 1994. The evolutionary history of whales and dolphins. *Annual Review of Earth and Planetary Sciences,* 22(1), 419-455.

Foster, H.J., 2016. A Zooarchaeological Study of Changing Meat Supply and Butchery Practices at Medieval Castles in England. Unpublished PhD Thesis. University of Exeter.

Frank, S.M., 1993. Scrimshaw: An Introduction and Overview, A.D. 800-1960*. In: Basberg, B.L., Ringstad, J.E. and Wexelsen, E., eds. *Whaling & History: Pespectives on the evolution of the Industry.* Sandefjord, Norway: Kdr. Chr. Christensens Hvalfangstmuseum. Pp.203-212.

Fraser, F.C., 1970. An early 17th century record of the California grey whale in Icelandic waters. *Investigations Cetacea*, 2, pp.13-20.

Gaborit-Chopin, D. and Bardoz M.-C., 2005. *La France romane: au temps des premiers capétiens, 987-1152, exposition Paris Musée du Louvre, 10 mars, 6 juin 2005*, Paris: Musée du Louvre Ed. Hazan. p.407.

Gabriel, I., 1988. Hof-und Sakralkultur sowie Gebrauchs- und Handelsgut im Spiegel der Kleinfunde von Starigard/Oldenburg. *Bericht der Römischen Kommission*, 69, pp.240-242.

Galik, A. and Kunst, G.K., 2004. Dietary habits of a monastic community as indicated by animal bone remains from Early Modern Age in Austra. In: O'Day, S.J., 2004. *Behaviour Behind Bones. The Zooarchaeology of ritual, religion, status and identity. Proceedings of the 9th ICAZ conference, Durham 2002.* Oxford: Oxbow Books. pp.224-232.

Garcia, A.C.A., 2020. Pollution and marine debris in Portuguese Atlantic harbours during the Early Modern age: Administration and functionality problems. In: Roque, A.C., Brito, C. and Veracini, C., eds. *Peoples, Nature and Environments Learning to Live Together*. Newcastle: Cambridge Scholars Publishing. Pp.83-96.

Gardiner, M., 1997. The Exploitation of Sea-Mammals in Medieval England: Bones and their Social Context. *Archaeological Journal*, 154(1), pp.173-195.

Gardiner, M., Stewart, J. and Priestley-Bell, G., 1998. Anglo-Saxon Whale exploitation: some evidence from Dengemarsh, Lydd, Kent. *Medieval Archaeology*, 42, pp.96-101.

Gardiner, M.F., 1988. Medieval settlement and society in the Broomhill area and excavations at Broomhill church. In: Eddison, J. and Green, C., eds. *Romney Marsh: Evolution, Occupation and Reclamation.* Oxford: Oxford University Committee for Archaeology. Pp.112-127.

Garrison, E.G., McFall, G., Cherkinsky, A. and Noakes, S.E., 2012. Discovery of a Pleistocene mysticete whale, Georgia Bight (USA). *Palaeontologica Electronica*, 15(3), 10. DOI 10.26879/323.

Garrison, E.G., Morgan, G.S., McGrath, K., Speller, C. and Cherkinsky, A., 2019. Recent dating of extinct Atlantic grey whale fossil, (*Eschrichtius robustus*), Georgia Bight and Florida, western Atlantic Ocean. *PeerJ*, 7:e6381. DOI 10.7717/peerj.6381.

Gatesy, J., Geisler, J.H., Chang, J., Buell, C., Berta, A., Meredith, R.W., Springer, M.S. and McGowen, M.R. 2013. A phylogenetic blueprint for a modern whale. *Molecular Phylogenetics and Evolution*, 66, pp.479-506.

Geraci, J.R. and Smith, T.G., 1979. Vitamin C in the Diet of Inuit Hunters from Holman, Northwest Territories. *Arctic*, 32(2), pp.135-139.

Gifford, D. and Seidmann, R.J., 1974. *Ulysses Annotated; Notes for James Joyce's Ulysses*. 2nd edition. New York: E.P. Dutton.

Gilbert, M.T.P., Kivisild, T., Gronnow, B., Andersen, P.K., Metspalu, E., Reidla, M., Tamm, E., Axelsson,

E., Götherström, A., Campos, P.F., Rasmussen, M., Metspalu, M., Higham, T.F.G., Schwenninger, J.-L., Nathan, R., De Hoog, C.-J., Koch, A., Moller, L.N., Andreasen, C., Meldgaard, M., Villems, R., Bendixen, C., Willerslev, E., 2008. Paleo-Eskimo mtDNA genome reveals matrilineal discontinuity in Greenland. *Science*, 320, pp.1787-1789.

Gimeno, G.-L., 1978. Hallazgo de un broche altomedieval trabajado en hueso. *Boletín del Seminario de Arte y Arqueología*, 44, pp.430-434.

Giry, A. and Prou, M., 1952, eds. *Receuil des Actes de Charles II le Chauve*, 2. Paris: Imprimerie Nationale.

Goley, P.D. and Straley, J.M., 1994. Attack on grey whales (*Eschrichtius robustus*) in Monterey Bay, California by killer whales (*Orcinus orca*) previously identified in Glacier Bay, Alaska. *Canadian Journal of Zoology*, 72(8), pp.1528-1530.

Gordon, R.L., 2015. Feeding the city: zooarchaeological perspectives on urban provisioning and consumption behaviours in post-medieval England (AD1500 – AD1900). Unpublished PhD thesis. University of Leicester.

Goyheneche Farnie, E., 1984. La peche a la baleine. *Cuader. Sec. Hist.-Geo.*, 2, pp.8-24.

Grant, A. 1992. Animal resources. In: Astill, G. and Grant, A., eds. *The Countryside of Medieval England*. Oxford: Blackwell Press. 14987.

Grau-Sologestoa, I. and García-García, M., 2018. Zooarqueología y Edad Media en la península ibérica. In: Castillo, J.A.Q., ed. *Treinta años de Arqueología Medieval España*. Oxford: Archaeopress Publishing Ltd.

Grau-Sologestoa, I., 2016. Socio-economic status and religious identity in medieval Iberia: The Zooarchaeological evidence. *Environmental Archaeology*. DOI: 10.1080/14614103.2016.1153818.

Grau-Sologestoa, I., 2018. Pots, chicken and building deposits: The archaeology of folk and official religion during the High Middle Ages in the Basque Country. *Journal of Anthropological Archaeology*, 49, pp.8-18.

Grau-Sologestoa, I., Albarella, U. and Castillo, J.A.Q., 2016. Urban medieval and post-medieval zooarchaeology in the Basque Country: Meat supply and consumption. *Quaternary International*, 399, pp.1-12.

Gray, J.E., 1864. Notes on the whale-bone whale: with a synopsis of the species. *Annals and Magazine of Natural History (ser. 3),* 14, pp.345-353. DOI 10.1080/00222936408681724.

Gray, J.E., 1866. *Catalogue of seals and whales in the British museum*. Second Edition. London: Taylor and Francis.

Gray, N.-M., Kainec, K., Madar, S., Tomko, L. and Wolfe, S. 2006. Sink or swim? Bone density as a mechanism for buoyancy control in early cetaceans. *Anat. Rec.: Ad. Integr. Anat. Evol. Biol.*, 29, pp.638-653.

Greenwood, J., 2011. *The changing technology of post-medieval sea-salt production in England.* Available at: <http://www.citcem.org/encontro/pdf/new_01/TEXTO%20- %20Jeremy%20Greenwood.pdf> [Accessed 20-11-2019].

Grenier, R., Stevens, W. and Bernier, M.-A., 2007. *L'archéologie subaquatique de Red Bay: La construction navale et la pêche de la baleine basque au XVIe siècle*. 5 vols. Ottawa, Ontario: Parcs Canada.

Groenman-van Waateringe, W. and Wijngaarden-Bakker, L.H., n.d. *Medieval archaeology and environmental research in the Netherlands*. pp.283-297.

Guizard, F., 2011. Retour sur un monster marin au Moyen Âge: la baleine. In: Gautier, A. and Martin, C., eds. *Échanges, communications et reseaux dans le Moyen Âge*. Turnhout: Brepols. pp.261-276.

Guizard, F., 2018. Delfines nec non et ballenae… Les cétacés de l'Atlantique nord au haut Moyen Âge: representation, identifidation et consummation. *Anthropozoologica*, 53(10), pp.115-123.

Gustavsson, R., Hennius, A. and Ljungkvist, J., 2015. Are many Vendel and Viking Period gaming pieces made of whale bone? *Forn Vännen, Journal of Swedish Antiquarian Research,* 110(1), pp.51-54.

Haelters, J., 2018. De narwhal van Bornem. *Zoogdier*, 29(3), pp.10-11.

Hagen, A., 1994. *A Handbook of Anglo-Saxon Food: Processing and Consumption*. Pinner: Anglo-Saxon Books.

Hallen, Y., 1994. The use of bone and antler at Foshigarry and Bac Mhic Connain, two Iron Age sites on North Uist, Western Isles. *Proceedings of the Society of Antiquaries of Scotland*, 124, pp.189-231.

Hambrecht, G. and Gibbons, K.S., 2018. Whale bone as fuel at an Iceland farm in Early Modern Iceland. *Arctic Anthropology*, 55(1), pp.63-72.

Hamilakis, Y., 1999. Food technologies/technologies of the body: The social context of wine and oil production and consumption in Bronze Age Crete. *World Archaeology*, 31, pp.38-54.

Hamilakis, Y., 2003. The sacred geography of hunting: wild animals, social power and gender in early farming societies. In: Kotjabopoulou, E., Hamilakis, Y., Halstead, P., Gamble, C. and Elefanti, P., eds. 2003. *Zooarchaeology in Greence: recent advances*. London (BSA Studies Volume 9), pp.239-246.

Hammond, P.S., Berggren, P., Benke, H., Borchers, D.L. Collet, A., Heide-Jorgensen, M.P., Heimlich, S. Hiby, A.R., Leopold, M.F. and Oien, N., 2002. Abundance of harbour porpoise and other cetaceans in the North Sea and adjacent waters. *Journal of Applied Biology,* 39, pp.361-376.

Hammond, P.W., 1998. *Food and Feast in Medieval England.* Sutton Publishing.

Hansen, L.I. and Olsen, B., 2004. *Samenes historie fram til 1750.* Oslo: Cappelen.

Harland, J.F., 2012. Animal Husbandry: the Mammal Bone. In: Barrett, J.H., ed. 2012. *Being an Islander.* Cambridge: McDonald Institute for Archaeological Research. Ch.8.

Harwood, J., 2001. Marine Mammals and their environment in the twenty-first century. *Journal of Mammalogy*, 82, pp.630-640.

Häussermann, V., Gutstein, C.S., Bedington, M., Cassis, D., Olavarria, C., Dale, A.C., Valenzuela-Tora, A.M., Perez-Alvarez, M.J., Sepúlveda, H.H., McConnoll, K.M., Horwitz, F.E., Försterra, G., 2017. Largest baleen whale mass mortality during strong El Niño event is likely related to harmful toxic algal bloom. *PeerJ*, 5:e3123.

Hedges, J.W., 1987. *Bu, Gurness and the Brochs of Orkney, Part II: Gurness, BAR British Series 164.* Oxford: BAR.

Heers, J., 1955. Le commerce des Basques en Méditerranée au XVe siècle (d'après les archives de Gênes). *Bulletin Hispanique*, tome 57, 3, pp.292-324.

Hennius, A., Gustavsson, R., Ljungkvist, J. and Spindler, L., 2018. Whalebone Gaming Pieces: Aspects of Marine Mammal Exploitation in Vendel and Viking Scandinavia. *European Journal of Archaeology*, pp.1-20.

Henriksen, S. and Roll Valen, C., 2013. *Skjærvika og Fjellvika, Hammerfest Kommune. Rapport fra de arkeologiske undersøkelsene 2009 og 2010 (Tromura Kulturvitenskap, Rapport 43).* Tromsø: Tromsø Museum, University of Tromsø.

Herman, O., 1887. *A Magyar halászat könyve*. Budapest: Királyi Magyar Természettudományi Társulat.

Heron, C. and Craig, O.E., 2015. Aquatic resources in foodcrusts: Identifcation and Implication. *Radiocarbon,* 57(4), pp.707-719.

Higgs, N.D., Little, C.T.S. and Glover, A.G., 2011. Bones as biofuel: a review of whale bone composition with implications for deep-sea biology and paleoanthropology. *Proceedings of the Royal Society of Biological Sciences*, 278, pp.9-17.

Hillson, S., 1992. *Mammal bones and teeth. An introductory guide to methods of identification.* Walnut Creek: Left Coast Press, Inc.

Hohn, A.A., Scott, M.D., Wells, R.S., Sweeney, J.C. and Irvine, A.B., 1989. Growth layers in teeth from known-age, free-ranging bottlenose dolphins. *Marine Mammals Science*, 5(4), pp.315-342.

Holm, P., Ludlow, F., Scherer, C., Travis, C., Allaire, B., Brito, C., Hayes, P.W., Matthews, J.A., Rankin, K.J., Breen, R.J., Legg, R., Lougheed, K. and Nicholls, J., 2019. The North Atlantic Fish Revolution (ca. AD 1500). *Quaternary* Research, pp.1-15. doi:10.1017/qua.2018.153

Hooker, S.K., 2009. Overview Toothed Whales. In: W.F. Perrin, B. Würsig and J.G.M. Thewissen., eds. *Encyclopaedia of Marine Mammals.* 2nd ed. London: Academic Press. pp.1173-1179.

Hufthammer, A.K., Arntsen, L., Kitchener, A.C. and Buckley, M., 2018. Grey whale (*Eschrichtius robustus*) in Norwegian waters 2000 years ago. *Palaeogeography, Palaeoclimatology, Palaeoecology*, 495, pp.42-47.

Hybel, N. and Poulsen, B., 2007. *The Danish Resources c. 1000-1550; Growth and Recession.* Leiden: Brill.

IJsseldijk, L.L., van Neer, A., Deaville, R., Begeman, L., van de Bildt, M., van den Brand, J.M.A., Brownlow, A., Czeck, R., Dabin, W., ten Doeschate, M., Herder, V., Herr, H., IJzer, J., Jauniaux, T., Fast Jensen, L., Jepson, P.D., Jo, W.K., Lakemeyer, J., Lehnert, K., Leopold, M., Leopold, M.F., Osterhaus, A., Perkins, M.W., Piatkowski, U., Prenger-Berninghoff, E., Pund, R., Wholsein, P., Gröne, A. and Siebert, U., 2018. Beached bachelors: an extensive study on the largest recorded sperm whale *Physeter macrocephalus* mortality event in the North Sea. *PLoS ONE,* 13(8), e0201221.

IJssennager-van der Pluijm, N.L., 2018. De kat in het bot vinden. In: Nieuwhof, A., Knol, E. and Schokker, J., eds. *Fragmenten uit de rijke wereld van de archeologie: Jaarverslagen van Vereniging voor Terpenonderzoek 99*, pp.169-172.

IUCN, 2015. *The IUCN Red List of Threatened Species.* [online] available at: <http://www.iucnredlist.org/> [Accessed 23 January 2016].

Iversen, F., 2013. The name of the game! The changing role of hunting on royal land in Norway during the Middle Ages. In: Grimm, O. and Schmölcke, U., eds. *Hunting in northern Europe until 1500 AD Old traditions and regional developments, continental sources and continental influences*. Neumünster: Wachholtz Verlag. pp.465-484.

Iverson, S.J. and Koopman, H.N., 2018. Blubber. In: Würsig, B., Thewissen, J.G.M. and Kovacs, K.M., 2018, eds. *Encyclopaedia of Marine Mammals.* 3rd Edition. London: Academic Press. pp.107-110.

Iverson, S.J., 2009. Blubber. In: W.F. Perrin, B. Würsig and J.G.M. Thewissen., eds. 2009. *Encyclopaedia of Marine Mammals.* 2nd ed. London: Academic Press. pp.115-120.

Jackson, J.A., Carroll, E.L., Smith, T.D., Zerbini, A.N., Patenaude, N.J., Baker, C.S., 2016. An integrated approach to historical population assessment of the great whales: case of the New Zealand southern right whale. *Royal Society of Open Science*, 3, 150669.

Jamieson, B.G.M. and Miller, D.L., 2007. *Reproductive Biology and Phylogeny of Cetaceans. Reproductive Biology and Phylogeny* 7. Boca Raton, FL: CRC Press.

Japan Whaling Association, 1987. *Whales and Traditions of Diet*. Japan Whaling Association.

Jarnut, J., 1985. Die frühmittelalterliche Jagd under rechts- und socialgeschichtlichen Aspekten. *Settimane di studio del Centro Italiano di studi sull'Alto Medioevo*, 31(1), pp.765-798.

Jefferson, T.A., Leatherwood, S. and Webber, M.A., 1993. *Marine Mammals of the World*. Rome: Food and Agriculture Organization of the United Nations.

Jefferson, T.A., Webber, M.A. and Pitman, R.L., 2008. *Marine mammals of the world; a comprehensive guide to their identification*. Academic Press.

Jenkins, J.T., 1921. A *history of the whale fisheries: from the Basque fisheries of the tenth century to the hunting of the finner whale at the present date*. London: H.F. & G. Witherby.

Johnston, R.A., 2011. *All things medieval: an encyclopaedia of the medieval world*. Santa Barbara, California: ABC-CLIO LLC.

Jones, M.L. and Swartz, S.L., 2009. *Grey Whale*. In: Perrin, W.F., Würsig, B. and Thewissen, J.G.M., eds. *Encyclopedia of marine mammals*. 2nd edition. Burlington, San Diego, New York, London: Academic Press. pp.503-511.

Jones, R.T., 2013. Running into whales: the history of the North Pacific from below the Waves. *The American Historical Review*, 118(2), pp.349-377. https://doi.org/10.1093/ahr/118.2.349

Karlsen, K., 1962. Development of tooth germs and adjacent structures in the whalebone whale (Balaenoptera physalus (L.)). *Hvalradets Skrifter*, 45, pp.1-56.

Kelleher, H. 2003 Archaeological excavation at Trench 309 and Trench 310 Washington Street. (Unpublished excavation report in the Office of the Cork City Archaeologist).

Kent, S., 1989. Cross-cultural perceptions of farmers as hunters and the value of meat. In: Kent, S., ed. *Farmers as Hunters*. Cambridge: Cambridge University Press.

King Alfred the Great, 1855. *A description of Europe, and the Voyages of Ohthere and Wulfstan*. Translated by J. Bosworth. London: Longman, Brown, Green, and Longmans, Paternoster-row.

Kiszka, J., Macleod, K., van Canneyt, O., Walker, D. and Ridoux, V., 2007. Distribution, encounter rates, and habitat characteristics of toothed cetaceans in the Bay of Biscay and adjacent waters from platform-of-opportunity data. *ICES Journal of Marine Science*, 65, pp.1033-1043.

Knag, N., 1938. *Matricul oc beshrifuelse ofuer Findmarchen for anno 1694*. UTNE and SOLBERG. pp.1-32.

Krauwer, M., 1982. *Hot botmateriaal uit de ijzertijd, Romeinse tijd en middeleeuwen van de nederzettingen bij Den Burg, Texel (Campagnes 1967 en 1971-5)*. Dissertation, IPP, University of Amsterdam.

Krivogorskaya, Y., Perdikaris, S. and McGovern, T.H., 2005. Cleaning up the farm: a later Medieval Archaeofauna from Gjögur, a fishing farm of NW Iceland. *Archaeological Islandica*, 5, pp.381-392.

Kruyff, L., 2009. *De keuken van Katharina van Kleef*. Available at: https://www.foodlog.nl/artikel/de_keuken_van_katharina_van_kleef/Hertogin/. Accessed: [27-03-2019].

Kurlansky, M., 1991. *The Basque History of the World*. New York: Walker & Company.

Kurlansky, M., 1999. *The Basque History of the World*. New York: Penguin Group USA.

Laist, D.W., 2017. *North Atlantic right whales: from hunted leviathan to Conservation Icon*. Baltimore, Maryland: John Hopkins University Press.

Lambert, E., MacLeod, C.D., Hall, K., Brereton, T., Dunn, T.E., Wall, D., Jepson, P.D., Deaville, R. and Pierce, G.J., 2011. Quantifying likely cetacean range shifts in response to global climatic change: implications for conservation strategies in a changing world. *Endangered Species Research*, 15(3), pp.205-222.

Lambert, E., Pierce, G.J., Hall, K., Brereton, T., Dunn, T.E., Wall, D., Jepson, P.D., Deaville, R. and MacLeod, C.D., 2014. Cetacean range and climate in the eastern North Atlantic: future predictions and implications for conservation. *Global Change Biology*, 20(6), pp.1782-1793.

Larson, G., Albarella, U., Dobney, K., Rowley-Conwy, P., Schibler, J., Tresset, A., Vigne, J-D., Edwards, C.J., Schulmbaum, A., Dinu, A., Bălăçsescu, A, Dolman, G., Tagliacozza, A., Manaseryan, N., Miracle, P., Wijngaarden-Bakker, L. van, Masseti, M., Bradley, D.G. and Cooper, A., 2007. Ancient DNA, pig domestication, and the spread of the Neolithic into Europe. *Proceedings of the National Academy of Sciences*, 104(39), pp.15276–15281.

Larsson, L., 1990. Dogs in Fraction - Symbols in Action. *Studia Praehistorica Belgica*, 5. Leuven: Leuven University Press. pp.153-160.

Lazarus, S., 2006. *Troubled Waters; The Changing Fortunes of Whales and Dolphins*. London: Natural History Museum.

Le Huray, J.D. and Schutkowski, H., 2005. Diet and social status during the La Tene period in Bohemia: Carbon and nitrogen stable isotope analysis of bone collagen from Kutna Hora-Karlov and Radovesice. *Journal of Anthropological Archaeology*, 25, pp.135-147.

Leatherwood, S., Reeves, R.R., Perrin, W.F. and Evans, W.E., 1988. *Whales, Dolphins, and Porpoises of Eastern North Pacific and Adjacent Arctic Waters. A Guide to Their Identification*. New York: Dover Publications.

Lebecq, S., 1989. Frisons et Vikings: Remarques sur les relations entre Frisons et Scandinaves aux VIe-IXe siècles. In. *Les mondes normands (VIIIe-XIIe s.) Actes du deuxième congrès international d'archéologie médiévale (Caen, 2-4 octobre 1987; Actes des congrès de la Société d'archéologie médiévale, 2)*. Caen: Société d'Archéologie Médiévale. pp.45-59.

Lebecq, S., 1997. Scènes de chasse aux mammifères marins (mers du Nord, VIe-XIIe siècles). In: Mornet, E. and Morenzoni, F., eds. *Milieux naturels, espaces sociaux. Etudes offertes à Robert Delort*. Paris. pp.243-246.

Lepiksaar, J., 1975. Über die Tierknochenfunde aus den mittelalterlichen Siedlungen Südschwedens. In: Clason, A.T., ed. *Archaeozoological studies*. Amsterdam. pp.230-239.

Lestocquoy, J., 1948. Baleine et ravitaillement au Moyen Age. *Revue du Nord*, 117, pp. 39-43.

Lilljeborg, W., 1861. *Hvalben, Funna ij orden, paa, Gräsön i Roslagen i Sverige*. Copenhagen: Skandinavske Naturforskeres, Ottende Møde. pp.599-661.

Linacre, A. and Tobe, S.S., 2011. An overview to the investigative approach to species testing in wildlife forensic science. *Investigative Genetics,* 2(2), pp.1-9.

Lindquist, O., 1995a. *Whales, Dolphins and Porpoises in the Economy and Culture of Peasant Fishermen in Norway, Orkney, Shetland, Faroe Islands, and Iceland ca. 900-1900 AD and Norse Greenland ca. 1000-1500 AD. Vol 1*. PhD dissertation, University of St. Andrews.

Lindquist, O., 1995b. Whaling by Peasant Fishermen in Norway, Orkney, Shetland, the Faroe Islands, Iceland and Norse Greenland: Medieval and Early Modern Whaling Methods and Inshore Legal Régimes. In: Basber, B.L., Ringstad, J.E. and Wexelsen, E., eds. *Whaling and History Perspectives on the Evolution of the Industry*. Sandefjord: Kommmandor Chr. Christenses. pp.17-54.

Lindquist, O., 1997. *Peasant fisherman whaling in the Northern Atlantic area, ca 900-1900 AD*. Háskólinn á Akureyri.

Lindquist, O., 2000. *The North Atlantic grey whale (Eschrichtius robustus): An historical outline based on Icelandic, Danish-Icelandic, English and Swedish sources dating from ca 1000 AD to 1792*. St Andrews: University of St Andrews and Stirling.

Loeng, H., 1989. Ecological features of the Barents Sea. In: L. Rey and V. Alexander, eds. *Proceedings of the 6th Conference Comité Arct. Internat.* 15 May 1985. Leiden: E. J. Brill. pp.327-365.

Loeng, H., 1991. Features of the physical oceanographic conditions of the Barents Sea. In: Sakshaug, E., Hopkins, C.C. and Britsland, N.A., eds. *Proceedings of the Pro Mare Symposium on Polar Marine Ecology, Trondheim, 12-16 May 1990. Polar Research* (10/1), pp.5-18.

Loewen, B., 2009. Historical Data on the Impact of 16th-Century Basque Whaling on Right and Bowhead Whales in the Western North Atlantic. *Canadian Zooarchaeology*, 26, 3-24.

Lombardo, D., 2010. *Cape Cod National Seashore: The first 50 years*. Arcadia Publishing.

Losey, R.J., and Yang, D.Y., 2007. Opportunistic whale hunting on the southern northwest coast: ancient DNA, artefact, and ethnographic evidence. *American Antiquity*, 72(4), pp.657-676.

Loveluck, C., 2001. Wealth, waste and conspicuous consumption. Flixborough and its importance for mid and late Saxon settlement studies. In: Hamerow, H. and MacGregor, A., eds. *Image and power in the archaeology of early medieval Britain: essays in honour of Rosemary Cramp*. Oxford: Oxbow Books. Pp.79-130.

Lowe, C., 2012. Tere Tohora Tere Tangata: Where Whales Journey, People Follow. Unpublished Bachelors Thesis. Harvard College.

Lyman, R.L., 1987. Zooarchaeology and Taphonomy: a general consideration. *Journal of Ethnobiology*, 7, pp.93-117.

Lyman, R.L., 1994. *Vertebrate Taphonomy*. Cambridge: Cambridge University Press.

Lyman, R.L., 2008. *Quantitative Paleozoology*. Cambridge: Cambridge University Press.

Macé, M., 2003. Did the gray whale, *Eschrichtius robustus*, calve in the Mediterranean? *Lattara,* 16, pp.153-164.

MacGarvin, M., 1990. *The North Sea*. London: Collins & Brown Limited.

MacGregor, A., 1974. The Broch of Burrain, North Ronaldsay, Orkney. *Proceedings of the Society of Antiquaries of Scotland*, 105, pp.63-118.

MacGregor, A., 1989. Bone, antler and horn industries in the urban context. In: Serjeantson, D. and Waldron T., eds. *Diet and crafts in towns. The evidence of animal remains from the roman to the post-Medieval Periods*. BAR British Series 199. BAR British Series 499. Oxford: BAR Publishing.

MacLaren, A., 1974. A Norse House on Drimore Machair, South Uist. *Glasgow Archaeological Journal*, 3. pp.9-18.

MacLeod, C.D., Bannon, S.M., Pierce, G.J., Schweder, C., Learmonth, J.A., Herman, J.S. and Reid, R.J., 2005. Climate change and the cetacean community of north-west Scotland. *Biological Conservation*, 124(4), pp.477-483.

Makowiecka, M. and Makowiecki, D., 2007. Gospodarka zwietzętami we wczesnomiejskim centrum osadnic zym w Kołobtzega Budzistowie w świetle nowych badań at cheozoologicznych. In: Leciejewicz, L. and Rębkowski, M., eds. *Kołobrzeg. Wczesne miasto nad Bałtykiem*. Warszawa: Trio.

Makowiecki, D., and Makowiecka, M., 2014. The Character of Animal Exploitation and the Environment at the Polish/Prussian Frontier in the Medieval Period: A Case Study. *Archaeologia Baltica*, 20, pp.91-116.

Mann, M.E., Zhang, Z., Rutherford, S., Bradley, R.S., Hughes, M.K., Shindell, D., Ammann, C., Faluvegi, G. and Ni, F., 2009. *Science*, 326, pp.1256-1260.

Mannino, M.A., Talamo, S., Tagliacozzo, A., Fiore, I., Nehlich, O., Piperno, M., Tusa, S., Collina, C., Di Salvo, R., Schimmenti, V. and Richards, M.P., 2015. Climate-driven environmental changes around 8,200 years ago favoured increases in cetacean strandings and Mediterranean hunter-gatherers exploited them. *Scientific Reports*, 5.

Martin, M., 1716. *A description of the Western Isles of Scotland*. Bell: London.

Martín, V., Tejedor, M., Pérez-Gil, M., Dalebout, M.L., Arbelo, M. and Fernández, A., 2011. A Sowerby's beaked whale (*Mesoplodon bidens*) stranded in the Canary Islands: The most southern record in the eastern North Atlantic. *Aquatic mammals*, 37(4).

Marx, F.G. and Fordyce, R.W., 2015. Baleen boom and bust: a synthesis of mysticete phylogeny, diversity and disparity. *Royal Society Open Science*, 2, pp.1-14.

McKenna, M.C., and Bell, S.K. 1997. *Classification of mammals above the species level*. New York: Columbia University Press.

McLeod, B.A., Brown, M.W., Moore, M.J., Stevens, W., Barkham, S.H., Barkham, M., and White, B.N., 2008. Bowhead Whales, and Not Right Whales, Were the Primary Target of 16th- to 17th-Century Basque Whalers in the Western North Atlantic. *Arctic*, 61, pp.61-75.

McMillan, A.D., 2019. Non-human whalers in Nuu-chah-nulth art and ritual: Reappraising orca in archaeological context. *Cambridge Archaeological Journal*, 29(2), pp.309-326.

Mead, J.G. and Mitchell, E., 1984. Atlantic grey whales. In: Jones, M.L., Swartz, S.L., Leatherwood, S., eds. *The grey whale, Eschrichtius robusutus*. New York: Academic Press, Inc,. pp. 33-53.

Mehler, N., 2007. Viking age and medieval craft in Iceland: Adaptation to extraordinary living conditions on the edge of the Old World. In: Klápste, J. and Sommer, P., eds. 2007. *Arts and Crafts in Medieval Rural Environment. Ruralia VI*. Turnhout, Belgium Brepols Publishers. pp.227-224

Melnikov, V.V. and Zagrebin, I.A., 2005. Killer whale predation in coastal waters of the Chukotka Peninsula. *Marine Mammal Science*, 21(3), pp.550-556.

Melville, H., 1851. *Moby Dick; or, The Whale*. ed. 1922. Boston: C.H. Simonds Company.

Méniel, P. and Arbogast, R.-M., 1989. Les restes de mammifères de la cour Napoléon du Louvre (Paris) du XIVe au XVIIIe siècle. *Revue de Paléobiologie, Museum d'Histoire Naturelle de la Ville de Geneve*, 8(2), pp.407-466.

Merples, B., 1976. The animal bones. Excavations at Oxford Castle, 1965-1973. *Oxoniensia*, 41, pp.302-304.

Middelfart museum, 2017. *Marsvinjagt og marsvin*. Available at: https://www.middelfart-museum.dk/marsvinelaug [Accessed 11-01-2019].

Middleton, N., 2005. Early medieval post customs, tolls and controls on foreign trade. *Early Medieval Europe*, 13(4), 313-358.

Mitchell, R.J. and Leys, M.D.R., 1950. *A history of the English People*. London: Longmans.

Moesker, T. and Cavallo, C., 2016. Dierlijk bot. In: Dijkstra, M.F.P., Verhoeven, A.A.A. and van Straten, K.C.J., 2016. eds. *Nieuw licht op Leithon. Archeologisch onderzoek naar de vroegmiddeleeuwse bewoning in plangebied Leiderdorp-Plantage*. Themata 8. Amsterdam: Universiteit van Amsterdam/Diachron UvA bv. pp.575-624.

Moffat, R., Spriggs, J. and O'Connor, S., 2008. The Use of Baleen for Arms, Armour and Heraldic Crests in Medieval Britain. *Antiquaries Journal*, 88, pp.207-215.

Møhl, U., 1961. Oversigt over dyreknogler fra Naesholm. In: La Cour, V., ed. *Naesholm*. Kobenhavn: Nationalmuseet. pp.358-424

Mol, D., Post, K., Reumer, J.W.F., van der Plicht, J., de Vos, J., van Geel, B., van Reenen, G., Pals, J.P. and Glimmerveen, J., 2006. The Eurogeul – first report of the palaeontological, palynological and archaeological investigations of this part of the North Sea. *Quaternary International*, pp.142-143.

Monks, G.G., 2001. Quit blubbering: An examination of Nuu'chah'nulth (Nootkan) whale butchery. *International Journal of Osteoarchaeology*, 11, pp.136–149.

Monks, G.G., 2005. An oil utility index for whale bones. In: Monks, G.G., ed. *The Exploitation and Cultural Importance of Sea Mammals*. Oxford: Oxbow Books. pp.138-155.

Moore, S.A., 1888. *A history of the foreshore and the law relating thereto*. London: Stevens & Haynes.

Moore, S.E., 2008. Marine Mammals as Ecosystem Sentinels. *Journal of Mammalogy*, 89(3), pp.534-540.

Mortier, R., 2016. Wat eten we vandaag? Een analyse van de laatmiddeleeuwse voedingsgeschiedenis van de stedelijke burgerij en de stedelijke abdijen op basis van kookboeken en rekeningen. Unpublished Masters thesis. University of Ghent.

Motteau, J., 1991. *Catalogue des objets des fouilles 1973-1977, Recherches sur Tours, n° 5, Tours, Association pour le développement des études d'archéologie urbaine à Tours*. p.138.

Mulville, J., 2002a. A Whale of a Problem? Zooarchaeology and Modern Whaling. In: Monks, G.G., ed. 2002. *The Exploitation and Cultural Importance of Sea Mammals*. Oxford: Oxbow Books, pp.154-166.

Mulville, J., 2002b. The Role of Cetacea in Prehistoric and Historic Atlantic Scotland. *International Journal of Osteoarchaeology*, 13, pp.34-48.

Murray, M.S., 2008. Zooarchaeology and arctic marine mammal biogeography, conservation, and management. *Ecolcological Applications*, 18(2), S41-S55.

Musset, L., 1964. Quelques notes les baleiniers Normands du Xe au XIIIe siècle. *Revue d'histoire économique et sociale*, 42(2), pp.147-161.

NADNuis, 2018. *Weefzwaard van walvisbeen*. Available at: http://nadnuis.nl/. [Accessed 05-08-2018].

NADNuis, 2019. *Noordelijk Archeologisch Depot Nuis*. Available at: <http://www.nadnuis.nl/>. [Accessed 29-05-2019].

Naismith, R., 2019. The Laws of London? IV Æthelred in Context. *The London Journal*, 44(1), pp.1-16.

Neer, W. van, Lentacker, A. and Ervyck, A., 2013. Dierenresten uit veenwinningsput spoornr.27. In: De Smeldt, I. and in 't Ven, I., eds. *Het Archeologische onderzoek in Raversijde (Oostende) in de periode 1992-2005*. Agentschap Onroerend Erfgoed. pp.469-470.

Nichols, C., Herman, J., Gaggiotti, O.E., Dobney, K.M., Parsons, K. and Hoelzel, A.R., 2007. Genetic isolation of a now extinct population of bottlenose dolphins (*Tursiops truncatus*). *Proceedings of the Royal Society B*. doi:10.1098/rspb.2007.0176.

Nilsen, A.C., 2015. The Slettnes type: Proto-Sámi dwellings in Northern Norway 0-1050 AD. Master's Thesis in Archaeology Department of Social Anthropology and Archaeology University of Tromsø.

Nilsen, G., 2017. Surplus Production and Marine Resource Use in the North Norwegian Iron Age. *The International Journal of Nautical Archaeology*, 46(2), pp.231–252.

NMS (National Museum of Scotland), 2019a. *Casket*. Available at: <http://nms.scran.ac.uk/database/record.php?usi=000-100-002-043-C> [Accessed 16-5-2019].

NMS (National Museum of Scotland), 2019b. *Casket of a carved whalebone*. Available at: <http://nms.scran.ac.uk/database/record.php?usi=000-100-049-339-C> [Accessed 16-5-2019].

Noakes, S.E., Garrison, E.G. and McFall, G., 2009. Underwater paleontology: recovery of a prehistoric whale mandible offshore Georgia. In: Pollock, N.W., ed. *Diving for Science 2009*. Atlanta: Proceedings of the American Academy of Underwater Sciences 28th Symposium. pp.245-251.

Noakes, S.E., Pyenson, N.D. and McFall, G., 2013. Late Pleistocene grey whales (*Eschrichtius robustus*) offshore Georgia, U.S.A., and the antiquity of grey whale migration in the North Atlantic Ocean. *Palaeogeography, Palaeoclimatology, Palaeoecology*, 392, pp.502-509.

Ó Ríordáin, B., 1973. The High Street Excavations. In: Almqvist B. and Greene, D. eds. *Seventh viking congress*. London: Royal Irish Academy. pp.135-140.

Ó Ríordáin, B., 1975. Dublin City: Christchurch Place. In: Delaney, T.G., ed. *Excavations 1974, Association of Young Irish Archaeologists, Ulster Archaeological Society and Group for the Study of Irish Historic Settlement*. pp.14-15.

O'Connor, S., 2013. Exotic materials used in the construction of Iron Age sword handles from South Cave, UK. In: Choyke, A. and O'Connor, S., eds. 2012. *From these bare bones. Raw Materials and the study of worked osseous objects*. Oxford and Oakville: Oxbow Books. pp.188-200.

O'Connor, S., Solazzo, C. and Collins, M., 2015. Advances in identifying archaeological traces of horn and other keratinous hard tissues. *Studies in Conservation*, 60(6), pp.393-417. DOI: 10.1179/2047058414Y.0000000134

O'Connor, T., 2000. *The Archaeology of Animal Bones*. Phoenix Mill: Sutton Publishing Limited.

O'Connor, T.P., 1989. *Bones from Anglo-Scandinavian Levels at 16-22 Coppergate*. London: Council for Britisch Archaeology.

O'Dell, A.C., Stevenson, R.B.K., Brown, T.N., Plenderleith, H.J. and Bruce-Mitford, R.L.S., 1959. The St Ninian's Isle Silver Hoard. *Antiquity*, 33, pp.241-268.

Odner, K., 1992. *The Varanger Saami; Habitation and Economy AD 1200-1900*. Oslo: The Institute of Comparative Research in Human Culture.

O'Hara, 2010. An Iron Age and Early Medieval cemetery at Collierstown 1, Co. Meath. Interpreting the changing character of a burial ground. Unpublished Masters Thesis

Olaus Magnus, 2010. *Olaus Magnus, A Description of the Northern Peoples, 1555: Volume III*. Translated by P. Foote. Farnham, Surrey, England: Routledge.

Omura, H., Ohsumi, S., Nemoto, R., Nasu, K. and Kasuya, T., 1969. Black right whales in the North Pacific. *The Scientific Reports of the Whales Research Institute*, 21, pp.1–78.

Oosterbaan, A., 2018. De Walvisbottencollectie Vonk: Nieuwe Gegevens over het Voorkomen van Walvisachtigen in de Noordzee. *Cranium*, 35(2), pp.43-46.

Ørsted Brandt, L., Haase, K. and Collins, M.J., 2018. Species identification using ZooMS, with reference to the exploitation of animal resources in the medieval town of Odense. *Danish Journal of Archaeology*.

Orton, D.C., Morris, J., Locker, A. and Barrett, J.H., 2014. Fish for the city: meta-analysis of archaeological cod remains and the growth of London's northern trade. *Antiquity*, 88(344), pp. 516-530.

OSPAR Commission, 2016. *OSPAR Commission, Protecting and Conserving the North-East Atlantic and its Resources*. [online] Available at: <http://www.ospar.org/convention/the-north-east-atlantic> [Accessed 17 January 2016].

O'Sullivan, D. and Young, R., 1995. *Book of Lindisfarne, Holy Island*. London: B.T. Batsford Ltd/English Heritage.

Owen, O. and Dalland, M., 1999. *Scar: A Viking Boat Burial on Sanday, Orkney*. East Linton: Tuckwell Press.

Pales, L. and Lambert, C., 1971. *Atlas osteologique pour server a l'identification des mammiferes du Quaternaire*. Paris: Editions du Centre national de la recherche scientifique.

Pálsdóttir, L.B. and Sveinbjarnarson, Ó.G., 2011. *Under the Glacier: 2011 Archaeological investigations on the fishing station at Gufuskálar, Snæfellsnes*. Reykjavik.

Parks Canada, 2017. *Archaeology of a Sixteenth-Century Basque Whaling Boat*. [online] available at: <https://www.pc.gc.ca/en/lhn-nhs/nl/redbay/decouvrir-discover/basque> [Accessed 8 August 2018].

Patton, A. and Savelle, J., 2006. The symbolic dimensions of whale bone use in Thule winter dwellings. *Études/Inuit Studies*, 30(2), pp.137–161.

Peltier, H., Dabin, W., Daniel, P., Van Canneyt, O., Dorémus, G., Huon, M. and Ridoux, V., 2012. The significance of stranding data as indicators of cetacean populations at sea: Modelling the drift of cetacean carcasses. *Ecological Indicators*, 18, pp.278-290.

Pengelly, W., 1865. On cetacean remains washed ashore at Babacombe, South Devon. *Transactions of the Devonshire Association*, 1, pp.86-89.

Pengelly, W., 1878. Remains of whales found on the coast of Devonshire, part II. *Transactions of the Devonshire Association*, 10, pp.630-635.

Perdikaris, S., Hambrecht, G., Brewington, S. and McGovern, T., 2007. Across the fish event horizon: a comparative approach. In: Hüster Plogmann, H., ed. The role of fish in ancient time. *Proceedings of the 13th Meeting of the ICAZ Fish Remains Working Group in October 4th – 9th, Basil, August 2005*. pp.51-62.

Perrin, W.F. and Geraci, J.R., 2009. Stranding. In: W.F. Perrin, B. Würsig and J.G.M. Thewissen., eds. 2009. *Encyclopaedia of Marine Mammals*. 2nd ed. London: Academic Press. pp.1118-1122.

Perrin, W.F., Würsig, B. and Thewissen, J.G.M., eds. 2009. *Encyclopaedia of Marine Mammals*. 2nd ed. London: Academic Press.

Persson, C., 1986. *Description to the quaternary maps*. Uppsala: Sveriges Geologiska Undersbkning, 71.

Pétillon, J.-M., 2013. Circulation of whale-bone artifacts in the northern Pyrenees during the Late Upper Paleolithic. *Journal of Human Evolution*, 65(5), pp.525-543.

Pétillon, J-M., 2008. First evidence of a whale-bone industry in the western European Upper Paleolithic: Magdalanian artifacts from Isturitz (Pyrénées-Atlantiques, France). *Journal of Human Evolution* 54(5), pp.720-726.

Pierce, G.J., Ward, N., Brownlow, A. and Begoña Santos, M., 2018. Analysis of historical and recent diet and strandings of sperm whales (*Physeter microcephalus*) in the North Sea. *Lutra*, 61(1), pp.71-86.

Pigière, F., Boone, I., Udrescu, M., van Neer, W. and Vanpoucke, S., 2004. Status as reflected in food refuse of late medieval noble and urban household at Namur (Belgium). In: O'Day, S.J., ed. *Behaviour Behind Bones. The Zooarchaeology of ritual, religion, status and identity. Proceedings of the 9th ICAZ conference, Durham 2002*. Oxford: Oxbow Books. pp.233-243.

Pike, G.C., 1962. Migration and feeding of the grey whale (*Eschrichtius robustus*). *Journal of the Fisheries Research Board of Canada*, 19(5), pp.815-838.

Pingree, R. and Le Cann, B., 1990. Structure, strength and seasonality of the slope currents in the Bay of Biscay region. *Journal of the Marine Biological Association of the United Kingdom*, 70(4). pp.857-885.

Pipe, A., 1995. Environmental Evidence. In: Mills, P., 1995. *Excavations at the Dorter Undercroft, Westminster Abbey. Transactions of the London and Middlesex Archaeological Society*. pp.69-124.

Pipe, A., 2007. The Vertebrate Remains. Miller, P. and Saxby, D., 2007. ed. *The Augustinian priory of St Mary Merton, Surrey*. MOLAS Monograph 34. Museum of London Archaeology Service. pp.246-250.

Pipe, A., Rielly, K. and Ainsley, C., 2011. Animal Bone. In: T. Dyson, M. Samuel, A. Steele and S.M. Wright, 2011. *The Cluniac Priory and abbey of St Saviour Bermondsey, Surrey; Excavations 1984-15*. MOLA Monograph 50. London: Museum of London Archaeology.

Pluskowski, A., 2004. Narwhals or Unicorns? Exotic Animals as Material Culture in Medieval Europe. *European Journal of Archaeology* 7(3), pp.291-313.

Poplin, F., 1983. La dent de Cachalot sculptée du Mas d'Azil, avec remarques sur les autres restes de cétacés de la Préhistoire française. In: Poplin, F., ed. *La faune et l'Homme Préhistoriques. Société préhistoriques française*. Paris. pp.81-99.

Post, K. and Bosselaers, M., 2005. Late Pliocene occurrence of Hemisyntrachelus (Odontoceti, Delphinidae) in the southern North Sea. *Deinsea*, 11, pp.29-45.

Post, K., 2005. A Weichselian marine mammal assemblage from the southern Norse Sea. *DEINSEA*, 11, pp.21-27.

Posturas, 1873. *Posturas da camara municipal da Cidadedes S. Salvador, Capital da Provincia da Bahia*. Typ. Constitucional-AO Aljube.

Poulard, J-C. and Blanchard, F., 2005. The impact of climate change on the fish community structure of the eastern continental shelf of the Bay of Biscay. *ICES Journal of Marine Science*, 7(1). pp.1436-1443.

Price, S.A., Bininda-Emonds, O.R.P. and Gittleman, J.L., 2005. A complete phylogeny of the whales, dolphins and even-toed hoofed mammals (Cetartiodactyla). *Biological Review*, 80, pp.445-43.

Prilloff, R.-J., 2002. Archäozoologische Beiträge zur Geschichte der Stadt Erfurt. *Erfurter Beiträge*, 3, pp.71-102.

Proulx, J.-P., 1986. *Whaling in the North Atlantic From Earliest Times to the Mid-19th Century*. Ottawa.

Prummel, W., 1987. Atlas for identification of foetal skeletal elements of cattle, horse, sheep and goat. Part 2. *Archaeozoloogia*, 1/2, pp.11-41.

Prummel, W., de Vries, L., Laarman, F. and van den Hurk, Y., 2018. Een gehoorkapsel van een grijze walvis (*Eschrichtius robustus*) uit Wijster (Dr.). *Paleo-Aktueel*, 29, pp.43-49.

Prummel, W., van Gent, J.T. and Kampanje, E.J.O., 2012. Walvisbotten uit Friese en Groninger terpen *Paleo-aktueel*, 23, pp.41-48.

Pyenson, N.D., Gutstein, C.S., Parham, J.F., Le Roux, J.P., Carreno Chavarria, C., Little, H., Metallo, A., Rossi, V, Valenzuela-Toro, A.M., Velez-Juarbe, J., Santelli, C.M., Rubilar Rogers, D., Cozzuol, M.A., Suarez, M.E., 2014. Repeated mass strandings of Miocene marine mammals from Atacama Region of Chile point to sudden death at sea. *Proceedings of the Royal Society B*, 281.

Rastogi, T., Brown, M.W., McLeod, B.A., Frasier, T.R., Grenier, R., Cumbaa, S.L., Nadarajah, J. and White, B.N., 2004. Genetic analysis of 16th-century whale bones prompts a revision of the impact of Basque whaling on right and bowhead whales in the western North Atlantic. *Canadian Journal of Zoology*, 82, pp.1647-1654.

Redman, N., 2004. *Whales' Bones of the British Isles*. Great Britain: Redman Publishing

Redman, N., 2009. *Whales' Bones of Germany, Austria, Czech Republic and Switzerland*. Great Britain: Redman Publishing.

Redman, N., 2010a. *Whales' Bones of the British Isles, Supplement 2004-2010*. Great Britain: Redman Publishing.

Redman, N., 2010b. *Whales' Bones of the Netherlands and Belgium*. Great Britain: Redman Publishing.

Redman, N., 2013. *Whales' Bones of the Nordic Countries, Central and Eastern Europe*. Great Britain: Redman Publishing.

Redman, N., 2014. *Whales' Bones of France, Southern Europe, Middle East and North Africa*. Great Britain: Redman Publishing.

Reeves, R.R., Smeenk, C., Kinze, C.C., Brownell, R.L. and Lien, J., 1999a. *White beaked dolphin Lagenorhynchus albirostris (Gray, 1846)*. In: Ridgway, S.H. and Harrison, R., 1999, eds. *Handbook of Marine Mammals. Volume 6: The second book of dolphins and porpoises*. London: Academic Press. pp.1-30.

Reid, J.B., Evans, P.G.H. and Northridge, S.P., 2003. *Atlas of Cetacean distribution in north-west European waters*. Petersborough: Joint Nature Conservation Committee.

Reitz, E.J. and Wing, E.S., 2008. *Zooarchaeology 2nd edition. Cambridge Manuals in Archaeology*. Cambridge: Cambridge University Press.

Rice, D.W., 2009a. Ambergris. In: Perrin, W.F., Würsig, B. and Thewissen, J.G.M., eds. *Encyclopaedia of Marine Mammals*. 2nd ed. London: Academic Press. pp.28-29.

Rice, D.W., 2009b. Spermaceti. In: Perrin, W.F., Würsig, B. and Thewissen, J.G.M., eds. *Encyclopaedia of Marine Mammals*. 2nd ed. London: Academic Press. pp.1098-1099.

Rice., D.W., 2009c. Classification (Overall). In: Perrin, W.F., Würsig, B. and Thewissen, J.G.M., eds. *Encyclopaedia of Marine Mammals*. 2nd ed. London: Academic Press. pp.234-238.

Rice., D.W., 2009d. Baleen. In: Perrin, W.F., Würsig, B. and Thewissen, J.G.M., eds. *Encyclopaedia of Marine Mammals*. 2nd ed. London: Academic Press. pp.78-80.

Riddler, I. and Trzaska-Nartowski, N., 2014. Hamwic et l'artisanat de l'os de baleine aux VIIe-IXe siecles. *Cahiers LandArc 2014 – N°6.*

Riddler, I.D. and Trzaska-Nartowski, N., 2013. Lundenwic and the Middle Saxon Worked Bone Interlude. *Anglo-Saxon Studies in Archaeology and History*, 18, pp.75-98.

Riddler, I.D., 2014. The Archaeology of the Anglo-Saxon Whale. In: Klein, S.S., Schipper, W. and Lewis-Simpson, S.M., eds. 2014. *The Maritime world of the Anglo-Saxons. Medieval and Renaissance texts and studies 448. Essays in Anglo-Saxon Studies 5*. Tempe: Ariz. pp.337-354

Ridgeway, V., 2000. A medieval saltern mound at Millfields Caravan Park, Bramber, West Sussex. *Sussex Archaeological Collections*, 138, pp.135-152.

Rielly, K., 2003. The animal and fish bone. In: Malcolm, G., Bowsher D. and Cowie R., eds. 2003. *Excavations at the Royal Opera House 1989-99.* London: Museum of London Archaeology Service. Ch.7.20.

Rielly, K., 2015. APPENDIX 10: Animal bone Assessment. In: Killock, D., Simonson, J., Brown, J. and Butler, J., 2015. *The Adelphi Buidling John Adam Street London WC2 City of Westminster. Assessment of an archaeological excavation.* London: Pre-construct Archaeology.

Rijksdienst voor het Cultureel Erfgoed, 2019. *BoneInfo.* [online]. Available at: <http://livelink.archis.nl/livelink/livelink.exe?func=ll&objId=3118333&objAction=browse&sort=name> [Accessed 12 February 2019].

Rodrigues, A.S.L., Charpentier, A., Bernal-Casasola, D., Gardeisen, A., Nores, C., Millán, J.A.P., McGrath, K. and Speller, C.F., 2018. Forgotten Mediterranean calving grounds of gray and North Atlantic right whales: evidence from Roman archaeological records. *Proceedings of the Royal Society B.*

Rodrigues, A.S.L., Kolska Horwitz, L., Monsarrat, S. and Charpentier, A., 2016. Ancient whale exploitation in the Mediterranean: species matters. *Antiquity*, 90(352), pp.928-938.

Roman, J. and Palumbi, S.R., 2003. Whales Before Whaling in the North Atlantic. *Science*, 301, pp.508-510.

Romero, A., 2006. More private gain than public good: whale and ambergris exploitation in 17th century Bermuda. *Bermuda Journal of Archaeology and Maritime History*, 17. P.6.

Ropelewski, A., 1957. Morswin jako przylow w Polskim rybolowstwie Baltyckim (the common porpoise as a by-catch in Polish Baltic Fishersies) - Prace Marsk. *Inst. Rybackiego w Gdyni,* 9, pp.427-437.

Russell, D., 2011. *Eye of the whale; Epic passage from Baja to Siberia.* Washington, Covelo, London: Island Press/Shearwater Books.

Sabin, R., Bendrey, R. and Riddler, I., 1999. Twelfth-century porpoise remains from Dover and Canterbury. *Archaeological Journal,* 156, pp.363-369.

Sabin, R.C., 2002. From the Palaeolithic to the Present-Day: The research value of marine mammal remains from archaeological contexts and the use of contemporary museum reference collections. Ed. Monks, G.G., 2002. *The Exploitation and Cultural Importance of Sea Mammals. Proceedings of the 9th ICAZ Conference, Durham 2002*. Oxford: Oxbow Books. pp.1-5.

Savelle J.M. and McCartney, A.P., 1991. Thule Eskimo Subsistence and Bowhead Whale Procurement. In: Stiner, M.C., ed. *Human Predator and Prey Mortality.* Colorado, Boulder: Westview Press. pp.201-216.

Savelle, J.M. and Friesen, T.M., 1996. An odontocete (Cetacea) meat utility index. *Journal of Archaeological Science* 23, pp.713–721.

Savelle, J.M. and Kishigami, N., 2013. Anthropological research on whaling: Prehistoric, Historic and Current Contexts. *Senri Ethnological Studies*, 84, pp.1-48.

Savelle, J.M. and Vadnais, A., 2011. Releasing the soul: Zooarchaeological evidence for a whale cult among prehistoric Thule Inuit in Canada. *Bulletin of the National Museum of Ethnology,* 36(1), pp.93-112.

Savelle, J.M., 1997. The Role of Architectural Utility in the Formation of Zooarchaeological Whale Bone Assemblages. *Journal of Archaeological Science*, 24(10), pp.869–885.

Savelle, J.M., 2000. Information systems and Thule Eskimo bowhead whaling. In: Rowley-Cowny, P. ed. *Animal Bones: Human Societies.* Oxford: Oxbow Books. pp.74-86.

Scheinin, A.P., Karem, D., MacLeod, C.D., Gazo, M., Chicote, C.A., Castellote, M., 2011. Grey whale (*Eschrichtius robustus*) in the Mediterranean Sea: anomalous event or early sign of climate driven distribution change? *Marine Biodiversity Records*, 4, e28.

Schmid, E., 1972. *Atlas of animal bones for prehistorians, archaeologists and Quaternary geologists*. Amsterdam-London-New York.

Schnall, U., 1992. Der Kampf um die "Gabe Gottes". *Deutsches Schiffahrtsarchiv*, 15, pp.209-222.

Schnall, U., 1993. Medieval Scandinavian Laws as Sources for the History of Whaling. In: Basberg, B.L., Ringstad, J.E. and Wexelsen, E., 1993, eds. *Whaling & History: Pespectives on the evolution of the Industry*. Sandefjord, Norway: Kdr. Chr. Christensens Hvalfangstmuseum. Pp.11-15.

Schumann, N., Gales, N.J., Harcourt, R.G., and Arnould, J.P.Y., 2013. Impacts of climate change on Australian marine mammals. *Australian Journal of Zoology*, 61(2), pp.146-159.

Scottish Marine Mammal Stranding Scheme, 2014. *Scottish Marine Mammal Strandings*. Available at: <http://www.strandings.org/> [Accessed 28 Feb 2016].

SeaOnScreen, 2013. *Cetaceans.* [online] Available at: <http://www.zeeinzicht.nl/vleet/index.php?id=4185&template=template-vleeteng&language=2&item=Cetaceans> [Accessed 28 October 2013].

Seersholm, F.V., Cole, T.L., Grealy, A., Rawlence, N.J., Greig, K., Knapp, M., Stat, M., Hansen, A.J., Easton, L.J., Shepherd, L., Tennyson, A.J.D., Scofield, R.P., Walter, R. and Bunce, M., 2018. Subsistence practices, past biodiversity, and anthropogenic impacts revealed by New Zealand-wide ancient DNA survey. *PNAS*, 115(30), pp.7771-7776.

Seersholm, F.V., Winter Pedersen, M., Jensen Søe, M., Shokry, H., Mak, S.S.T., Ruter, A., Raghaven, M., Fitzhugh, W., Kjær, K.H., Willerslev, E., Meldgaard, M., Kapel, C.M.O., Hansen, A.J., 2016. DNA evidence of bowhead whale exploitation by Greenlandic Paleo-Inuit 4,000 years ago. *Nature Communications*, 7, 13389.

Shetelig, H. 1917. Skibet. IBrøgger, A.W., Falk, H.J. and Shetelig, H., eds. Osebergsfundet. Bind 1. Kristiania: Universitetets Oldsaksamling. pp.283–364.

Shirihai, H. and Jarrett, B., 2011. *Whales Dolphins and Seals; a Field Guide to the marine mammals of the World.* London: A & C Black.

Sinding, M.-H.S., Gilbert, M.T., Gronnow, B., Gullov, H.C., Toft, P.A. and Foote, A.D., 2012. Minimally destructive DNA extraction from archaeological artefacts made from baleen. *Journal of Archaeological Science*, 39, pp.3750-3753.

Sjövold, T., 1974. *The Iron Age settlement of Arctic Norway. A study in the expansion of European Iron Age culture within the Arctic Circle. Part II. Late Iron Age (Merovingian and Viking periods).* Tromsö, Oslo, Bergen: Norwegian Universities Press.

Skomsvoll, J., 2012. Nordnorske spillsaker fra jernalder og middelalder. Unpublished Master's Thesis at the arkeologi Institutt for arkeologi og sosialantropologi Fakultet for humaniora, samfunnsvitenskap og lærerutdanning Universitetet i Tromsø Våren.

Skóra, K.E., 1991. Notes on cetacea observed in the Polish Baltic Sea: 1979-1990. *Aquatic Mammals*, 17(2), pp.67-70.

Slijper, E.J., 1962. *Whales*. New York: Basic Books.

Smeenk, C. and Evans, P.G.H., 2018. Review of sperm whale (*Physeter microcephalus*) strandings around the North Sea. *Lutra,* 61(1), pp.29-70.

Smeenk, C., 1997. Stranding of sperm whales Physeter microcephalus in the North Sea: history and patterns. *Bulletin de l'institut royal des sciences naturelles de Belgique/Bulletin van het koninklijk Belgisch instituut voor natuurwetenschappen. Biologie*, 67-SUPPL, pp.15-28.

Smith, A.B. and Kinahan, J., 1984. The invisible whale. *World archaeology*, 16(1), pp.89-97.

Smyth, M., 1996. *Studies in Celtic History; Understanding the Universe in Seventh-Century Ireland.* Woodbridge: Boydell Press.

Snoeren, P., 1981. Jongste kick op culinair gebied: ridderlijk schransen. *Leidse Courant*, 12 September. Available from: <https://leiden.courant.nu/issue/LLC/1981-09-12/edition/0/page/17>. Accessed [27-03-2019]

Solazzo, C., Fitzhugh, W., Kaplan, S., Potter, C. and Dyer, J.M., 2017. Molecular markers in keratins from Mysticeti whales for species identification of baleen in museum and archaeological collections. *PloS one*, 12(8), e0183053.

Sousa, A. and Brito, C., 2011. Historical strandings of cetacean on the Portuguese coast: anecdotes, people and naturalists. *Marine Biodiversity Records*, 4, pp.1-8.

Speller, C., van den Hurk, Y., Charpentier, A., Rodrigues, A., Gardeisen, A., Wilkens, B., McGrath, K., Rowsell, K., Spindler, L., Collins, M. and Hofreiter, M., 2016. Barcoding the largest animals on Earth: ongoing challenges and molecular solutions in the taxonomic identification of ancient cetaceans *Philosophical Transactions B,* 371.

Steevens, I., 2014. Walvisvangst en -verwerking rond het jaar 1000. In: Lehouck, A., van Acker, J. and Stockelynck, S., eds. *Koksijde Golf ter Hille; Van abdijhoeve tot golf.* Oostkamp, Koksijde: Stichting Kunstboek. pp.22-23.

Stern, A.F., 2000. *Slag bij Woeringen* (1288). Available at: <http://www.graafschap-middeleeuwen.nl/joomla/index.php/oorlogvoering/oorlogen/196-slag-bij-woeringen-1288.html>. Accessed 27-03-2019.

Stige, L.C., Ottersen, G., Brander, K., Chan, K.-S. and Stenseth, N.C., 2006. Cod and climate: effect of the North Atlantic Oscillation on recruitment in the North Atlantic. *Marine Ecology Progress Series*, 325, pp.227-241.

Storå, N., 1968. *Massfångst av sjöfågel I Nordeurasien. En etnologisk undersökning av fångstmetoderna.* Abo: Abo Tidnings och Tryckeri Ab.

Stratford, N., 1997. *The Lewis Chessmen; and the enigma of the hoard.* London: British Museum Press.

Strohalm, M., Hassman, M., Kosata, B. and Kodícek, M., 2008. mMass data miner: an open source alternative for mass spectrometric data analysis. *Rapid Commun Mass Spectrom.*, 22(6), pp.905-908.

Suzuki, T., 1993. The Nutritional Characteristics of Minke Whale Meat. *ISANA,* 8.

Svanberg, I., 2008. *Thet war storth wndher: En valstrandning i Roslagen 1489.* Uppland: Årsbok.

Svanberg, I., 2018. Strandade valar och valben i Sverige. *Forum Navale*, 74, pp.148-179.

Swanton, M., 1975. Ed. *Anglo-Saxon Prose*. Letchworth: Aldine Press.

Swift, J.H., Aagaard, K., and Malmberg, S.-A., 1980. The contribution of the Denmark strait overflow to the deep North Atlantic. *Deep Sea Research Part A. Oceanographic Research Papers*, 27(1), pp.29-42.

Sydney, N.V., Machado, F.A. and Hingst-Zaher, E., 2012. Timing of ontogenetic changes of two cranial regions in *Sotalia guianesis* (Delphindae). *Mammalian Biology – Zeitschrift für Säugetierkunde*, 77(6), pp.397-403.

Sykes, N., 2004. The dynamics of status symbols; Wildfowl exploitation in England AD 410-1550. *Archaeological Journal*, 161, pp.82-105.

Szabo, V., 2012. Subsistence whaling and the Norse Diaspora: Norsemen, Basques, and whale use in the western North Atlantic, ca. AD 900-1640. In: Hudson, B., 2012, ed. *Studies in the Medieval Atlantic*. New York, NY: Palgrave Macmillan.

Szabo, V.E., 2005. "Bad to the bone"? The Unnatural History of Monstrous Medieval Whales. *The Heroic Age. A Journal of Early Medieval Northwestern Europe*, 8, pp.1-30.

Szabo, V.E., 2008. *Monstrous Fishes and the Mead-Dark Sea: Whaling in the Medieval North Atlantic*. The Netherlands: Brill.

Tamm, D. and Vogt, H., 2016. *The Danish Medieval Laws. The Laws of Scania, Zealand and Jutland*. London and New York: Routledge.

Tardiff, J., 1866, ed. *Monuments Historiques: Archives Nationales de l'Empire, Inventaires et Documents*. Paris: J. Claye.

Tate, J., Reiche, I. and Pinzari, F., 2014. The Lewis Chessmen: what can examination of the surfaces tell us? In: Caldwell, D.H. and Hall, M.A., eds. *The Lewis Chessmen; new perspectives*. Edinburgh: NMS Enterprises Limited. pp.11-28.

Taylor, B.L., Rojas-Bracho, L., Moore, J., Jaramillo-Legorreta, A., Ver Hoef, J.M., Cadenas-Hinojosa, G., Nieto-Garcia, E., Barlow, J., Gerrodette, T., Tregenza, N., Thomas, L. and Hammond, P.S., 2016. Extinction is imminent for Mexico's endemic porpoise unless fishery bycatch is eliminated. *Conservation Letters*.

Tedford, R.H., Barnes, L.G. and Ray, C.E., 1994. The Early Miocene Littoral Ursoid Carnivoran Kolponomos: Systematics and Mode of Life. In: Berta, A. and Deméré, A., eds. *Contributions in Marine Mammal Paleontology Honoring Frank C. Whitmore, JR.* San Diego: San Diego Natural History Museum. Ch.1.

Teixeira, A., Venâncio, R. and Brito, C., 2014. Archaeological Remains Accounting for the Presence and Exploitation of the North Atlantic Right Whale Eubalaena glacialis on the Portuguese Coast (Peniche, West Iberia), 16th to 17th century, *PLOS ONE*, 9(2), e85971, pp.1-12.

Telfer, A., 2003. Medieval drainage near Smithfield Market: excavations at Hosier Lane, EC1. *London Archaeologist*, 10(5), pp.115–120.

Thewissen J.G.M., 2009. Archaeocetes, Archaic. In: W.F. Perrin, B. Würsig and J.G.M. Thewissen., eds. 2009. *Encyclopaedia of Marine Mammals*. 2nd ed. London: Academic Press. pp.46-48.

Thewissen, J.G.M. and Fish, F.E., 1997. Locomotor evolution in the earliest cetaceans: Functional model, modern analogues, and paleontological evidence. *Paleobiology*, 23, pp.482-490.

Thomas, R., 2007. They were what they ate: Maintaining social boundaries through the consumption of food in medieval England. In: Twiss, K.C., ed. *The archaeology of Food and Identity*, Occasional Paper No. 34. Carbondale: Centre for Archaeological Investigations, Southern Illinois University. pp.130-151.

Tinker, S.W., 1988. *Whales of the World*. New York: E.J. Brill.

Tulloch, V.J.D., Plagányl, É.E., Matear, R., Brown, C.J. and Richardson, A.J., 2018. Ecosystem modelling to quantify the impact of historical whaling on Southern Hemisphere baleen whales. *Fish and Fisheries*, 19, pp.117-137.

Turvey, S.T., Pitman, R.L., Taylor, B.L., Barlow, J., Akamatsu, T., Barrett, L.A., Zhao, X., Reeves, R.R., Stewart, B.S., Wang, K., Wei, Z., Zhang, X., Pusser, L.T., Richlen, M., Brandon, J.R. and Wang, D., 2007. First human-caused extinction of a cetacean species? *Biology Letters*, 3, pp.557-540.

Twiss, K., 2012. The archaeology of food and social diversity. *Journal of Archaeological Research*, 20, pp.357-395.

Uhen, M.D., 2009a. Basilosaurids. In: Perrin, W.F., Würsig, B. and Thewissen, J.G.M., eds. *Encyclopaedia of Marine Mammals*. 2nd ed. London: Academic Press. pp.91-94.

Uhen, M.D., 2009b. Evolution of Dental Morphology. In: Perrin, W.F., Würsig, B. and Thewissen, J.G.M., eds. 2009. *Encyclopaedia of Marine Mammals*. 2nd ed. London: Academic Press. pp.302-307.

UK Cetacean Strandings Investigation Programme, 2014. *UK Cetacean Strandings Investigation Programme*. [online] Available at: <http://ukstrandings.org/> [24 October 2017].

Valdés, L., López-Urrutia, A., Cabal, J., Alvarez-Ossorio, M., Bode, A., Miranda, A., Cabanas, M., Huskin, I., Anadón, R., Alvarez-Marqués, F., Llope, M. and Rodríguez, N., 2007. A decade of sampling in the Bay of Biscay: What are the zooplankton time series telling us? *Progress in Oceanography*, 74(2-3), pp.98-114.

Van Beneden, P.J. and Gervais, P., 1868. *Ostéographie des Cétacés Vivants et Fossils: Comprenant la description et l'iconographie du squelette et du système dentaire de ces animaux*. Paris: Arthus Bertrand.

Van Dam, P.J.E.M., 2003. Feestvissen en vastenvissen. Culturele, ecologische en economische aspecten van de visconsumptie in de Nederlanden in de Late Middeleeuwen. *Tijdschrift voor Sociale Geschiedenis*, 29(1), pp.467-496.

Van Dam, P.J.E.M., 2009. Fish for feast and fast fish consumption in the Netherlands in the late middle ages. In: Sicking, L. and Abreu-Ferreira, eds., 2009. *Beyond the catch: fisheries of the North Atlantic, the North Sea and the Baltic, 900-1850*. Pp.309-336.

Van den Hurk, Y., forthcoming. Osteological Reference for Cetaceans in Archaeology (ORCA) Manual.

Van den Hurk, Y., Rielly, K. and Buckley, M., forthcoming. *Medieval Cetacean Exploitation in London.*

Van Deinse, A.B. and Junge, G.C.A., 1937. Recent and older finds of the California grey whale in the Atlantic. *Temminckia*, 2, pp.161-188.

van Dijk, J., 2018. Dierlijk bot. In: Meijlink, B.H.F.M. and Silkens, B., 2018. *Archeologisch proefsleuvenonderzoek en opgraving aan de Derringmoerweg, plangebied Hazenburg II te Arnemuiden, gemeente Middelburg.* Middelburg: Walcherse Archeologische Dienst. pp.87-91.

Van Es, W.A., 2018. Dierritueel in de Frankische Nederzetting bij Wijster (Dr.). In: Nieuwhof, A., Knol, E. and Schokker, J., eds. *Fragmenten uit de rijke wereld van de archeologie: Jaarverslagen van Vereniging voor Terpenonderzoek* 99, pp.95-106.

Van Geel, B. and Borger, G.J., 2005. Evidence for medieval salt-making by burning Eel-grass (*Zostera marina* L.) in the Netherlands. *Netherlands Journal of Geosciences*, Geologie en Mijnbouw, 84(1), pp.43-49.

Van Giffen, A.E, 1914. Die Faune der Wurten. *Tijdschrift der Nederlansche Dierkundige Vereeniging*, 13 (2), pp.1-166.

Van Neer, W. and Ervynck, A., 2004. Remains of traded fish in archaeological sites: indicators of status, or bulk food? In: O'Day, S.J., 2004. *Behaviour Behind Bones. The Zooarchaeology of ritual, religion, status and identity. Proceedings of the 9th ICAZ conference, Durham 2002*. Oxford: Oxbow Books. pp.203-214.

Van Winter, J.M., 2013. Suiker in de middeleeuwse keuken. *Jaarboek 'de Oranjeboom'*, 66, 44-71

Vanselow, K.H., Jacobsen, S., Hall, C. and Garthe, S., 2017. Solar storms may trigger sperm whale strandings: explanation approaches for multiple strandings in the North Sea in 2016. *International Journal of Astrobiology.*

Viaene, A., 1971. *Biekorf. Jaargang 72*. St.-Andries: E. Vercruysse en Zoon.

Victoria and Albert Museum, 2017. *The adoration of the Magi*. Available at: <http://collections.vam.ac.uk/item/O92687/the-adoration-of-the-magi-panel-unknown/> [Accessed 23-05-2019.

Voipio, A., 1981. *The Baltic Sea*. Amsterdam/Oxford/New York: Elsevier Scientific Publishing Company.

Waldick, R.C., Kraus, S., Brown, M. and White, B.N., 2002. Evaluating the effects of historic bottleneck events: an assessment of microsatellite variability in the endangered, North Atlantic right whale. *Molecular Ecology*, 11, pp.2241-2249.

Walvisstrandingen.nl, 2014. *Walvisstrandingen.nl*. [online] Available at: <http://www.walvisstrandingen.nl/> [Accessed 24 October 2017].

Walvisstrandingen.nl, 2019. *Walvisstrandingen.nl*. [online] Available at: <http://www.walvisstrandingen.nl/> [Accessed 01 February 2019].

Webster, L. and Backhouse, J., 1991. *The Making of England: Anglo-Saxon Art and Culture, A.D. 600-900*. London: British Museum Press.

Webster, L., 2011. The Prittlewell (Essex) burial: a comparison with other Anglo-Saxon princely graves. In: Panhuysen, T.A.S.M. and Ludowici, B., eds. *Transformations in North-Western Europe (AD 300-1000). Proceedings of the 60th Sachsensymposion 19-23 September 2009 Maastricht. Neue Studien zur Sachsenforschung 3.* Hannover: Die Publishing Company. pp.266-276.

Wellman, H.P., Torben, C.R., Rodrigues, A.T. and Yang, D.Y., 2016. Evaluating Ancient Whale Exploitation on the Northern Oregon Coast Through Ancient DNA and Zooarchaeological Analysis. *The Journal of Island and Coastal Archaeology*, 12(2), pp.255-275. DOI: 10.1080/15564894.2016.1172382

Went, A.E.J., 1968. Whaling from Ireland. *The Journal of the Royal Society of Antiquaries of Ireland*, 98(1), pp.31-36.

Whitmore, F.C. and Sanders, A.E., 1977. Review of the Oligocene Cetacea. *Systematic Zoology*, 25(4), pp.304-320.

Whittaker, H., 2006. Game-boards and gaming-pieces in funerary contexts in the Northern European Iron Age. *Nordlit: Tidsskrift I litteratur og kultur*, 10(2), pp.103-112.

Williams, E.M., 1998. Synopsis of the earliest cetaceans: Pakicetidae, Ambulocetidae, Remingtonocetidae, and Protocetidae. In: Thewissen, J.G.M., ed. 1998. *Emergence of whales, Evolutionary Patters in the Origin of Cetacea*. New York: Plenum Press. pp.1-28.

Wilson, C., Sastre, A.V., Hoffmeyer, M., Rowntree, V.J., Fire, S.E., Santinelli, N.H., Díaz Ovejero, S., D'Agostino, V., Marón, C.F., Doucette, G.J., Broadwater, M.H., Wang, Z., Montoya, N., Seger, J., Adler, F.R., Sironi, M. and Uhart, M.M., 2016. Southern right whale (*Eubalaena australis*) calf mortality at Península Valdés, Argentina: Are harmful algal blooms to blame? *Marine Mammal Science*, 32(2), pp.423-451.

Wilson, C.A., 1973. *Food and Drink in Britain: From the Stone Age to Recent Times*. London: Constable.

Wolff, W.J., 2000. The south-eastern North Sea: losses of vertebrate fauna during the past 2000 years. *Biol Conserv*, 95, pp.209-217.

Wood, I.N., 1990. Ripon, Francia and the Franks Casket in the Early Middle Ages. *Northern History*, 26, pp 1-19.

Wylie, J., 1987. *The Faroe Islands: Interpretation of History*. Lexington: University of Kentucky.

Yarborough, L.F., 1995. Prehistoric Uses of Cetacean Species in the Northern Gulf of Alaska. In: McCartney, A.P., ed. *Hunting the Largest Animals: Native Whaling in the Western Arctic and Subarctic*. Calgary: University of Alberta, the Canadian Circumpolar Institute. pp.63-81.

Young, S., Deméré, T.A., Ekdale, E.G., Berta, A. and Zellmer, N., 2015. Morphometrics and Structure of Complete Baleen Racks in Gray Whales (Eschrichtius robustus) From the Eastern North Pacific Ocean. *The Anatomical Record*, 298(4), pp.703-719.

Ypma, Y.N., 1962. Geschiedenis van de Zuiderzeevisserij. *Publicaties van de stichting voor het bevolkingsonderzoek in de drooggelegde Zuiderzeepolders*, no.27. Amsterdam.

Zeiler, J.T., 2018. *Vee, vogels, huisdieren en wild; Archeozoölogisch onderzoek van dierlijke resten uit de Hof ter Hille te Koksijde, België (laatste kwart 9e - eerste kwart 12e eeuw AD). ArchaeoBone rapport nr.154.* Haren: ArchaeoBone.

Zhou, K., 2009. Baiji. In: Perrin, W.F., Würsig, B. and Thewissen, J.G.M., eds. 2009. *Encyclopaedia of Marine Mammals*. 2nd ed. London: Academic Press. pp.71-76.

Zvelebil, M., 1992. Hunting in farming societies – the prehistoric perspective. In: Grant, A., ed. *Animals and Their Products in Trade and Exchange*. Paris: HASRI (Anthropozoologica 16). pp.7-18.

Appendix

Appendix table 1. Zooarchaeological data assessed as part of this PhD study. Toponym (name of site), Geographic region (province, county, state, etc.), site type (ecclesiastical, high-status, urban, rural, grave, or other), Species (Uc: Unknown cetacean, Lc: Large cetacean, L/Mc: Large/Medium cetacean, Mc: Medium cetacean, M/Sc: Medium/Small cetacean, Sc: Small cetacean, FHGRw: Fin, Humpback, Grey, or Right whale, SRHw: Sperm, Right or Humpback whale, Uo: Unknown Odontoceti, Sw: Sperm whale, Sw?: Sperm whale?, Cbw: Cuvier's beaked whale, Nbw: Northern bottlenose whale, Sbw: Sowerby's beaked whale, Ubw: Unknown beaked whale, Awsd: Atlantic white-sided dolphin, Cbd: Common bottlenose dolphin, Kw: Killer whale, Lfpw: Long-finned pilot whale, Rd: Risso's dolphin, Glob: Globicephalinae, Sbcd: Short beaked common dolphin, Wbd: White-beaked dolphin, Cbd?: Common bottlenose dolphin?, Cbd/Wbd: Common bottlenose dolphin/White beaked dolphin, Cbd/Lfpw: Common bottlenose dolphin/Long finned pilot whale, Lfpw?: Long finned pilot whale?, Ud: Unknown dolphin, Ud/p: Unknown dolphin/porpoise, Be: Beluga, Hp: Harbour porpoise, Up: Unknown porpoise, Ubw: Unknown baleen whale, Boww: Bowhead whale, NArw: North Atlantic right whale, Boww/NArw: Bowhead whale/North Atlantic right whale, Cmw: Common minke whale, Fw: Fin whale, Hw: Humpback whale, Seiw?: Sei whale?, Cmw?: Common minke whale?, Hw?: Humpback whale?, Ur: Unknown rorqual, Gw: Grey whale), details in regard to the identified specimens, NISP, X (no NISP data was acquired, so only the presence of a species was determined), TOTAL (NISP + X), Min. No. Spec. (Minimal Number of Species) present at a site, and references

Toponym	Geographic region	Start date	End date	Site type	Species	Details	NISP	X	TOTAL	Min. No. Spec.	Reference
Scotland											
Ardnave	Hebrides	2000BC	600	Rural	Lc: 1	Lc: vertebral epiphyses	1	0	1	1	Ritchie and Welfare, 1983
Bernera	Outer Hebrides	600	1100	Rural	Uc: 1	Uc: whale bone plaque	0	1	1	1	Owen and Dalland, 1999, 83; Isaksen, 2012
Bornais (The House floors)	Outer Hebrides	1210	1470	Rural	Mc: 1	Mc: rib	1	0	1	1	Ingrem, Mulville and Carledge, 2005
Bornais (abandonment kiln/barn)	Outer Hebrides	1400	1480	Rural	Lc: 1	Lc: Burnt whale vertebra (butchered and the lateral processes were removed and the bone scorched)	1	0	1	1	Ingrem, Mulville and Carledge, 2005
Bornais	Outer Hebrides	800	1000	Rural	NArw: 1	NArw: Worked fragment, identified using ZooMS	1	0	1	1	Buckley *et al.*, 2014
Bornais	Outer Hebrides	600	1100	Rural	NArw: 1, Fw: 1	NArw: Worked fragment. Fw: worked fragment. Both identified using ZooMS	2	0	2	2	Buckley *et al.*, 2014
Bornais	Outer Hebrides	600	1000	Rural	Sw: 1	Sw: fragment identified using ZooMS	1	0	1	1	Buckley *et al.*, 2014
Bornais	Outer Hebrides	600	1100	Rural	Uc: 1	Uc: whale bone plaque (3 fragments)	1	0	1	1	Owen and Dalland, 1999; Isaksen, 2012
Buckquoy 1	Orkney	1150	1272	Rural	Uc: 7		7	0	7	1	Ritchie, 1976
Buckquoy 2	Orkney	1100	1200	Rural	Uc: 1		1	0	1	1	Ritchie, 1976

Toponym	Geographic region	Start date	End date	Site type	Species	Details	NISP	X	TOTAL	Min. No. Spec.	Reference
Buckquoy 3	Orkney	200	800	Rural	Uc: 1		1	0	1	1	Ritchie, 1976
Buckquoy 4	Orkney	200	800	Rural	Uc: 1		1	0	1	1	Ritchie, 1976
Buckquoy 5	Orkney	600	699	Rural	Uc: 1		1	0	1	1	Ritch, 1976
Burrian	Orkney	50BC	799	Rural	Uc: 1, Lc: 17, Mc: 1, Kw: 3, Rd: 8	Mc: intervertebral disc. Lc: 1 cancellous bone. Kw: 3 teeth. Rd: 8 teeth.	30	0	30	5	MacGregor, 1974
Barvas	Outer Hebrides	900	1099	Rural	Uc: X		0	1	1	1	Armit, 1996
Brough of Birsay	Orkney	800	1000	Rural	Uc: 11	Uc: unperforated large pin, 1 vice or clamp with a flat base the top curved and decorated with four rows of dot-in-circle, 1 large pin, notched peg, line stretcher, large whale bone implement, 2 blocks of whale bone, 2 whale bone blocks with perforations, and 1 rib fragment.	11	0	11	1	Curle, 1982
Castle Bar Dunbar	East Lothian	654	954	High Status	Uc: 1		1	0	1	1	Smith, 2000
Castle Bar Dunbar 2	East Lothian	500	1500	High Status	Uc: 1		1	0	1	1	Smith, 2000
Castle Hill Pumping Station	Aberdeenshire	1000	1300	Rural	Sw: 3	Sw: vertebral epiphysis, a fragment of a possible rib and a large fragment of rib, which has been significantly modified. The rib showed signs of chopping and cutting and got holes, and could have been used for architectural features. Smaller rib has signs of burning.	3	0	3	1	Bailey *et al.*, 2015
Chapal and Burial Ground on St. Ninian's Isle III	Shetland	500BC	1200	Burial	Uc: 2	1 worked cancellous whale bone fragment with large drilled central perforation maybe used for fishing, or wooden handle, or the whole perhaps acting as a mallet. And another shaped whale bone vertebral centrum.	2	0	2	1	Gidney, 2011
Chapal and Burial Ground on St. Ninian's Isle VII	Shetland	1200BC	2000	Rural	Uc: 2	Amorphous lumps	2	0	2	1	Gidney, 2011
Chapal and Burial Ground on St. Ninian's Isle VII (2)	Shetland	1200BC	2000	Rural	Uc: 1	Amorphous lumps	1	0	1	1	Gidney, 2011
Clickhimin	Shetland	200	700	Rural	Uc: 8	Scapula, caudal vertebra and 6 pieces of cancellous bone	8	0	8	1	Grahame, 1968

Toponym	Geographic region	Start date	End date	Site type	Species	Details	NISP	X	TOTAL	Min. No. Spec.	Reference
Constantine's Cave	Fife	0	800	Rural	Lfpw?: 1		1	0	1	1	Wace and Jehu, 1915
Drimore	Outer Hebrides	800	1100	Rural	Uc: 4	1 whale bone cleaver found, a fragment of a flat whale bone with two broken perforations, another worked fragment of whale bone, and a vertebra (120 by 135 mm and 55m thick), which have been used as sources of raw material.	4	0	4	1	MacLaren, 1974
Eglinton Castle	North Ayrshire	1400	1550	NON Archaeological	Uc: 1	Casket (made of 6 bone plates with decoration)	1	0	1	1	NMS, 2019a
Fife (unknown site)	Fife	1400	1550	NON Archaeological	Uc: 1	Casket (made of 6 bone plates with decoration)	1	0	1	1	NMS, 2019b
Freswick	Highland	700	1499	Rural	Uc: 1	Object made of cetacean bone	1	0	1	1	Curle, 1938
Freswick Links	Highland	0	1299	Rural	Uc: 15	2 whale bone whorls, 7 whale bone snecks, 2 worked whale bones as gaming pieces, and 4 other worked whale bones.	15	0	15	1	Batey, 1987
Iona (Vallum ditch)	Hebrides	563	1000	Ecclesiastical	M/Sc: 3	2 Ribs and 1 vertebra	3	0	3	1	Murray, McCormick and Plunkett, 2004
Iona (Guest house)	Hebrides	563	1000	Ecclesiastical	Uc: 2		2	0	2	1	Murray, McCormick and Plunkett, 2004
Jarlshof 1	Shetland	800	850	Rural	Uc: X, Lc: X	Couple of whale bones	0	2	2	2	Hamilton, 1956
Jarlshof 2	Shetland	900	1099	Rural	Lc: X	Lots of grey seal, also bone from sperm whale and other large whales	0	2	2	2	Hamilton, 1956
Jarlshof 3	Shetland	1000	1299	Rural	Uc: 1	Several whale bones	0	1	1	1	Hamilton, 1956
King's Cross Point	North Ayrshire	850	900	Rural	Uc: 1	Whale bone plaque	1	0	1	1	Grieg, 1940; Isaksen, 2012
Kirkwall	Orkney	1100	1850	Rural	Cbd/ Lfpw: 1, Narw/ Boww: 2	2 fragments of a baleen whale jaw and 1 vertebra from a bottlenose dolphin/young pilot whale	3	0	3	2	McGavin, 1983
Ladyhill	Moray	1000	1699	High Status	Lc: 1		1	0	1	1	Hall *et al.*, 1998
Law Ting Holm	Shetland	45	630	Rural	Lc: 5	5 whale bone pieces	5	0	5	1	Kunst, 2014
Lewis	Outer Hebrides	1100	1199	Other	Sw: 6	4 warders and 2 pawns (chess pieces) made of sperm whale teeth	6	0	6	1	Stratford, 1997
Northton	Isle of Harris	500	1799	Rural	Ur: 1	whale fragment of a rorqual	1	0	1	1	Simpson, Murphy and Gregory, 2006

Toponym	Geographic region	Start date	End date	Site type	Species	Details	NISP	X	TOTAL	Min. No. Spec.	Reference
Perth	Perth and Kinross	1300	1499	High Status	Ubw: X	Remains of baleen	1	0	1	1	Moffat, Spriggs and O'Connor, 2008
Pool 1	Orkney	100	800	Rural	L/Mc: 62, Sc: 17		79	0	79	2	Hunter, 2007
Pool 2	Orkney	800	1099	Rural	L/Mc: 40, Sc: 5		45	0	45	2	Hunter, 2007
Quoygrew 1	Orkney	900	999	Rural	Uc: 4	1 worked whale bone	4	0	4	1	Harland, 2012
Quoygrew 2 to 3	Orkney	1000	1299	Rural	Uc: 11	3 vertebral discs (perforated) and another vertebral disc (unperforated), a spindle whorl and another worked whale bone	11	0	11	1	Harland, 2012
Quoygrew 2	Orkney	1000	1199	Rural	Uc: 1	Weaving sword whale bone	1	0	1	1	Harland, 2012
Quoygrew 4	Orkney	1400	1599	Rural	Uc: 6	Rope shortener or swivel	6	0	6	1	Harland, 2012
Quoygrew 2, 4	Orkney	1100	1299	Rural	Uc: 3		3	0	3	1	Harland, 2012
Quoygrew 4 to 5	Orkney	1400	1799	Rural	Uc: 1	Whale bone used as pivot	1	0	1	1	Harland, 2012
Quoygrew 7	Orkney	900	2000	Rural	Uc: 1	Spindle whorl	1	0	1	1	Harland, 2012
Rattray	Aberdeenshire	1150	1550	High Status	Uc: 1, Sc: 1, Sw?: 1, Sbw: 1	Whale bone used as toggle. Objectno. 281	4	0	4	4	Hamilton-Dyer *et al.*, 1993
Scar	Orkney	895	1030	Burial	Uc: 23	22 whale bone gaming pieces and 1 whale bone plaque	23	0	23	1	Owen and Dalland, 1999
Foshigarry & Bac Mhic Connain	Orkney	200BC	800	Rural	Uc: 99, Cbw: 1	3 unid. bone plaques, 15 cetacean vertebra containers and lids, 4 implements, 2 handles, 10 implements of bone and antler (of which one is a scapula), 1 long handled comb, 1 mirror handle, 7 misc. cetacean bone objects, 1 modelling tool, 35 notched implements and related types (of which 5 are ribs), 2 pendants, 2 perforated plates, 1 points, 1 pinheads, 2 points/ pins, 1 polishers/ hide-working tools, 3 turned objects, 1 wedges, 2 worked bones (1 ulna and 1 humerus), 10 working debris (2 mp/ phalanges). 1 vertebra of a Cuvier's beaked whale	100	0	100	2	Hallen, 1994

Toponym	Geographic region	Start date	End date	Site type	Species	Details	NISP	X	TOTAL	Min. No. Spec.	Reference
Foshigarry & Bac Mhic Connain	Orkney	200BC	800	Rural	Uc: X		0	1	1	1	Hallen, 1994
St Ninian's Isle	Shetland	750	825	Other	Hp: 1	Harbour porpoise mandible Part of the St Ninian Treasure	1	0	1	1	O'Dell *et al.*, 1959
Saevar Howe	Orkney	700	800	Rural	Uc: 1	Piece of whale bone plaque	1	0	1	1	Graham-Campbell, 1994, 216-217; Isaksen, 2012
England											
Bishopstone 1	East-Sussex	800	899	Rural	Uc: 1, Lc: 1		2	0	2	2	Poole, 2010
Bishopstone 2	East-Sussex	900	999	Rural	Lc: 1		1	0	1	1	Poole, 2010
Bishopstone 3	East-Sussex	800	999	Rural	Lc: 1, Mc: 1		2	0	2	2	Poole, 2010
Blythburgh	Suffolk	700	799	Ecclesiastical	Lc: 1	Whale bone plaque	1	0	1	1	Pestell, 2004
Botolphs	East-Sussex	450	550	Rural	Lc:1	A small part of the skeleton of a whale, not identifiable to species or element, consists, in the main of cancellous tissue. It is of triangular cross section and across the outer surface numerous knife marks and depressions are visible	1	0	1	1	Gardiner, 1990
Castle Rising Castle	Norfolk	1100	1499	High Status	Sbcd: 1		1	0	1	1	Jones, Reilly and Pipe, 1997
Aula Nova	Kent	1160	1165	Urban	Lc: 1, Hp: 9	5th caudal vertebra, 12th or 13th caudal vertebra, 4th or 5th thoracic vertebra, partial scapula, near complete rostrum, 10th lumbar vertebra, cranial fragment, 5th or 6th caudal vertebra and another partial scapula.	10	0	10	2	Sabin, Bendrey and Riddler, 1999
Canterbury, Linacre Garden	Kent	1100	1349	Urban	Uc: 1, Hp: 1	A vertebra from a porpoise and a vertebra from a whale	2	0	2	2	Gardiner, 1998
Carisbrooke Castle	Isle of Wight	1100	1199	High Status	Uc: 1	Chopping board	1	0	1	1	Young, 2000
Carlton Colville	Suffolk	400	1199	Rural	Uc: 5	5 Vertebral blocks	5	0	5	3	Riddler and Sabin, 2009
Cathedral Refectory	Norfolk	1094	1538	Ecclesiastical	Hp: 1	Porpoise vertebra	1	0	1	1	Curl, 2006
Chalk Pit Field (Trench 14)	Norfolk	410	1066	Rural	Uc: 4	Huge pieces of bone	4	0	4	1	Anonymous, n.d.
Clarendon Centre	Oxfordshire	1301	1400	Urban	Sc: 7	7 small cetacean remains hand collected	7	0	7	1	Douglas *et al.*, 2015

Toponym	Geographic region	Start date	End date	Site type	Species	Details	NISP	X	TOTAL	Min. No. Spec.	Reference
Clarendon Centre	Oxfordshire	1301	1400	Urban	Sc: 3	3 small cetacean remains sieved	3	0	3	1	Rielly, 2015
Dengemarsh	Kent	840	1043	Other	NArw: 2	2 partial skeletons	2	0	2	1	Gardiner, Stewart and Priestley-Bell, 1988
Duke's Warf	Norfolk	1000	1500	Urban	Ud/p: X		0	1	1	1	Curl, 2007
Ely	Cambridgeshire	?	?	Rural	Uc: 1	Whale bone plaque	1	0	1	1	Anathon and Shetelig, 1940
Jarrow	Tyne and Wear	700	899	Ecclesiastical	Uc: 2	A sub-rectangular block of cetacean bone tissue, sawn at one end, with some traces of faceting by knife, and a ring made of whale bone	2	0	2	1	Riddler, 2006
Lion Walk	Essex	1400	1699	Rural	Cbd: 1	vertebra with butchery marks	1	0	1	1	Crummy, 1984
Isle of Portland	Dorset	10000BC	1500	Rural	Sc: 1	Atlas of small cetacean	1	0	1	1	Maltby, 2009
Townwall Street	Kent	1150	1250	Urban	Uc: 1, Ud: 1, Hp: 5	5 caudal remains of harbour porpoise. Unknown dolphin rostrum. Tabula pieces made of whale bone	7	0	7	3	Sabin, Bendrey and Riddler, 1999
Flixborough 1	Lincolnshire	677	750	Rural	Uc: 1, Cbd: 7, Cbd?:4		12	0	12	3	Dobney *et al.*, 2007
Flixborough 2	Lincolnshire	750	833	Rural	Uc: 4, Cbd: 22, Cbd?: 17	All from dump Maybe Risso's Dolphin and/or White beaked dolphin remains.	43	0	43	3	Dobney *et al.*, 2007
Flixborough 3	Lincolnshire	800	899	Rural	Cbd: 14, Cbd?: 14, Cmw: 1, Cmw?: 1	All from dump Juvenile minke whale	30	0	30	4	Dobney *et al.*, 2007
Flixborough 4	Lincolnshire	900	999	Rural	Uc: 6, Cbd: 55, Cbd?: 12, Kw: 1, Cmw?: 2	dump. Juvenile minke whale	76	0	76	5	Dobney *et al.*, 2007
Flixborough 5	Lincolnshire	977	1099	Rural	Uc: 4, Cbd: 17, Cbd?: 3, Cmw: 2	dump	26	0	26	4	Dobney *et al.*, 2007
Hamwic SOU 14	Hampshire	700	899	Urban	Lc: 1	Bone working place. Pit. lots of bone working material and saw marks on most bones	1	0	1	1	Morton, 1992
Hamwic SOU 32	Hampshire	700	899	Urban	Uc: 5	Bone working place. Lots of bone working material.	5	0	5	1	Morton, 1992

Toponym	Geographic region	Start date	End date	Site type	Species	Details	NISP	X	TOTAL	Min. No. Spec.	Reference
Hamwic Six Dials	Hampshire	700	900	Urban	Uc: X		0	1	1	1	Andrews, 1997
Hamwic Ascupart Street	Hampshire	670	800	Urban	Uc: 240	Large quantity of worked whale bones, working debris	240	0	240	1	Riddler and Trzaska-Nartowski, 2014
Hinxton	Cambridgeshire	1000	1150	Rural	Uc: X	A casket made of whale bone.	0	1	1	1	Gardiner, 1997
Ipswich	Suffolk	500	1500	Urban	Uc: 10, Lc: 1	Eleven pieces of whale bone from 5 different sites in Ipswich. Most of the fragment of whale bone waste consist of strips of rectangular or triangular section. One vertebral disc of large cetacean, and a spinous process from a vertebra.	11	0	11	2	Riddler, Trzaska-Nartowski and Hatton, in press
Larling	Norfolk	750	799	Ecclesiastical	Uc: 1	A carved panel fragment of whale bone dating to the late eighth century, perhaps from a book-cover	1	0	1	1	Webster and Backhouse, 1991, 179
Launceston Castle 1	Cornwall	1104	1175	High Status	Lc: 1	vertebra	1	0	1	1	Alberella and Davis, 1996
Launceston Castle 2	Cornwall	1227	1266	High Status	Lc: 1	vertebra	1	0	1	1	Alberella and Davis, 1996
Launceston Castle 3	Cornwall	1266	1299	High Status	Lc: 7, Sc: 5, Sbcd: 1	Mandible of common dolphin, small cetacean vertebrae, large cetacean vertebrae	13	0	13	3	Alberella and Davis, 1996
Launceston Castle 4	Cornwall	1400	1499	High Status	Lc: 6, Sc: 9, Sbcd: 1	Small cetacean vertebrae, large cetacean vertebrae	16	0	16	3	Alberella and Davis, 1996
Lewes Priory	East Sussex	900	1099	Ecclesiastical	Hp: 6	Five porpoise vertebrae and one skull from an immature individual	6	0	6	1	Gardiner, 1997
Green Shiel	Northumberland	850	900	Rural	Uc: 1		1	0	1	1	O'Sullivan and Young, 1995, 86
Lindisfarne	Northumberland	1200	1500	Ecclesiastical	Hp: 1, Ubw: 1	Whale bone plaque, probably piece of mandible of large baleen whale and a vertebra of a harbour porpoise	2	0	2	2	Wilkins, Petts and Dave, 2016
London (Calvert's Buildings, 15-23 Southwark Street)	Greater London	1200	1500	Urban	Hp: 1	Cervical vertebra transversely chopped indicating that the animal had been decapitated	1	0	1	1	Gardiner, 1998
London (Hare Court)	Greater London	600	799	Urban	Uc: 2	2 whale mid-blade rib fragments, both chopped trough transversely at each end	2	0	2	1	Bendrey, 2005
London (Royal Opera House) (phase 5)	Greater London	730	770	Urban	Uc: 1, Lfpw?: 1	Vertebra	2	0	2	2	Rielly, 2003

Toponym	Geographic region	Start date	End date	Site type	Species	Details	NISP	X	TOTAL	Min. No. Spec.	Reference
London, Westminster Abbey (First half of eleventh century)	Greater London	1000	1049	Ecclesiastical	Cbd/ Wbd: 1	Maxilla	1	0	1	1	Gardiner, 1998
London (Globe House)	Greater London	750	850	Urban	Lc: 1	Vertebra	1	0	1	1	Browsher, 1999
London - Bermondsey Abbey	Greater London	900	1066	Ecclesiastical	Uc: 2, Lfpw: 12, NArw: 1, Fw: 1		16	0	16	4	Personal communication Kevin Rielly, 2018
London - Bermondsey Abbey 2 (Period M6)	Greater London	1200	1250	Ecclesiastical	Hp: 1	Caudal vertebra	1	0	1	1	Pipe, Rielly and Ainsley, 2011
London - Westminster Abbey (Cellarium)	Greater London	1150	1350	Ecclesiastical	Hp: 2, Cmw: 1	Skull fragment and vertebra of a harbour porpoise + Skull fragment of a common minke whale	3	0	3	2	Personal communication Kevin Rielly, 2018
London - Vintry, 68-69 Upper Thames Street	Greater London	0	1600	Urban	Lc: 1	Proximal end of a large whale mandible	1	0	1	1	Personal communication Kevin Rielly, 2018
London - 2-12 Hosier Lane	Greater London	1200	1400	Urban	Lc: 1	Long bone or mandible fragment of a large cetacean	1	0	1	1	Telfer, 2003
London - 8-10 Old Jewry	Greater London	1050	1150	Urban	Mc: 1	Chopped and sawn vertebra	1	0	1	1	Personal communication Kevin Rielly, 2018
London - Deans Yard, Westminster Abbey	Greater London	1300	1500	Ecclesiastical	Sc: 1	Caudal vertebra	1	0	1	1	Personal communication Kevin Rielly, 2018
London - Augustinian Priory of St Mary Merton (Merton High Street)	Greater London	1500	1700	Ecclesiastical	Sc: 1	Vertebra	1	0	1	1	Pipe, 2007
London - Winchester Palace	Greater London	1066	1500	High Status	Mc: 1	Rib	1	0	1	1	Personal communication Kevin Rielly, 2018
London - Althrope Grove	Greater London	660	1300	Urban	Mc: 1	Small whale vertebra with signs of butchery	2	0	2	2	Blackmore and Cowie, 2001
Billingsgate	Greater London	1000	1200	Urban	Kw: 4	Rib and 3 caudal vertebrae	4	0	4	1	Personal communication Kevin Rielly, 2018
Westminster Sub-Vault of the Misericorde	Greater London	1100	1300	Ecclesiastical	Hp: 1	Tooth	1	0	1	1	Pipe, 1995
St Peters Hill	Greater London	1200	1500	Urban	Lc: 1	Rib	1	0	1	1	Personal communication Kevin Rielly, 2018
Trig Lane	Greater London	1200	1500	Rural	Cmw?: 1	Rostrum	1	0	1	1	Personal communication Kevin Rielly, 2018

Toponym	Geographic region	Start date	End date	Site type	Species	Details	NISP	X	TOTAL	Min. No. Spec.	Reference
Westminster songschool	Greater London	1570	1600	Ecclesiastical	Hp: 2	Vertebra and mandible	2	0	2	1	Personal communication Kevin Rielly, 2018
Adelphi building	Greater London	775	850	Urban	Hp: 1	Vertebra with chop/butchery mark	1	0	1	1	Personal communication Kevin Rielly, 2018
Lundenwic (15-16 Bedford Street)	Greater London	660	899	Urban	Uc: 1	Single offcut of whale bone consists of an unfinished and fragmentary tooth segment for a composite comb	1	0	1	1	Riddler and Trzaska-Nartowski, 2013
Lurk lane, Beverly	Yorkshire and the Humber	1070	1188	Ecclesiastical	Uc: 13		13	0	13	1	Scott, 1991
Norwich Castle	Norfolk	1450	1550	High Status	Ud: 2		2	0	2	1	Garcia, 2009
Oxford, Oxfordshire: Oxford Castle (thirteenth - mid fifteenth century)	Oxfordshire	1200	1499	High Status	Hp: 1	Vertebra that lacks the epiphyses, transverse processes and neural spine. The depth of the centrum is 52.5 mm.	1	0	1	1	Merples, 1976
Queenhithe 1 - Upper Thames Street	Greater London	899	1066	Urban	Sw: 1	Thoracic vertebra	1	0	1	1	Sidell, 2000
Queenhithe 2 - Bull Wharf (Building 5)	Greater London	1100	1199	Urban	Uc: 1		1	0	1	1	Sidell, 2000
Ramsgate	Kent	500	699	Rural	Uc: 1	Vertebral block	1	0	1	1	Riddler, 2014
Ripon	North Yorkshire	700	750	NON Archaeological	Uc: 1	Franks Casket	1	0	1	1	Wood, 1990
Sandtun	Kent	660	899	Rural	Lc: 1	Vertebra with cutmarks	1	0	1	1	Riddler, 1998
Seaford, Church Street	East-Sussex	1300	1399	Rural	Uc: 2	2 fragments of a baleen whale jaw and 1 vertebra from a bottlenose dolphin/young pilot whale	2	0	2	1	Brothwell, 1979
Southampton - Melbourne Street	Hampshire	660	899	Urban	Uc: 5	The compact bone wall is removed and are therefore mostly of porous, trabecular bone is left	5	0	5	1	Bourdillon and Coy, 1980
Southampton	Hampshire	450	1066	Urban	Cmw: 1	Vertebra used as a chopping block	1	0	1	1	Holsworth, 1976
Whitby Abbey	North Yorkshire	400	1066	Ecclesiastical	Uc: 1	Vertebra	1	0	1	1	Cramp, 1976
Witchampton	Dorset	900	1199	Urban	Uc: X	Multiple chessmen carved from the flippers (pectoral fin) of whales	0	1	1	1	Dalton, 1928
16-22 Coppergate 1	North Yorkshire	875	975	Urban	Uc: 1	Sword pommel	1	0	1	1	Bond and O'Connor, 1999
16-22 Coppergate 2	North Yorkshire	975	1099	Urban	Uc: 1	Double sided comb	1	0	1	1	O'Connor, 1989

Toponym	Geographic region	Start date	End date	Site type	Species	Details	NISP	X	TOTAL	Min. No. Spec.	Reference
Guersney											
Chapelle Dom Hue	Guersney	1300	1399	Ecclesiastical	Hp: 1	Harbour porpoise burial	1	0	1	1	Personal communication Phillip de Jersey, February 2019
Wales							0	0			
Ringwork in South Glamorgan	South Glamorgan	500	1500	High Status	Uc: 2	2 skull fragments	2	0	2	1	Noddle, Bramwell and Jones, 1977
Cardigan Castle	Dyfed	1050	1500	High Status	Ud: 1	Dolphin skull	1	0	1	1	Cardigan Castle, n.d.
Northern Ireland											
Downpatrick	County Down	600	1000	Ecclesiastical	Uc: 1	Part of a vertebra of whale, a massive disc of bone which found in final use as a slab in an Early Christian period pavement.	1	0	1	1	Hamlin and Lynn, 1988
Rathmullan	County Down	600	1000	Rural	Lc: 2	Lc: Multiple bones that were mostly long bone portions, which showed clear signs of being cut, though not clear whether these are butchery marks or marks that were made to create an artefact or tool	2	0	2	1	Lynn et al, 1981/1982
Rathmullan 2	County Down	900	1200	Rural	Lc: 1	Lc: scapula	1	0	1	1	Lynn et al, 1981/1982
Clough Castle	County Down	1250	1350	High Status	Lc: 1	Lc: fragment with 5 perforations that was used as a stamp	1	0	1	1	Waterman and Jope, 1954
Dunideny Castle	County Antrim	1400	1650	High Status	Lc: 1	Lc: vertebra with chopping signs, made into a square	1	0	1	1	McNeill, 2004
Lough Faughan	County Down	650	975	Rural	Sw: 1	Sp: tooth fragment	1	0	1	1	Collins et al, 1955
Ireland											
Iniskea North	Connacht	400	999	Ecclesiastical	Uc: 7	Whale bone disc (perforated, probably used for the creation of a basket). Scapula of a small whale. Fragments and 2 ribs of whale. 2 fragments of bones of whale	7	0	7	1	Henry, 1945; McCormick *et al.*, 2011
Cherrywood	Dublin	850	899	Rural	Uc: 1	whale bone plaque (which is often associated with women in Ireland and the Hebrides)	1	0	1	1	Ó'Néill, 2005
Christchurch Place	Dublin	1150	1350	Urban	Uc: 1	Double sided comb	1	0	1	1	Ó Ríordáin, 1975

Toponym	Geographic region	Start date	End date	Site type	Species	Details	NISP	X	TOTAL	Min. No. Spec.	Reference
Christ Church	Munster	1225	1275	Urban	Mc: 1	Vertebra, unfused at both ends. Chopped dorso-ventrally on the posterior and anterior sided and also on the lateral side	1	0	1	1	McCarthy, 1997
Clochan-na-carraige	Connacht	400	800	Rural	Uc: 1	Whale bone found under loose stones, possible being incorporated into wall structure	1	0	1	1	Leask, 1943
Fishamble Street	Leinster	?	?	Urban	Uc: X		0	1	1	1	McCormick, 1987
Grattan Street	Munster	1200	1400	Urban	Uc: 1	Vertebra	1	0	1	1	McCarthy, 2003
Grand Parade	Munster	900	1600	Urban	Uc: X		0	1	1	1	Personal communication McCarthy, 2016
South Great George's St./Stephen St.	Leinster	800	900	Burial	Uc: 1	Possible whale bone artefact	1	0	1	1	Personal communication Cathy Daly, April 2016
High Street	Leinster	850	1350	Urban	Uc: 1	Whale bone cleaver	1	0	1	1	Ó Ríordáin, 1973
Marshes Upper 5	Leinster	400	1000	High Status	Uc: X		0	1	1	1	Gosling, 1991
Marshes Upper 3	Leinster	600	799	Rural	Uc: 11		11	0	11	1	Gowen 1992; McCormick 1992
Collierstown	Leinster	427	606	Burial	Uc: 1	Whale bone sword pommel	1	0	1	1	O'Hara, 2010; McCormick *et al.*, 2011
Keel West	Connacht	500	1500	Rural	Hw?: 1	Cervical vertebra built into the west whale of a ruined stone building, the bone was possibly from a humpback whale.	1	0	1	1	Barton, 1943
Kilmainham	Leinster	750	1100	Rural	Uc: 1	3 fragments of whale bone plaque	1	0	1	1	Isaksen, 2012
Knowth Site M	Leinster	500	999	Rural	Sw: 1	Toggle like shaped object made of a sperm whale tooth	1	0	1	1	Stout & Stout 2008; McCormick *et al.*, 2011
Illaunloughan	Munster	773	970	Ecclesiastical	Sw: 1		1	0	1	1	Murray, McCormick and Plunkett, 2004; McCormick *et al.*, 2011
Illaunloughan (mid seventh to the mid tenth century deposit	Munster	650	899	Ecclesiastical	Kw: 1	Tooth of a young killer whale	1	0	1	1	Murray, McCormick and Plunkett, 2004
Patrick Street Site D	Munster	1200	1300	Urban	Lfpw: 2	A skull of a immature pilot whale with butchery marks	2	0	2	1	McCormick and Murphy, 1977
Patrick Street Site C	Munster	1200	1300	Urban	Uc: 2		2	0	2	1	McCormick and Murphy, 1977

Toponym	Geographic region	Start date	End date	Site type	Species	Details	NISP	X	TOTAL	Min. No. Spec.	Reference
Rosepark	Leinster	400	599	Rural	Uc: 3		3	0	3	1	Carroll 2008; McCarthy 2008
Washingtonstreet	Munster	1100	1199	Urban	Uc: 1	Anglo-Normanstyle gaming piece	1	0	1	1	Kelleher, 2003
Tobin Street	Munster	1225	1275	Urban	Ud: 1	Vertebra of an immature dolphin with a clean chop mark on the lateral side	1	0	1	1	McCarthy, 2003
North Gate	Munster	1250	1500	Urban	M/Sc	Candle holder	1	0	1	1	Hurley, 1995
Tory Island (Round Tower)	Ulster	1100	1199	Ecclesiastical	Uc: 1		0	1	1	1	Crumlish, 1993
Shandon	Munster	800	1199	High Status	Lc: 1	Proximal humerus	1	0	1	1	Elder *et al.*, 2007
Skiddy's Lane	Munster	1200	1400	Urban	Ud: 1	A single vertebra of an immature dolphin	1	0	1	1	McCarthy, 2003
Raheens	Munster	500	1500	Rural	Mc: 1	Cetacean vertebra, fused at both ends, fine dorso-ventral chop mark on the lateral side	1	0	1	1	McCarthy, 1994, 65
Wood Quay	Leinster	750	1100	Urban	Uc: X		0	1	1	1	Wallace, 2016
Faroe Islands											
Kvívík	Streymoy	793	1066	Rural	Lfpw: X		0	1	1	1	Dahl, 1971
Toftanes	Eysturoy	800	1000	Rural	Lc: 1		1	0	1	1	Hansen, 2013
Undir Junkarinsfløtti	Sandoy	800	1400	Rural	Uc: 3	Uc: tool debris	3	0	3	1	McGovern *et al.*, 2004
Poland											
Gdansk	Pomerian	950	1308	Urban	Hp: 1	not mentioned which elements	1	0	1	1	Makowiecki and Makowiecka, 2014
Gdansk 2	Pomerian	1308	1466	Urban	Hp: 7	not mentioned which elements	7	0	7	1	Makowiecki and Makowiecka, 2014
Gdansk 3	Pomerian	1308	1793	Urban	Hp: 4	not mentioned which elements	4	0	4	1	Makowiecki and Makowiecka, 2014
Gdansk 4	Pomerian	1308	1793	Urban	Hp: 5	not mentioned which elements	5	0	5	1	Makowiecki and Makowiecka, 2014
Gdansk 5	Pomerian	1466	1793	Urban	Hp: 5	not mentioned which elements	5	0	5	1	Makowiecki and Makowiecka, 2014
Gdansk 6	Pomerian	950	1797	Urban	Hp: 1	not mentioned which elements	1	0	1	1	Makowiecki and Makowiecka, 2014
Kołobrzeg	West Pomeranian Voivodeship	950	1000	Urban	Kw: 1	Kw: mandible piece with traces of chopping	1	0	1	1	Makowiecka and Makowiecki, 2007

Toponym	Geographic region	Start date	End date	Site type	Species	Details	NISP	X	TOTAL	Min. No. Spec.	Reference
Russia											
Novgorod	Novgorod Oblast	1000	1500	Urban	Uc: X	Whale bone was used for the production of knifehandles + 2-3 whale scapulae	0	1	1	1	Personal communication Dr. Liubov Holden, May 2016
Estonia							0	0			
Salme	Saaremaa	550	799	Burial/High Status	Uc: 50, Ud: 1	hnefatafl gaming pieces created with a bow lathe. Ud: 4 dices made of teeth	54	0	54	2	Peets and Maldre, 2010
Salme II	Saaremaa	550	799	Burial/High Status	Uc: 328	hnefatafl gaming pieces	328	0	328	1	Personal communication J. Peets, 2016
France											
Barreau Saint Georges	Nord-Pas-de-Calais	1000	1050	Rural	Uc: 2, Fw: 2	1 rib with two perforated holes, signs of boiling. + another fragment, and 2 fin whale remains (identified using aDNA analysis)	4	0	4	2	UBAAL, n.d.; Personal communication Dr Tarek Oueslati, July 2018
Cavalaire-sur-Mer	Provence-Alpes-Côte d'Azur	1479	1550	Other	Uc: X	Several whale bones. Including a vertebra of a young whale, which bears several chopping marks.	0	1	1	1	Delhaye, 1998
Compiègne	Picardy	1100	1199	Urban	Uc: X	tabula pieces of whale bone	0	1	1	1	Riddler, 2014
La Tour des Salves	Île-de-France	1500	1599	High Status	Ud: 3	Butchery signs	3	0	3	1	Clavel, 2001
Saint-Georges-de-Boscherville	Upper Normandy	1100	1199	Ecclesiastical	Sbcd: 1, Hp: 1		2	0	2	2	Clavel, 2001
Saint-Urnel-en-Plomeur	Brittany	230	1000	Burial	Uc: 3	Two adult graves with lateral furnishings carved out of large cetacean bones	3	0	3	1	Giot and Corgné, 1951
CHÂTEAU DE SUSCINIO (dépotoir (DEP_1045))	Brittany	1300	1399	High Status	Hp: 1	Lumbar vertebra	1	0	1	1	Borvon, 2017
Chateau de Suscinio (l'US2184 (Cour/Espace 4))	Brittany	1300	1399	High Status	Hp: 1		1	0	1	1	Borvon, 2017
Saint-Georges-sur-l'Aa	Hauts-de-France	900	1099	Rural	Uc: X		0	1	1	1	Personal communication Dr Tarek Oueslati, April 2019
la cour Napoléon du Louvre	Ile-de-France	1300	1699	Urban	Hp: 3	Vertebra, humerus and mandible with cutmarks	3	0	3	1	Méniel and Arbogast, 1989
Hôtel de Beauvais	Ile-de-France	1400	1499	Ecclesiastical	Sbcd: 1		1	0	1	1	Du Bouëtiez, Clavel, and Ravoire, 2013

Toponym	Geographic region	Start date	End date	Site type	Species	Details	NISP	X	TOTAL	Min. No. Spec.	Reference
Faculté de Sciences Sociales	Midi-Pyrénées	400	1499	Unknown	Ud: 1		1	0	1	1	Catalo, 2010
n° 80 quai du Châtelet	Centre-Val de Loire	1400	1550	High Status	Cbd: 1	Vertebra	1	0	1	1	Du Bouëtiez, Clavel, and Ravoire, 2013
Château de Talmont Saint-Hilaire	Pays de la Loire	1300	1499	High Status	Cbd: 1	Thoracic vertebra bottlenose dolphin with 2 cutmarks	1	0	1	1	Borvon, unpublished
Château de Talmont Saint-Hilaire	Pays de la Loire	1000	1399	High Status	Hp: 1	Mandible rostrum	1	0	1	1	Borvon, unpublished
La Moutte	Provence-Alpes-Côte d'Azur	950	1050	High Status	Uc: 2	2 Trictrac gaming pieces. One depicting two birds and one a four-legged creature.	2	0	2	1	Mouton, 2005
Saint-Gilles	Occitanie	1100	1299	Ecclesiastical	Uc: 1	Unworked whale bone found within tomb within cloister	1	0	1	1	Chazottes, 2017
Rue Saint Wulfran	Hauts-de-France	1000	1100	Unknown	Uc: 2	2 Trictrac gaming pieces with animals depicted	2	0	2	1	Gaborit-Chopin and Bardoz, 2005
Quai de la point	Hauts-de-France	1050	1100	Unknown	Uc: 3	3 Trictrac gaming pieces	3	0	3	1	Gaborit-Chopin and Bardoz, 2005
Tours	Centre-Val de Loire	1000	1200	Unknown	Uc: 1	1 Trictrac gaming piece	1	0	1	1	Motteau, 1991
Belgium											
Brugge	West Flanders	800	999	High Status	Uc: 2	Worked fragments	2	0	2	1	Ervynck, 1991
Gravensteen	East Flanders	1000	1199	Urban	Uc: 1	Scapula with multiple cutmarks	1	0	1	1	Van der Plaetsen, 1987
Sint-Veerleplein 11, Gravensteen	East Flanders	900	1250	High Status	Uc: X	Originally identified as mammoth bones. Chopmarks present on bone	0	1	1	1	Ervynck, Laleman and Berkers, 2012
Walraversijde	West Flanders	1270	1500	Rural	Uc: X, Wbd: 1, Hp: 4	Bone fragments	5	1	6	3	Van Neer, Lentacker and Ervyck, 2013
Walraversijde	West Flanders	1270	1500	Rural	Sw: 1	One young sperm whale. Multiple bones (atlas, others not known). Multiple chop and cutmarks	1	0	1	1	Van Neer and Ervynck, 1993
Koksijde (Hof ter Hille)	West Flanders	875	1125	Ecclesiastical	Uc: 1, Fw: 1, Cmw?: 1	1 bone with chopmarks, 1 thoracic vertebra of a minke whale? 6 bones from sieved material	12	0	12	3	Zeiler, 2018
De Motte van Veurne	West Flanders	850	1600	High Status	Uc: 1	Unknown whale bone	1	0	1	1	Van Neer and Ervynck, 1993

Toponym	Geographic region	Start date	End date	Site type	Species	Details	NISP	X	TOTAL	Min. No. Spec.	Reference
The Netherlands											
Achlum	Friesland	1000	1200	Terp	Sw: 1	14th or 15th caudal vertebra with 2/3 holes in it and burning marks. Also chop and cut marks present	1	0	1	1	Prummel, Van Gent and Kampanje, 2012
Achlum	Friesland	400BC	1400	Terp	Sw: 1, Nbw: 1	Worked piece of whale (medium to large) bone with perforation and a vertebral body (thoracic, lumbar or one of the first caudal)	2	0	2	2	NADNuis, 2019
Arnemuiden Hazenburg II	Zeeland	1000	1199	Rural	Uc: 1	Gaming piece of whale bone	1	0	1	1	van Dijk, 2018
Bleekveld	Gelderland	900	1550	Rural	NArw: 1	Vertebral spinous process of right whale (thoracic)	1	0	1	1	Van Doesburg and Bakker, 1999
Brederode	Noord-Holland	1275	1573	High Status	Kw: 1, Lfpw: 1	Lumbar and caudal vertebrae	2	0	2	2	Zeiler, 2007
Caetshage	Gelderland	1100	1299	High Status	Uc: 1	Rib fragment with saw marks	1	0	1	1	Halici, 2004
Beatrixlaan	Noord-Holland	1000	1200	Rural	NArw: 1	16 fragments (two butchery signs visible on one fragment)	1	0	1	1	Groenman van Waateringe and Wijngaarden-Bakker, n.d.
Beatrixlaan	Noord-Holland	750	1000	Rural	NArw: 1, Fw: 1, Gw: 1	Fin whale (10 unidentified fragments), right whale (transverse process of vertebra), grey whale (7 unid. fragments)	3	0	3	3	Groenman van Waateringe and Wijngaarden-Bakker, n.d.
Dorestad (Hoogstraat I (river-bed))	Utrecht	700	850	Urban	Ur: 1		1	0	1	1	Prummel, 1983
Dorestad (Northern section of the settlement)	Utrecht	700	850	Urban	Lc: 1, NArw: 1	1 rib + unidentified specimen	2	0	2	2	Prummel, 1983
Dorestad (Veilingterrein)	Utrecht	750	900	Urban	Uc: 1	Vertebral chopping block	1	0	1	1	Dijkstra, 2012
Dorpsheuvel	Zuid-Holland	42	899	High Status	Uc: 5, Hp: 1		6	0	6	2	Clason, 1961
Firdgum (Early Middle Ages)	Friesland	400	999	Terp	NArw: 1, Hp: 2	Humerus of right whale and two epiphyses of porpoise from thoracic and caudal vertebrae	3	0	3	2	Prummel, Gent and Kam
Frederik-Hendriklaan	Zuid-Holland	500	900	Rural	Hp: 1	Skull fragment, unfused so probably young individual	1	0	1	1	Esser, 2009

Toponym	Geographic region	Start date	End date	Site type	Species	Details	NISP	X	TOTAL	Min. No. Spec.	Reference
Hoogmadeseweg	Zuid-Holland	529	899	Rural	Uc: 1	Vertebra of whale, signs that it has been worked.	1	0	1	1	Van Dijk, 1995
Huispolanen	Zuid-Holland	1300	1351	High Status	Uo: 1	Scapula	1	0	1	1	Bult, 1986
Graven van Gelre	Noord-Brabant	1175	1232	High Status	NArw: 1	Vertebral epiphyses	1	0	1	1	Esser *et al.*, 2014
Johan van Oldenbarneveltlaan	Zuid-Holland	500	700	Rural	Uc: 1, Hp: 5		6	0	6	2	Magendans and Waasdorp, 1989
Katwijk, De Zanderij	Zuid-Holland	700	1000	Rural	Lc: 1, FHGw: 1	Fragments	2	0	2	2	Laarman, 2006
Koornmarkt	Gelderland	900	1049	Rural	NArw: 2	vertebral epiphyses used as chopping surface and vertebral arch	2	0	2	1	Holthuis, Smeenk and Laarman, 1998
Oldehoofsterkerkhof	Friesland	400	900	Terp	NArw: 1	Mandible	1	0	1	1	Thilderkvist, 2013
Maarland-Zuidzijde	Zuid-Holland	1250	1500	Urban	Sw: 1	epiphyses	1	0	1	1	Esser, 2004
Oost-Souburg	Zeeland	900	975	High Status	NArw: 1	Three pieces of cranium	1	0	1	1	Lauwerier, 1995
Oosterbeintum	Friesland	300	1099	Terp	-	Specimen identified to not be cetacean	0	0	0	0	Knol *et al.*, 1996
Oosterbuurt-Albertshoeve 1	Noord-Holland	650	899	Rural	Uc: 4	Bone with chopmarks	4	0	4	1	Lauwerier and Laarman, 1999
Oosterbuurt-Albertshoeve 2	Noord-Holland	850	1200	Rural	Uc: 1		1	0	1	1	Lauwerier and Laarman, 1999
Rijksweg57	Zeeland	1250	1950	Rural	NArw: 1	Mandible with holes drilled	1	0	1	1	Van Dijk et al, 2011
Rijnsburg (Na)	Zuid-Holland	725	1249	Ecclesiastical	Uc: 3	Three fragments	3	0	3	1	Clason, 1965
Rijnsburg (Nw)	Zuid-Holland	725	1249	Ecclesiastical	Uc: 2	Fragments	2	0	2	1	Clason, 1965
Tzummarum	Friesland	525	1000	Terp	Uc: X		0	1	1	1	Hopman, 1993
Tzummarum	Friesland	700	1000	Ecclesiastical/ Terp	Fw: 1	Worked bone with perforation	1	0	1	1	NADNuis, 2019
De Woerd	Zuid-Holland	800	999	Rural	Bd: 1		1	0	1	1	Sablerolles, 1990
Waldijk II	Noord-Holland	12BC	1999	Rural	Sw: 1	Transverse process of vertebra	1	0	1	1	Van Dijk, 2012
Wijnaldum-Tjistsma (Carolingian periode)	Friesland	525	900	Terp	Hp: 2	Mandible and cranium	2	1	1	1	Prummel, Gent and Kampanje, 2012
Wijnaldum-Tjistsma (Merovingian period)	Friesland	400	750	Terp	Uc: 1	Pointy stick made of whale bone	1	0	1	1	Prummel, Gent and Kampanje, 2012

Toponym	Geographic region	Start date	End date	Site type	Species	Details	NISP	X	TOTAL	Min. No. Spec.	Reference
Slot Op Den Hoef	Noord-Holland	1285	1573	High Status	Cbd: 2, Sw?: 1	2 lumbar vertebrae common bottlenose dolphin, 1 rib fragment of sperm whale with butchery sign on side	3	0	1	1	Zeiler, 2007
Golfslag	Zeeland	1000	1000	Rural	Uc: 1		1	0	1	1	van Dijk, 2012
Hoogeland	Zuid-Holland	450	1050	Rural	Ubw: 1, NArw: 1	Rib and central body of a vertebra with chopmarks	2	0	1	1	Goosens, 2012
Hoogeland	Zuid-Holland	450	1500	Rural	Uc: 1		1	0	1	1	Goosens, 2012
Mussentiend	Gelderland	450	1500	Rural	NArc: 1	Vertebra fragment	1	0	1	1	Médard, 2013
Aalmarktschool	Zuid-Holland	1200	1275	Urban	Nbw: 7	Fragments	7	0	7	1	Esser, Beerenhout and Kootker, 2010
Dokkershaven Zuidzijde	Zeeland	1575	1600	Urban	Lc: 1	whale vertebra, used as chopping block	2	0	1	1	Claeys, Jaspers and Ostkamp, 2010
Het Groenje	Zeeland	850	1000	Urban	Gw: 1	Cancellous bone fragment	1	0	1	1	Buitenhuis, 2011
Hallum	Friesland	100BC	1200	Terp	Gw: 1	Cervical vertebra	1	0	1	1	NADNuis, 2019
Rottum	Friesland	600BC	1200	Terp	Gw: 1	Weaving sword	1	0	1	1	NADNuis, 2019
Leens	Groningen	600BC	1200	Terp	NArw: 1	Weaving sword	1	0	1	1	NADNuis, 2019
Burseplein	Overijssel	900	1200	Urban	Sw: 1	Mandible fragment of sperm whale	1	0	1	1	Ijzereef and Laarman, 1986
Plantage	Zuid-Holland	760	850	Rural	Uc: 3	3 unidentified bones with 1 containing saw marks	3	0	1	1	Moesker and Cavallo, 2016
Plantage	Zuid-Holland	650	850	Rural	Uc: 5	Rib fragment with chop, burn and cutmarks + 1 scapula (mammoth), other 4 are not provided	5	0	5	1	Moesker and Cavallo, 2016
Frankenslag	Zuid-Holland	450	1050	Rural	Hp: 2	Cervical vertebrae	2	0	1	1	Esser, 2018
Molenslag	Zuid-Holland	400	800	Rural	Cbd?: 1	Lumbar vertebra	1	0	1	1	Esser and Dütting, 2018
Burgh-Haamstede	Zeeland	800	900	Rural	Fw: 1	Vertebra with chopmarks (Cranial and caudal side fused)	1	0	1	1	Personal communication Joost van den Berg and Aagje Feldbrugge
Burgh-Haamstede	Zeeland	50BC	1600	Rural	Sw: 1, Fw: 1	Sperm whale vertebra (no chopmarks), fin whale vertebra (lots of chopmarks)	2	0	2	2	Personal communication Joost van den Berg and Aagje Feldbrugge
Englum	Groningen	600BC	1200	Terp	Lfpw: 1	humerus of old individual with chop and cutmarks present	1	0	1	1	Nieuwhof, 2007

Toponym	Geographic region	Start date	End date	Site type	Species	Details	NISP	X	TOTAL	Min. No. Spec.	Reference
Unknown Terp (Unknown)	Unknown	500BC	1900	Terp	Sw: 1, Kw: 1, Nbw: 3, NArw: 1	Right whale caudal vertebra with chopmarks from adult. Caudal vertebra killer whale. Sperm whale right rib with cutmarks and probably worked. Bottlenose whale 2 caudal vertebrae and piece of another vertebrae.	6	0	6	4	Prummel, Van Gent and Kampanje, 2012
Germany											
Borgsumborg	Schleswig-Holstein	100	999	Rural	Hp: 4		4	0	4	1	Schmölcke, 2009
Dänische Wiek	Mecklenburg-Vorpommern	830	1020	Rural	Ubw: 1	Cervical vertebra	1	0	1	1	Aaris-Sorensen *et al.*, 2010
Dominikanerkloster Norden	Weser-Ems	1200	1599	Ecclesiastical	Hp: 4		4	0	4	1	Küchelmann, 2010
Elisenhof	Schleswig-Holstein	1200	1300	Rural	Lc: 1, Hp: 1	Porpoise vertebra and a whale rib	2	0	2	2	Reichstein and Heinrick, 1994
Emden (Kirchstrasse)	Weser-Ems	900	1699	Urban	Cbd: 1		1	0	1	1	Grimm, 2006
Erfurt	Thuringia	500	1500	Urban	Hp: 1	Thoracic vertebra from a harbour porpoise	1	0	1	1	Priloff, 2002
Haithabu	Schleswig-Holstein	700	999	Urban	Wbd: 8		8	0	8	1	Reichstein, 1991
Oldenburg	Schleswig-Holstein	793	1066	High Status	Uc: 14	Tabula pieces of whale bone	14	0	14	1	Gabriel, 1988
Oldenburg 2	Schleswig-Holstein	1200	1300	High Status	Hp: 1		1	0	1	1	Prummel, 1993
Plessenstrasse	Schleswig-Holstein	1100	1199	Urban	Uc: X	Tabula pieces of whale bone	0	1	1	1	Ulbricht, 1984
Zingst	Mecklenburg-Vorpommern	260	530	Unknown	SRHw: 1	Lumbar vertebra	1	0	1	1	Aaris-Sorensen *et al.*, 2010
Ralswiek	Mecklenburg-Vorpommern	700	1200	Urban	Hp: 1	porpoise mandible fragment	1	0	1	1	Benecke, 1999
Rostock-Katharinenkloster	Mecklenburg-Vorpommern	1300	1399	Ecclesiastical	Hp: 2	2 porpoise vertebrae	2	0	2	1	Benecke, 1999
Wellinghusen	Schleswig-Holstein	600	1399	Rural	Hp: 1		1	0	1	1	Becker, 2012

Toponym	Geographic region	Start date	End date	Site type	Species	Details	NISP	X	TOTAL	Min. No. Spec.	Reference
Iceland											
Aðalstræti 14-18	Capital Region	1000	1500	Rural	Uc: 2	Piece of flat and rectangular piece of whale bone, and a whale bone sword.	2	0	2	1	Roberts, 2001; 2004a
Alþingisreitur	Capital Region	1500	1800	Rural	Uc: 65, Lc: 2, Nbw: 1, Glob: 2, Ud: 3, Ud/p: 4	Delphinidae: 1 mandible, 1 scorched skull fragment, 1 rib, and 1 articulated vertebra with chopmarks. Bottlenose whale: 1 vertebral epiphysis. Globicephalinae: 1 vertebral epiphysis.	77	0	77	6	Pálsdóttir, 2010
Alþingisreitur	Capital Region	1226	1500	Rural	Uc: 16, Fw: 1	Fin whale: 1 vertebra	17	0	17	2	Pálsdóttir, 2010
Alþingisreitur	Capital Region	871	1226	Rural	Glob: 3, Ud: 1, Hw: 1	Globicephalinae: 2 vertebrae with chopmarks and 1 skull fragment. Delphinidae: 1 fragment. Humpback whale: 1 fragment.	5	0	5	3	Pálsdóttir, 2010
Baldursheimur	Northeastern Region	900	999	Rural	Uc: 1		1	0	1	1	Roesdahl and Wilson, 1992
Fornusandar II	Southern Region	1400	1700	Rural	Uc: 3	Fragments. 2 rib fragments and one needle	3	0	3	1	Amorosi, 1996
Gasir 1	Northeastern Region	1250	1400	Rural	Uc: 8, Sc: 1		9	0	9	2	Woollett and McGovern, 2003
Gasir 2	Northeastern Region	1250	1400	Rural	Uc: 8, Lc: 1, Sc: 7	Fragments	16	0	16	3	Roberts, 2004b
Gasir 3	Northeastern Region	1250	1400	Rural	Uc: 4, Lc: 1, Sc: 2	1 small cetacean juvenile rib and 1 large cetacean vertebra chopping block.	7	0	7	3	Roberts, 2005
Gasir 4	Northeastern Region	1250	1400	Rural	Uc: 1		1	0	1	1	Pálsdóttir and Roberts, 2006
Gasir 5	Northeastern Region	1250	1400	Rural	Uc: 2, Sc: 1		3	0	3	2	Pálsdóttir and Roberts, 2007
Gasir Area B	Northeastern Region	1250	1400	Rural	Uc: 4	Working debris fragments	4	0	4	1	Vésteinsson, 2008
Gjögur	Westfjords	1390	1450	Rural	Uc: 20	Fragments	20	0	20	1	Krivogorskaya, Perdikaris and McGovern, 2005
Granastadir	Northeastern Region	950	1000	Rural	Uc: 1	Fragment with cut marks	1	0	1	1	Amorosi, 1996

Toponym	Geographic region	Start date	End date	Site type	Species	Details	NISP	X	TOTAL	Min. No. Spec.	Reference
Gufuskálar 1	Western Region	1300	1499	Rural	Uc: 3	3 pieces of worked whale bone of which one was a gaming piece	3	0	3	1	Feeley, 2012
Gufuskálar 2	Western Region	1300	1499	Rural	Uc: 5	5 pieces of worked whale bone	5	0	5	1	Pálsdóttir and Sveinbjarnarson, 2011
Gufuskálar 3	Western Region	1300	1499	Rural	Uc: 1		1	0	1	1	Freeley *et al.*, 2010
Herjolfsdalur	Vestmannaeyjar	870	1300	Rural	Uc: 3		3	0	3	1	Amorosi, 1996
Hjallholl Test Cut 1	Eastern Region	1100	1400	Rural	Uc: 90	working debris	90	0	90	1	Amorosi, 1996
Hjallholl AU1 Test Cut 1	Eastern Region	1100	1400	Rural	Uc: 74	working debris	74	0	74	1	Amorosi, 1996
HjallHoll AU2 Test Cut 1	Eastern Region	1100	1400	Rural	Uc: 1	working debris	1	0	1	1	Amorosi, 1996
HjallHoll AU3 Test Cut 1	Eastern Region	1100	1400	Rural	Uc: 15	working debris	15	0	15	1	Amorosi, 1996
Hofstaðir 1	Northwestern Region	1250	1350	Burial	Uc: 6, Lc: 3, Sc: 36	Fragments of artefacts and tool making debris. Many thoracic and caudal vertebrae from small cetaceans. And a toothless fragment of a lower jaw.	45	0	45	3	McGovern, Smiarowski and Harrison, 2013
Hofstaðir 2	Northwestern Region	940	980	High Status	Uc: 2, Sc: 1	Whale bone fragments include and a single rib from small dolphin or porpoise	3	0	3	2	McGovern, 2009
Hofstaðir 3	Northwestern Region	980	1030	High Status	Uc: 6		6	0	6	1	McGovern, 2009
Hofstaðir 4	Northwestern Region	1030	1070	High Status	Uc: 1	Vertebra used as a chopping block	1	0	1	1	McGovern, 2009
Hrísbrú Kirkjuhóll	Capital Region	900	1500	High Status	Uc: 1		1	0	1	1	Zori and Wake, n.d
Hrísbrú Tún	Capital Region	900	1500	High Status	Uc: 1		1	0	1	1	Zori and Wake, n.d
Hrísheimar 1	Northeastern Region	900	999	Rural	Uc: 1	Substantial piece of whale bone which has been worked to a tapering blunt end	1	0	1	1	Edvardsson, 2003
Hrísheimar 2	Northeastern Region	800	1066	Rural	Uc: 1	Whale bone dice of elongated form	1	0	1	1	Edvardsson, 2003
Akurvík 1	Westfjords	1030	1290	Rural	Uc: 67	Many fragments	67	0	67	1	Amundsen *et al.*, 2005

Toponym	Geographic region	Start date	End date	Site type	Species	Details	NISP	X	TOTAL	Min. No. Spec.	Reference
Akurvík 2	Westfjords	1420	1475	Rural	Uc: 1528, NArw: 1	Many fragmented pieces with cut and chop marks. 1 North Atlantic Right whale vertebra used as corner support for building	1529	0	1529	2	Krivogorskaya, Perdikaris and McGovern, 2005
Möðruvellir	Northeastern Region	1200	1450	Rural	Uc: 1, Sc: 1	Whale bone worked with hole in top, with piece broken off at the drilled hole.	2	0	2	2	Harrison, 2008
Saltvik	Northeastern Region	950	1300	Rural	Lc: 1	Fragment	1	0	1	1	Frideriksson *et al.*, 2004
Siglunes	Northeastern Region	871	1399	Rural	Uc: 26, Lc: 1, Sc: 3	Working debris, and a possible chopping board and some possible shovel blades.	30	0	30	3	Harrison, 2014
Sjóhúsavík	Northeastern Region	1300	1477	Rural	Uc: 1	Fragment	1	0	1	1	Gísladóttir, 2014
Skriðuklaustur Monastery	Eastern Region	1400	1600	Ecclesiastical	Uc: 6	Fragments	6	0	6	1	Hamilton-Dyer, 2010
Skuggi	Northeastern Region	950	1050	Rural	Uc: 1	One unidentifiable chopped whale fragment	1	0	1	1	Harrison, 2010
Skútustaðir 1	Northeastern Region	1262	1300	Rural	Uc: 1		1	0	1	1	Hicks, 2010
Skútustaðir 2	Northeastern Region	1300	1410	Rural	Uc: 1	Whale bone clamp	1	0	1	1	Hicks and Pálsdóttier, 2011
Steinbogi	Northeastern Region	1200	1299	Rural	Uc: 1	The single fragment of whale bone appears to be a small fragment of a worked piece, possibly part of a broken tool	1	0	1	1	Brewington *et al.*, 2004
Storaborg Hus 14	Suðurland	1450	1838	Rural	Uc: 2	Chopping block and at least one cranial fragment of a northern bottlenose whale	2	0	2	1	Amorosi, 1996
Storaborg Hus 9/14	Suðurland	1450	1838	Rural	Uc: 1	Chopping block and at least one cranial fragment of a northern bottlenose whale	1	0	1	1	Amorosi, 1996
Storaborg Hus 17	Suðurland	1450	1838	Rural	Uc: 2	Chopping block and at least one cranial fragment of a northern bottlenose whale	2	0	2	1	Amorosi, 1996
Storaborg Hus 18	Suðurland	1450	1838	Rural	Uc: 4, Nbw: 4	Chopping block and at least one cranial fragment of a northern bottlenose whale	5	0	5	1	Amorosi, 1996

Toponym	Geographic region	Start date	End date	Site type	Species	Details	NISP	X	TOTAL	Min. No. Spec.	Reference
Storaborg Hus 19	Suðurland	1450	1838	Rural	Uc: 1	Chopping block and at least one cranial fragment of a northern bottlenose whale	1	0	1	1	Amorosi, 1996
Sveigakot 1	Northeastern Region	910	1140	Rural	Uc: 1	Worked whale bone	1	0	1	1	Vésteinsson, 2001
Sveigakot 2	Northeastern Region	1000	1199	Rural	Uc: 1	Whale bone plaque	1	0	1	1	McGovern, 2003
Sveigakot 3	Northeastern Region	900	1199	Rural	Uc: 1	Fragment	1	0	1	1	Vésteinsson, 2004
Sveigakot 4	Northeastern Region	950	1199	Rural	Uc: 12	12 pieces of whale bone	12	0	12	1	Vésteinsson, 2005
Sveigakot 5	Northeastern Region	950	1199	Rural	Uc: 2		2	0	2	1	Gisladóttir and Vésteinsson, 2008
Tjarnargata 4 (1)	Capital Region	875	1000	Rural	Lc: 3		3	0	3	1	Amorosi, 1996
Tjarnargata 4 (2)	Capital Region	875	1000	Rural	Uc: 3, Seiw?: 1		4	0	4	2	Amorosi, 1996
Vatnsdalur	Westfjords	900	1000	Burial	Uc: 2	2 tool fragments	2	0	2	1	Magnússon, 1966
Vatnsfjörður	Westfjords	900	999	High Status	Uc: 3, Lc: 2	All whale bones are from large species and have cut marks, probably working waste	5	0	5	2	Edvardsson and McGovern, 2005
Videy	Capital Region	1300	1600	rural	Uc: 1, Lc: 2, Sc: 2	small cetacean 2 vertebrae	5	0	5	3	Amorosi and McGovern, 1993
Denmark											
Asaa	North Denmark Region	1300	1500	Rural	Cmw: 1		0	1	1	1	Aaris-Sorensen, 2009
Borrebjerg	Zealand	500BC	800	Rural	Kw: 2, Hp: 1	2 killer whale teeth and 1 porpoise vertebra	3	0	3	2	Degerböl, 1933
Egense	North Denmark Region	680	850	Rural	Fw: 1		0	1	1	1	Aaris-Sorensen, 2009
Lonstrup	North Denmark Region	560	680	Rural	Boww/ NArc: 1	Mandible	1	0	1	1	Aaris-Sorenson *et al.*, 2010
Naesholm	Zealand	1240	1340	High Status	Be: 1, Kw: 1, Hp: 12	Beluga cranium fragment, killer whale rib fragment, harbour porpoise 1 humerus and 11 vertebrae.	14	0	14	3	Mohl, 1961
ørkild	Region of Southern Denmark	1050	1500	High Status	Hp: 1		1	0	1	1	Jansen, 1987

Toponym	Geographic region	Start date	End date	Site type	Species	Details	NISP	X	TOTAL	Min. No. Spec.	Reference
Poulsker	Capital region of Denmark	900	1150	Rural	Ur: 1	Vertebra	1	0	1	1	Aaris-Sorensen, 2010
Dommerhaven	Region of Southern Denmark	700	800	Urban	Kw: 1	A badly preserved tooth	1	0	1	1	Hatting, 1991
Soborg Ruin	Capital region of Denmark	1230	1390	Rural	Hw: 1	Vertebra	1	0	1	1	Aaris-Sorensen *et al.*, 2010
Strandby	North Denmark Region	440	640	Rural	Boww: 1	Radius	1	0	1	1	Aaris-Sorenson *et al.*, 2010
Odense	Southern Denmark	1100	1399	Urban	Sw: 1, Narw: 1	1 whale bone comb plate from a sperm whale and 1 North Atlantic right whale bone created into an artefact	2	0	2	2	Ørsted Brandt *et al.*, 2018
Felt	Region of Southern Denmark	820	850	Urban	Uc: 2	2 whale bone plaques. (1 consisting out of 22 fragments and 1 other fragment)	2	0	2	1	Feveile, 2006, 297; Isaksen, 2012
Rønne	Capital region of Denmark	1255	1500	Rural	Uc: 1	1 whale bone disc. Suggested to have served as a plate for food	1	0	1	1	Staal, n.d.
Nibe	North Denmark Region	1300	1500	Rural	Lc: 1	1 vertebra of a large whale	1	0	1	1	Nordjyske, n.d.
Odense	Southern Denmark	800	1500	Urban	Uc: 2	1 whale bone disc decorated, and 1 plaque also decorated, but unlike those commonly found in Norway.	2	0	2	1	Eskildsen, 2013
Lejre	Zealand	700	1200	Rural	Uc: 1	Gaming pieces	1	0	1	1	Hennius *et al.* 2018
Norway											
Trondheim	Sør-Trøndelag	1000	1099?	Urban	Uc: X	Whale bone tool debris	0	1	1	1	Long, n.d.
Kongshavn	Finnmark	1290	1440	Rural	Uc: 1		1	0	1	1	Amundsen, 2008
Kongshavn Room 1	Finnmark	1250	1500	Rural	Uc: 29	2 with butchery signs	29	0	29	1	Amundsen, 2008
Kongshavn Room 2	Finnmark	1250	1500	Rural	Uc: 28	5 with butchery signs	28	0	28	1	Amundsen, 2008
Kongshavn Room 3	Finnmark	1250	1500	Rural	Uc: 7	2 with butchery signs	7	0	7	1	Amundsen, 2008
Kongshavn Room 4	Finnmark	1250	1500	Rural	Uc: 12	1 with butchery signs	12	0	12	1	Amundsen, 2008
Kongshavn Room 4b	Finnmark	1550	1800	Rural	Uc: 16	1 with knife mark	16	0	16	1	Amundsen, 2008
Kongshavn Room 5	Finnmark	1550	1800	Rural	Uc: 47	20 worked, 1 with chop mark, 1 with knife mark	47	0	47	1	Amundsen, 2008

Toponym	Geographic region	Start date	End date	Site type	Species	Details	NISP	X	TOTAL	Min. No. Spec.	Reference
Kongshavn Midden A	Finnmark	1250	1500	Rural	Uc: 81	16 worked, 2 sawn, 1 with knife marks, 4 chopped	81	0	81	1	Amundsen, 2008
Kongshavn Midden B	Finnmark	1250	1500	Rural	Uc: 42	9 chopped, 4 worked	42	0	42	1	Amundsen, 2008
Kongshavn Exterior Midden	Finnmark	1250	1500	Rural	Uc: 26		26	0	26	1	Amundsen, 2008
Laukvika	Finnmark	1430	1640	Rural	Uc: 3		3	0	3	1	Amundsen, 2008
Laukvika 2	Finnmark	1430	1640	Rural	Uc: 3		3	0	3	1	Amundsen, 2008
Gæccevaj'njar'ga	Finnmark	1400	1450	Rural	Uc: 21	Probably used for chopping blocks	21	0	21	1	Amundsen, 2008
Skonsvika	Finnmark	1270	1410	Rural	Uc: 9		9	0	9	1	Amundsen, 2008
Skonsvika 2	Finnmark	1280	1410	Rural	Uc: 4		4	0	4	1	Amundsen, 2008
Skonsvika SU 12	Finnmark	1200	1500	Rural	Uc: 51	3 slice fragments, 2 with chopping signs	51	0	51	1	Amundsen, 2008
Skonsvika SU 14	Finnmark	1200	1500	Rural	Uc: 32	2 worked, 3 chopped	32	0	32	1	Amundsen, 2008
Skonsvika SU 46	Finnmark	1200	1500	Rural	Uc: 9		9	0	9	1	Amundsen, 2008
Skonsvika Pit 7	Finnmark	1200	1500	Rural	Uc: 22	5 worked, 1 chopped	22	0	22	1	Amundsen, 2008
Skonsvika Pit 2 SU 46	Finnmark	1200	1500	Rural	Uc: 7		7	0	7	1	Amundsen, 2008
Skonsvika Pit 3 SU 46	Finnmark	1200	1500	Rural	Uc: 2		2	0	2	1	Amundsen, 2008
Skonsvika Pit 3 SU 14	Finnmark	1200	1500	Rural	Uc: 11		11	0	11	1	Amundsen, 2008
Skonsvika Pit 9	Finnmark	1200	1500	Rural	Uc: 1		1	0	1	1	Amundsen, 2008
Vadsoya	Finnmark	1450	1660	Rural	Uc: 1		1	0	1	1	Amundsen, 2008
Valderöia	More og Romsdal	500BC	800	Rural	Kw: X		0	1	1	1	Aaris-Sorenson *et al.*, 2010
Bjornerem	More og Romsdal	?	800	Rural	Hp: X		0	1	1	1	Clark, 1947
Dolm	Sor-Trondelag	?	800	Rural	Kw: X		0	1	1	1	Clark, 1947
S. Kaupang	Vestfold	600	1030	Unknown	Uc: 1	8-shaped tool. Line stretcher	1	0	1	1	Petersen, 1952, 522
Strandby	Rogaland	600	1030	Unknown	Uc: 1	8-shaped tool. Line stretcher	1	0	1	1	Petersen, 1952, 522
Ommestad	Hedmark	900	950	Grave	Uc: 1	Whale boneplaque	1	0	1	1	Petersen, 1952, 330
Toranger	Hordaland	780	1066	Unknown	Uc: 1	Whale boneplaque	1	0	1	1	Petersen, 1952, 330
Folkedal	Hordaland	780	1066	Unknown	Uc: 1	Whale boneplaque	1	0	1	1	Petersen, 1952, 330
Seim	Hordaland	780	1066	Unknown	Uc: 1	Whale boneplaque	1	0	1	1	Petersen, 1952, 330
Hopperstad	Sogn og Fjordane	780	1066	Unknown	Uc: 1	Whale boneplaque	1	0	1	1	Petersen, 1952, 330

Toponym	Geographic region	Start date	End date	Site type	Species	Details	NISP	X	TOTAL	Min. No. Spec.	Reference
Haugateig av Hopperstad	Sogn og Fjordane	780	1066	Unknown	Uc: 1	Whale boneplaque	1	0	1	1	Petersen, 1952, 330
Grande	Sør-Trøndelag	780	1066	grave	Uc: 2	Whale boneplaques	2	0	2	1	Isaksen, 2012
Dombu	Sør-Trøndelag	780	1066	Grave	Uc: 1	Whale boneplaque	1	0	1	1	Isaksen, 2012
Trondheim 2	Sør-Trøndelag	780	1066	urban	Uc: 1	Whale boneplaque	1	0	1	1	Isaksen, 2012
Vikestad	Nord-Trøndelag	780	1066	grave	Uc: 1	Whale boneplaque	1	0	1	1	Isaksen, 2012
Halsan	Nord-Trøndelag	780	1066	Unknown	Uc: 1	Whale boneplaque	1	0	1	1	Isaksen, 2012
Nedre Eggen	Nord-Trøndelag	780	1066	grave	Uc: 1	Whale boneplaque	1	0	1	1	Isaksen, 2012
Dun	Nord-Trøndelag	780	1066	Unknown	Uc: 1	Whale boneplaque	1	0	1	1	Isaksen, 2012
Stor-Skogmo	Nord-Trøndelag	780	1066	Unknown	Uc: 2	Whale boneplaques	2	0	2	1	Isaksen, 2012
Melhus	Nord-Trøndelag	700	750	grave	Uc: 1	Whale boneplaque	1	0	1	1	Isaksen, 2012
Klinga	Nord-Trøndelag	780	1066	Unknown	Uc: 1	Whale boneplaque	1	0	1	1	Isaksen, 2012
Hestmannen	Nordland	600	1050	Grave	Uc: 1	Whale boneplaque	1	0	1	1	Isaksen, 2012
Hov	Nordland	780	1066	grave	Uc: 1	Whale boneplaque	1	0	1	1	Isaksen, 2012
Meloy	Nordland	780	1066	grave	Uc: 1	Whale boneplaque	1	0	1	1	Isaksen, 2012
Akvik	Nordland	780	1066	Grave	Uc: 1	Whale boneplaque	1	0	1	1	Isaksen, 2012
Tommeide	Nordland	780	1066	Grave	Uc: 1	Whale boneplaque	1	0	1	1	Isaksen, 2012
Alsoy	Nordland	900	950	Grave	Uc: 1	Whale boneplaque	1	0	1	1	Isaksen, 2012
Myre	Nordland	780	1066	Unknown	Uc: 1	Whale boneplaque	1	0	1	1	Isaksen, 2012
Grytoy	Troms	780	1066	Unknown	Uc: 1	Whale boneplaque	1	0	1	1	Isaksen, 2012
Gare	Troms	800	800	rural	Uc: 4	Whale boneplaques	4	0	4	1	Isaksen, 2012
Slagstad	Troms	780	825	Grave	Uc: 1	Whale boneplaque	1	0	1	1	Isaksen, 2012
Krøttøy	Troms	780	1066	Grave	Uc: 1	Whale boneplaque	1	0	1	1	Isaksen, 2012
Tisnes	Troms	750	800	Grave	Uc: 1	Whale boneplaque	1	0	1	1	Isaksen, 2012
Akeroya	Nordland	780	1066	Unknown	Uc: 1	Whale boneplaque	1	0	1	1	Isaksen, 2012
Akerøy	Troms	780	1066	Grave	Uc: 1	Whale boneplaque	1	0	1	1	Isaksen, 2012
Engstad	Nord-Trøndelag	780	1066	Unknown	Uc: 1	Whale boneplaque	1	0	1	1	Isaksen, 2012
Staulosen	Nordland	700	800	grave	Uc: 1	Whale boneplaque	1	0	1	1	Isaksen, 2012
Loppasanden	Finnmark	780	900	Grave	Uc: 1	Whale boneplaque. Nine pieces glued together	1	0	1	1	Isaksen, 2012
Nordheim	Troms	700	800	Grave	Uc: 1	Whale boneplaque	1	0	1	1	Isaksen, 2012

Toponym	Geographic region	Start date	End date	Site type	Species	Details	NISP	X	TOTAL	Min. No. Spec.	Reference
Huseby	Troms	800	800	Rural	Uc: 1	Whale boneplaque	1	0	1	1	Isaksen, 2012
Soberg	Nordland	780	1066	Grave	Uc: 2	Whale boneplaques	2	0	2	1	Isaksen, 2012
Føre	Nordland	550	1066	grave	Uc: 1	Whale boneplaque	1	0	1	1	Isaksen, 2012
Vik	Nordland	780	1066	Grave	Uc: 2	Whale boneplaques	2	0	2	1	Isaksen, 2012
Mellom Steigen	Nordland	780	1066	Grave	Uc: 1	Whale boneplaque	1	0	1	1	Isaksen, 2012
Marnes	Nordland	780	1066	Grave	Uc: 1	Whale boneplaque	1	0	1	1	Isaksen, 2012
Enge	Nordland	780	1066	grave	Uc: 1	Whale boneplaque	1	0	1	1	Isaksen, 2012
Lilleberge	Nord-Trøndelag	780	1066	grave	Uc: 1	Whale boneplaque	1	0	1	1	Isaksen, 2012
Skei	Nord-Trøndelag	780	1066	grave	Uc: 1	Whale boneplaque	1	0	1	1	Isaksen, 2012
Villa	More og Romsdal	780	1066	Grave	Uc: 1	Whale boneplaque	1	0	1	1	Isaksen, 2012
Stuorrassi'da	Finnmark	1500	1600	Rural	Uc: 1		1	0	1	1	Odner, 1992
Saggusnjar'ga	Finnmark	1520	1650	Rural	Uc: 1		1	0	1	1	Odner, 1992
Gaeccevajnjar'ga (244B)	Finnmark	1200	1400	Rural	Lc: 7		7	0	7	1	Odner, 1992
Gaeccevajnjar'ga (in and front of 244B)	Finnmark	1200	1400	Rural	Uc: 1		1	0	1	1	Odner, 1992
Gaeccevajnjar'ga (5m NNW of 244B)	Finnmark	1200	1400	Rural	Uc: 1		1	0	1	1	Odner, 1992
Mellaneset	Finnmark	540	690	Rural	Uc: X		0	1	1	1	Universitetet, 2019
Klubbhelleren	Hordaland	400	600	Other	Hp: X		0	1	1	1	Universitetet, 2019
Steigen gårdshaug	Nordland	1100	1600	Unknown	Hp: X		0	1	1	1	Universitetet, 2019
Helgøygården	Troms	1100	1800	Unknown	Uc: 1	Gaming piece made of whale bone	1	1	2	2	Skomsvoll, 2012
Dreggsalmenningen 1980	Hordaland	1100	1600	Unknown	Uc: X, Ud: X, Hp: X		0	3	3	3	Universitetet, 2019
Televerkstomten	Sør-Trøndelag	1100	1600	Unknown	Kw: X, Hp: X		0	2	2	2	Universitetet, 2019
Soløy	Troms	800	1700	Rural	Hp: X		0	1	1	1	Universitetet, 2019
Storgt.33/ Tjømegt.1	Vestfold	1100	1600	Unknown	Hp: X		0	1	1	1	Universitetet, 2019
Erkebispegården	Sør-Trøndelag	1250	1532	Ecclesiastical	Uc: 1, Ud: 3, Hp: 5		9	0	9	3	Universitetet, 2019

Toponym	Geographic region	Start date	End date	Site type	Species	Details	NISP	X	TOTAL	Min. No. Spec.	Reference
Erkebispegården	Sør-Trøndelag	1532	1672	Ecclesiastical	Uc: 1, Ud: 1, Hp: 1		3	0	3	3	Universitetet, 2019
Blomsøy 1918+1920	Nordland	Unkown	Unknown	Unknown	Lfpw: X		0	1	1	1	Universitetet, 2019
Giske	Møre og Romsdal	550	1400	Rural	Uc: X, Kw: 1		3	0	3	2	Bratbak and Hufthammer, 2011
Skagen 3	Rogaland	1100	1100	Unknown	Wbd: X		0	1	1	1	Universitetet, 2019
Gamlebyen Søndre Felt	Oslo	1100	1500	Unknown	Awsd: X		0	1	1	1	Universitetet, 2019
Oseberg	Vestfold	793	1099	Other	Uc: X	Baleen found	0	1	1	1	Universitetet, 2019
Tysnes	Nordland	793	1099	Unknown	Uc: X		0	1	1	1	Universitetet, 2019
Andenes (Ts 4767)	Nordland	1200	1700	Unknown	Uc: X		0	1	1	1	Universitetet, 2019
Kaupang 1954	Møre og Romsdal	1000	1250	Urban	Uc: X		0	1	1	1	Universitetet, 2019
Bryggen 1958	Hordaland	1100	1600	Urban	Uc: X		0	1	1	1	Universitetet, 2019
Kaupang 1972	Vestfold	793	1099	Urban	Uc: X		0	1	1	1	Universitetet, 2019
Bryggen 1972	Hordaland	1100	1350	Unknown	Uc: X		0	1	1	1	Universitetet, 2019
Oslogt.	Oslo	1000	1500	Urban	Uc: X		0	1	1	1	Universitetet, 2019
Vadsøya	Finnmark	1500	1600	Unknown	Uc: X		0	1	1	1	Universitetet, 2019
Rosenkrantz gt. 4, I	Hordaland	1100	1600	High Status	Uc: X		0	1	1	1	Universitetet, 2019
Vesle Hjerkin 1985	Oppland	?	?	High Status	Uc: X		0	1	1	1	Universitetet, 2019
Bryggen Kjøpmannsstuen, Tyskebryggen	Hordaland	1100	1600	Urban	Uc: X		0	1	1	1	Universitetet, 2019
Kaupang Borgund	Møre og Romsdal	1025	1155	Unknown	Ubw: 1	Rib with multiple butchery marks	1	0	1	1	Personal communication Eirik Søyland Laskemoen Herdlevær, 20-06-2018
Tussøy	Troms	800BC	1600	Unknown	Kw: 1	killer whale tooth used as gaming piece or dice	1	0	1	1	Skomsvoll, 2012
Fjelldal	Nordland	800BC	1066	Unknown	Uc: 1	Weaving sword (107 cm long)	1	0	1	1	Petersen, 1952
Bleik	Nordland	600BC	800	Unknown	Uc: 1	Weaving sword (99 cm long)	1	2	3	3	Petersen, 1952
Tromsø	Troms	780	1066	Unknown	Uc: 1	Weaving sword (91,5 cm)	1	0	1	1	Petersen, 1952
Lurøy	Nordland	780	1066	Unknown	Uc: 1	Weaving sword (90 cm)	1	0	1	1	Petersen, 1952
Myklebostad	Sogn og Fjordane	780	1066	Unknown	Uc: 1	Weaving sword (64 cm long and 3.5 cm broad)	1	0	1	1	Petersen, 1952

Toponym	Geographic region	Start date	End date	Site type	Species	Details	NISP	X	TOTAL	Min. No. Spec.	Reference
Sande	Nordland	780	1066	Unknown	Uc: 1	Weaving sword (61 cm)	1	0	1	1	Petersen, 1952
Andsnes	Finnmark	780	1066	Unknown	Uc: 1	Weaving sword (55,5 cm)	1	0	1	1	Petersen, 1952
Vårberget	Finnmark	780	1066	Unknown	Uc: 1	Weaving sword (28 cm long and 2,5 cm broad)	1	0	1	1	Petersen, 1952
Hov	Nordland	780	1066	Unknown	Uc: 1	Weaving sword (6,5 cm broad)	1	0	1	1	Petersen, 1952
Hilde	Sogn og Fjordane	780	1066	Unknown	Uc: 1	Weaving sword (2.8 cm broad)	1	0	1	1	Petersen, 1952
Tisnes	Troms	780	1066	Unknown	Uc: 1	Weaving sword (handle of 16.7 cm (smallest)	1	0	1	1	Petersen, 1952
Sandstrand	Nordland	780	1066	Unknown	Uc: 1	Weaving sword (handle of 35 cm long)	1	0	1	1	Petersen, 1952
Reviken	Sogn og Fjordane	780	1066	Unknown	Uc: 2	2 whale bone cleavers	2	0	2	1	Petersen, 1952, 341
Austråt	Trøndelag	780	1066	Unknown	Uc: 1	1 whale bone cleaver	1	0	1	1	Petersen, 1952, 341
Lo	Trøndelag	780	1066	Unknown	Uc: 1	1 whale bone cleaver	1	0	1	1	Petersen, 1952, 341
Meløy	Nordland	850	900	Unknown	Uc: 1	1 whale bone cleaver	1	0	1	1	Sjövold, 1974
Alstahaug	Nordland	780	1066	Unknown	Uc: 1	1 whale bone cleaver	1	0	1	1	Sjövold, 1974
Tisnes	Troms	780	1066	Unknown	Uc: 1	1 whale bone cleaver	1	0	1	1	Petersen, 1952, 341
Skarstein	Nordland	800	850	Unknown	Uc: 1	1 whale bone cleaver	1	0	1	1	Sjövold, 1974
Sandvik	Troms	800	850	Unknown	Uc: 1	1 whale bone cleaver	1	0	1	1	Sjövold, 1974
Skjærvika	Finnmark	0	1000	Rural	Uc: 862	Many unidentified, but 3 cranial fragments, 1 mandible, 50 vertebrae, 38 ribs, 2 scapula, 1 humerus, and 764 unidentified	862	0	862	1	Hufthammer, 2010
Fjellvika	Finnmark	0	1000	Rural	Uc: 10	10 whale bones	10	0	10	1	Hufthammer, 2010
Sweden					Uc: 0						
Lödöse	Smaland and the Islands	1200	1499	Urban	Hp: X	Unspecified bones	0	1	1	1	Lepiskaar, 1975
Lund	South Sweden	1020	1400	Urban	Ur: 2	Unspecified bones	2	0	2	1	Clason, 1977; Bergquist and Lepiskaar, 1957
Lahibia Cave	South Sweden	650	1699	Other	Hp: X	Unspecified bones	0	1	1	1	Jennbert, 2011
Birka	Stockholm County	600	1050	Urban	Uc: 4	Whale bone plaque	4	0	4	1	Isaksen, 2012
Langgatan	Stockholm County	600	1050	Urban	Uc: 1	Whale bone plaque	1	0	1	1	Isaksen, 2012

Toponym	Geographic region	Start date	End date	Site type	Species	Details	NISP	X	TOTAL	Min. No. Spec.	Reference
Hjulsta	Stockholm County	900	1050	High-status grave	Uc: 1	Whale bone plaque	1	0	1	1	Isaksen, 2012
Gnista	Uppland	550	600	Grave	Uc: 42	42 gaming pieces	42	0	42	1	Gustavsson, Hennius, and Ljungkvist, 2015
Valsgärde 7	Uppland	550	1066	Grave	Uc: X	Gaming pieces	0	1	1	1	Gustavsson, Hennius, and Ljungkvist, 2015
Välsgarde 13	Uppland	550	1066	Grave	Uc: X	Gaming pieces	0	1	1	1	Gustavsson, Hennius, and Ljungkvist, 2015
Prästgarden 1	Uppland	550	1066	Grave	Uc: X	Gaming pieces	0	1	1	1	Gustavsson, Hennius, and Ljungkvist, 2015
Storby backe 512	Uppland	550	1066	Grave	Uc: X	Gaming pieces	0	1	1	1	Gustavsson, Hennius, and Ljungkvist, 2015
Storby backe 596	Uppland	550	1066	Grave	Uc: X	Gaming pieces	0	1	1	1	Gustavsson, Hennius, and Ljungkvist, 2015
Brista	Stockholm County	300	1210	Urban	Uc: 50	Gaming pieces	50	0	50	1	Hennius *et al.* 2018
Gamla	Uppland	875	950	Grave	Uc: 1	Gaming pieces	1	1	2	2	Hennius *et al.* 2018
Finland					Uc: 0						
Svartsmara	Åland	550	1066	Grave	Uc: X	Uc: gaming pieces	0	1	1	1	Gustavsson, Hennius, and Ljungkvist, 2015
Spain					Uc: 0						
La Solana	Catalonia	500	799	Rural	Uc: 2, Cbw: 1	Cranium and mandible piece of unidentified cetacean and a bone belonging to a Ziphius though not specified in text	3	0	3	2	Martin and Lorenzo, 2007
Gorliz	Basque country	1000	1500	Unknown	Uc: X		0	1	1	1	Personal communication Idoia Grau Sologestoa
Ponte do Burgo	Galicia	1400	1599	Rural	Ud: 1	Vertebra of dolphin with cutmark on the spinous process	1	0	1	1	Costa, 2012
Baelo Claudia	Cadiz	642	773	Rural	Lfpw: 1	Identified using aDNA and ZooMS research	1	0	1	1	Rodrigues *et al.*, 2018
Iulia Traducta	Cadiz	251	422	Urban	Gw: 1	Identified using aDNA and ZooMS research	1	0	1	1	Rodrigues *et al.*, 2018
Septem	Ceuta	720	896	Urban	Sbcd: 1	Identified using aDNA and ZooMS research	1	0	1	1	Rodrigues *et al.*, 2018

Toponym	Geographic region	Start date	End date	Site type	Species	Details	NISP	X	TOTAL	Min. No. Spec.	Reference
Septem	Ceuta	226	410	Urban	NArw: 1	Identified using aDNA and ZooMS research	1	0	1	1	Rodrigues *et al.*, 2018
Palais Zarautz	Basque country	1000	1100	High Status	Uc: X	Various whale bones	0	1	1	1	Aranzadi Zientzi Elkartea, 2004
Santa María de Hito	Cantabria	900	1000	Other	Sw: 1	Decorated belt buckler made of sperm whale rib	1	0	1	1	Gimeno, 1978
Portugal					Uc: 0						
Alcáçova de Santarém	Alentejo	800	1199	Rural	Ud: 1	1 vertebra (both sides unfused)	1	0	1	1	Davis, 2006
Silves	Algarve	1100	1299	Rural	Uc: 3	3 large fragments, probably parts of vertebral centrum. One has cut marks as was probably used as a chopping board	3	0	3	1	Davis, Gonçalves and Gabriel, 2008
Peniche	Centro Region	1450	1650	Other	NArw: 80	4 Occipital regions, 10 other skull bones, 3 sets of neck vertebrae, 1 ulna, 11 large vertebrae fragments, 1 vertebral disc, 8 rib fragments, 42 other small bone fragments	80	0	80	1	Teixeira, Vanâncio and Brito, 2014
Ribat da Arrifana	Algarve	850	1050	Ecclesiastical	Uc: 1		1	0	1	1	Antunes, 2007
Castelo de Paderne	Algarve	1100	1350	High Status	Ud: 1	Unidentified dolphin vertebra	1	0	1	1	Pereira, 2013
TOTAL					Uc:		5456	72	5528	18	
Number X (Only the presence of remains noted in report, not the actual numbers)					Uc:		72	-	72	-	
TOTAL+ NUMBER X (assuming that every "X" is at least 1 bone of the specific species)					Uc:		5528	-	5528	-	

References used in Appendix

Aaris-Sorensen, K., 2009. Diversity and dynamics of the mammalian fauna in Denmark throughout the last glacial-interglacial cycle, 115-0 kyr BP. *Fossils Strata*, 57, p.30.

Aaris-Sorensen, K., Rasmussen, K.L., Kinze, C. and Petersen, K.S., 2010. Late Pleistoceneand Holocene whale remains (Cetacea) from Denmark and adjacent countries: Species, distribution, chronology, and trace element concentrations. *Marine Mammal Science*, 26(2), pp.253-281.

Albarella, U. and Davis, S.J.M., 1996. Mammals and birds from Launceston Castle, Cornwall: decline in status and the rise of agriculture. Circaea, *The Journal of the Association for Environmental Archaeology*, 12(1), pp.1-156.

Amorosi, T. and McGovern, T.H., 1993. *The 1987-1988 Archaeofauna from Videy Iceland.* Hunter College CUNY: North Atlantic Biocultural Organization.

Amorosi, T., 1996. *Icelandic Zooarchaeology: New Data Applied to Issues of Historical Ecology, Paleoeconomy and Global Change*. Unpublish PhD Thesis.

Amundsen, C., Perdikaris, S., McGovern, T.H., Krivogorskaya, Y., Brown, M., Smiarowski, K., Storm, S., Modugno, S., Frik, M. and Koczela, M., 2005. Fishing Booths and Fishing Strategies in Medieval Iceland: an Archaeofuana from the of Akurvík, North-West Iceland. *Environmental Archaeology*, 10, pp.127-142.

Amundsen, C.P., 2008. Culture Contact, Ethnicity and Food Practices of Coastal Finnmark, Norway (1200 to 1600 A.D.). Unpublished PhD Thesis.

Anathon, B. and Shetelig, H., 1940. Part IV Viking Antiquities in England with a supplement of Viking Antiquities on the continent of western Europe. In: *Viking Antiquities in Great Britain and Ireland.* Oslo: H. Aschehoug & Co. pp.64-67.

Andrews, P., 1997. *Excavations at Hamwic Volume 2: excavations at Six Dials*. York: Council for British Archaeology.

Anonymous, n.d. *Chalkpit Field North; SHARPREPORT*. Historical and Archaeological Research Project Sedgeford.

Antunes, M.T., 2007. *Ribat da Arrifana: estudo arqueozoologico. Ribat da Arrifana: cultura material e espiritualidade*. Aljezur. pp.83-86.

Aranzadi Zientzi Elkartea, 2004. *Descurbierto un yacimiento arqueológico clave para la historia de getaria.* Available at: <http://www.aranzadi.eus/old/noticias/01yacimiento_getaria.htm>. [Accessed 10-05-2019].

Armit, I., 1996. *The archaeology of Skye and the Western Isl*es. Edinburgh: Edinburgh University Press.

Bailey, L., Holden, T., Franklin, J., Smith, C. and Cerón-Carrasco, R., 2015. An Archaeological Excavation of Medieval Fishermen's Midden at Castle Hill Pumping Station, Banff. *Scottish Archaeological Journal*, 36-7. pp.1-35.

Barton, D., 1943. A Whale Bone Built into a Wall. *Journal of the Royal Society of Antiquities of Ireland*, 73. p.298.

Batey, C.E., 1987. *Freswick Links, Caithness. A re-appraisal of the Late Norse site in its context*. Oxford: B.A.R.

Becker, C., 2012. Aus dem Dunkel eines Magazins ans Licht gebracht: Archäozoologische Untersuchungen zu Hatzum-Boomborg, einer Siedlung der Vorrömischen Eisenzeit in Ostfriesland. *Siedlungs- und Küstenforschung im südlichen Nordseegebiet*, 35, pp.201-294.

Bendrey, R., 2005. The Animal Bone. In: Butler, B., ed. 2005. *Saxons, Templars & Lawyers in the Inner Temple; Archaeological excavations in Church Court & Hare Court*. London: Pre-Construct Archaeology Limited. pp.64-67.

Benecke, N., 1999. Die jungpleistozäne und holozäne Tierwelt Mecklenburg-Vorpommern - Faunenhistorische und kulturgeschichtliche Befunde. *Documenta naturae*, 124. München.

Bergquist, H. and Lepiskaar, J., 1957. *Medieval animal bones found in Lund. Archaeology of Lund.*

Blackmore, L. and Cowie, R., 2001. Saxon and medieval Battersea excavations at Althorpe Grove, 1975-8. *Surrey Archaeological Collections*, 88, pp.67-92.

Bond, J.M. and O'Connor, T.P., 1999. *The Archaeology of York; the Animal Bones. Bones from medieval deposits at 16-22 Coppergate and other sites in York*. York: York Archaeological Trust.

Borvon, A., 2017. Étude des vestiges fauniques. In: Vincent, K. and Dubois, A., 2017. *Le Château de Suscinio (Morbihan - Sarzeau)*. Le Logis Nord. Tome II - Volume 1.

Borvon, A., unpublished. *Étude archéozoologique du château de Talmont Saint-Hilaire (Vendée), bâtiment nord secteur 1, XIe-XIVe siècles.*

Bourdillon, J. and Coy, J., 1980. The Animal Bones. In: Holdsworth, P., ed. 1980. *Excavations at Melbourne Street, Southampton, 19714-76. Southampton Archaeological Research Committee Report I CBA Research Report 33*. London: Council for British Archaeology.

Bratbak, O.F. and Hufthammer, A.K., 2011. Det osteologiske materialet fra utgravningene pa Giske, 2-13 mai 2011. In: Hatling, S., 2012. *Arkeologiske undersokelser ved kabeltrasé, gnr. 127, bnr, 2 og 23, Giske k., More og Romsdal. Spor etter Giskeaettens gard*. Bergen: Universitatet I Bergen.

Brewington, S., Harrison, R., Amundsen, C. and McGovern, T.H., 2004. An Early 13th c Archaeofauna from Steinbogi. In: Vésteinsson, O., ed. 2004. *Landscapes of settlement 2002; Report on investigations at five medieval sites in Mývatnssveit*. Reykjavík: Fornleifastofnun Íslands.

Brothwell, D.R., 1979. Notes on the Mammal remains in Medieval pits and well at Seaford Church Street. *Sussex Archaeological Collections*, 117, pp.231-233.

Browsher, J., 1999. Saxon foreshore at Temple Place. *London Archaeologist*, 9(3), pp.82-89.

Buckley, M., Fraser, S., Herman, J., Melton, N.D., Mulville, J. and Palsdottir, A.H., 2014. Species identification of archaeological marine mammals using collagen fingerprinting. *Journal of Archaeological Science*, 41, pp.631-641.

Buitenhuis, H., 2011. 10. Faunaresten. In: Ufkes, A. ed. 2011. *Een archeologische opgraving in de vroegmiddeleeuwse ringwalburg van Domburg, gem. Veere (Z.)*. Groningen: ARC bv. pp.149-160.

Bult, E.J., 1986. Polanen, Monster; Een analyse van het zoo-archeologisch materiaal. Scriptie project zoo-archeologie, A.E. van Giffen Instituut voor Prae- en Protohistorie, Universiteit van Amsterdam.

Cardigan Castle, n.d. *Conservation*. Available at: <https://www.cardigancastle.com/history/conservation/> [Accessed 04-06-2019]

Carroll, J., 2008. *Archaeological Excavations at Rosepark, Balrothery, Co. Dublin. Volume 1: Balrothery Excavations*. Dublin: Judith Carroll and Co. and the Heritage Council.

Catalo, J. 2010. *"Faculté de Sciences Sociales", Toulouse (Haute-Garonne), 07/12/2009 - 11/12/2009*. Rapport de diagnostic, INRAP, SRA Midi-Pyrénées.

Chazottes, M.-A., 2017. Matières du quatoidien, matières de luxe: os, bois de cerf, ivoire, corail, nacre, corne, fanon de baleine et écaille de tortue dans l'artisanat médiéval et postmédiéval en Provence à partir de l'étude conjointe des sources archéologiques, écrites et iconographiques. Unpublished PhD thesis.

Church, M.J., Arge, S.V., Brewinton, S., McGovern, T.H., Woollett, J.M., Perdikaris, S., Lawson, I.T., Cook, G.T., Amundsen, C., Harrison, R., Krivogorskaya, Y. and Dunbar, E., 2005. Puffins, Pigs, Cod and Barley: Palaeoeconomy at Undir Junkarinsflotti, Sandoy, Faroe Islands. *Environmental Archaeology*, 15, pp.179-197.

Claeys, J., Jaspers, N.L. and Ostkamp, S., 2010. *Vier eeuwen en sterven aan de Dokkershaven. Een archeologische opgraving van een postmiddeleeuwse stadswijk in het Scheldekwartier in Vlissingen*. ADC Rapport 1635/ ADC Monografie 9. Amersfoort: ADC ArcheoProjecten.

Clark, G., 1947. Whales as an Economic Factor in Prehistoric Europe. *Antiquity*, 21, pp.84-104.

Clason, A.T., 1961. Some remarks on the faunal remains from the Roman Castellum at Valkenburg, prov. of South Holland. *Palaeohistoria*, VIII. pp.139-147.

Clason, A.T., 1965. Animal and Man in Holland's past: An investigation of the animal world surrounding man in prehistoric and early historical times in the provinces of North and South Holland. In: Waterbolk, H.T., ed. 1965. *Palaeohistoria*, XIIIB, pp.24-27.

Clason, A.T., 1977. *Jacht en veeteelt van prehistorie tot middeleeuwen*. Haarlem: Ribula-van Dishoeck.

Clavel, B., 2001. L'Animal Dans L'Alimentation Médiévale et Moderne En France Du Nord (XIIIe – XVIIe siècles). *Revue Archaeologique de Picardie*, 19.

Collins, A.E.P., Patterson, R., Proudfoot, B., Morrison, M.E.S. and Jope, M., 1955. Excavations in Lough Faughan Crannog, Co. Down, 1951-52. *Ulster Journal of Archaeology*, Thirds series (18), 1955, pp.45-82.

Costa, L., 2012. Antropologia de los restos humanos de Galicia: estudio de la poblacion romana y medieval gallega. Unpublished PhD thesis. Universidad de Granada.

Cramp, 1976. Analysis of the fiends register and location plan of Whitby Abbey. In: Wilson, D.M., ed. 1976. *The Archaeology of Anglo-Saxon England*. Cambridge: Cambridge University Press. pp.453-457.

Crumlish, R., 1993. The Archaeology of Tory Island. *Archaeology Ireland*, 7(1), pp.22-25.

Crummy, P., 1984. *Colchester Archaeological Report 3: Excavations at Lion Walk, Balkerne Lane, and Middleborough, Colchester, Essex*. Colchester: Colchester Archaeological Trust Ltd.

Curl, J., 2006. Animal, bird and fish bone. In: Wallis, H., ed. 2006. *Excavations on the site of Norwich Cathedral Refectory 2001-3*. Norfolk: Archaeology and environment Division, Norfolk Museums and Archaeology Service.

Curl, J., 2007. Faunal Remains. In: Emery, G., ed. 2007. *An Archaeological evaluation at the former Eastern Electricity Offices, Duke Street, Norwich, in advance of the Dukes Warf re-development*. NAU Archaeology.

Curle, A.O., 1938. A Viking settlement at Freswick, Caithness. Report on the excavations carried out in 1937 and 1938. *Proceedings of the Society of Antiquaries of Scotland*, 73, pp.71-110.

Curle, C.L., 1982. Pictish and Norse Finds from the Brough of Birsay 1934-74. *Society of Antiquaries of Scotland*, Monograph Series 1, pp.1-141.

Dahl, S., 1971. The Norse settlement of the Faroe Islands. *Medieval Archaeology*, XIV, pp.60-73.

Dalton, O.M., 1928. Early chessmen of whale's bone excavated in Dorset. *Archaeologia*, 77, pp.77-86.

Davis, S.J.M., 2006. Faunal remains from Alcáçova de Santarém, Portugal. *Trabalhos de Arqueologia*, 43. Lisboa: Instituto Português de Arqueologia.

Davis, S.J.M., Gonçalves, M.J. and Gabriel, S., 2008. Animal remains from a Moslem period (12th/13th century AD) lixeira (garbage dump) in Silves, Algarve, Portugal. *Revista Portuguese de Arquelogia*, 11, pp.183-258.

Degerböl, M., 1933. Danmarks Pattedyr i Fortiden i sammenligning med recente Former. *Vidensk. Medd. Nat For*. 95. Copenhagen. p.396.

Delhaye, M., 1998. L'épave médiévale de Cavalaire: un exemple de l'évolution navale architecturale avant la Renaissance. *Itsas Memoria, Revista de estudios Marítimos de l País Vasco*, 2, pp.43-48.

Dijk, J. van, 2012. Archeozoölogisch Onderzoek. In: Meijlink B.H.F.M. and Silkens, B., ed. *Middeleeuwse bewoning bij de Duinburg, een archeologisch opgraving aan de Roosjesweg (plangebied De Golfslag) te Domburg*. Middelburg: Walcherse Archeologische Dienst.

Dijkstra, J., 2012. *Het domein van de boer en de ambachtsman. Een opgraving op het terrein van de voormalige fruitveiling te Wijk bij Duurstede: een deel van Dorestad en de villa Wijk archeologisch onderzoek*. ADC Monografie 12.

Dijkstra, M.F.P. and Horssen, J. van., 2010. In: Dijkstra, M.F.P. and Brandenburgh, C.R., eds. *Bodemonderzoek en Bouwhistorie in Leiden 1; Leiden-Aalmarktschool Archeologisch en bouwhistorisch onderzoek naar 800 jaar bewoning langs de Oude Rijn, ter plaatse van het voormalige St. Catharinagasthuis. Een walvis op het droge*. pp.232-237.

Dobney, K., Jaques, D., Barrett, J. and Johnstone, C., 2007. *Farmers, Monks and Aristocrats; The Environmental Archaeology of Anglo-Saxon Flixborough*. Oxford: Oxbow Books.

Douglas, A., Jarrett, C., Rielly, K., Riddler, I. and Trzaska-Nartowski, N., 2015. *Medieval Craft-Working in Oxford: Excavations at 15-17 Clarendon Centre*. Wallingford: Pre-Construct Archaeology. pp.64-83.

Du Bouëtiez, E., Clavel, B. and Ravoire, F., 2013. Vestiges médiévaux et modernes découverts sur le site de l'hôtel de Beauvais à Paris (4e arrondissement). *Revue Archéologique d'Île de France*, 5, pp.193-254.

Edvardsson, R. and McGovern, T.H., 2005. Archaeological Excavations at Vatnsfjördur 2003-04. *Archaeologica Islandica*, 4, pp.16-30.

Edvardsson, R., 2005. *Hrísheimar 2004: Interim Report*. Reykjavik: Fornleifastofnun Íslands. pp.1-31.

Elder, S., Dennehy, E., Tierney, J., Noonan, D., and Doolan, A., 2007. Archaeological Excavation Report, 00E0422 ext, 01E0327 ext, 02E0809 - Shandon td., Dungarvan, Co. Waterford. Medieval Moated Site. *Eachtra Journal*, 14. Cork.

Ervynck, A. 1991. De beer die woonde op de burg...: mens en dier in een vroegmiddeleeuwse versterking. In: de Witte, H. ed. 1991. *De Brugse Burg. Van grafelijke versterking tot moderne stadskern*, pp.170-181.

Ervynck, A., Laleman, M.C. and Berkers, M., 2012. Sint-Veerleplein 11, Gravensteen: Bewerkt bot en gewei. Stadsarcheologie. *Bodem en monument in Gent*, 2(6), pp.215-236.

Eskildsen, S., 2013. *Arkæologer finder mystiske benplader fra Middelalderen*. Available at: <https://ekstrabladet.dk/nyheder/samfund/article4816968.ece>. [Accessed 10-05-2019].

Esser, E. and Dütting, M.K., 2018. *Boeren en vissers in de duinen. Een archeozoölogische analyse van Ijzertijd en Merovingische dierlijke resten (Ossicle 338)*. Delft: Archeoplan Eco.

Esser, E., 2004. *Enkele dierlijke resten uit Brielle (object 09-214). Ossicle, 84*.

Esser, E., 2009. Bot. In: Pavlovic, A., ed. 2009. *Archeologisch onderzoek in het riooltraject op de Frederik Hendriklaan Gemeente Den Haag; Definitief Onderzoek*. Den Haag: Afdeling Archeologie Dienst Stadsbeheer gemeente Den Haag.

Esser, E., 2018. *Dieren in het vroegmiddeleeuwse Frankenslag II (Den Haag). Ossicle 334*. Delft: Ossicle.

Esser, E., Beerenhout, B. and Kootker, L.M., 2010. Dierlijk botmateriaal. In: Dijkstra, M.F.P. and Brandenburgh, C.r., eds. *Bodemonderzoek en Bouwhistorie in Leiden 1; Leiden-Aalmarktschool Archeologisch en bouwhistorisch onderzoek naar 800 jaar bewoning langs de Oude Rijn, ter plaatse van het voormalige St. Catharinagasthuis*. pp. 172-231.

Esser, E., Kootker, L., van der Sluis, M. and Beerenhout, B., 2014. Dineren in de burcht van Rode. Archeozoölogisch en Isotopenonderzoek naar de samenstelling, productie en distributie van voedsel uit de burcht van Sint-Oedenroden. In: *Ossiculum* 1.

Feeley, F., 2012. *Mammal Consumption at the Medieval Fishing Station at Gufuskálar; a Preliminary*

Zooarchaeology Report from the 2011 Excavation. New York: Cuny.

Feveile, C., 2006. Ribe Studier. Det ældste Ribe. Utgravninger pa nordsiden af ribe A 1984-2000. Bind 1.2. *Jysk Arkæologisk Slskabs Skrifter* 51.

Freeley, F.J., Perdikaris, S., Hicks, M. and Smiarowski, K., 2010. Preliminary Assessment of the faunal remains from the 2008 excavations at Gufuskálar, Snaefellsnes. *NORSEC*, 25, pp.1-12.

Fridriksson, A., Batey, C.E., Gestsdóttir, H., McGovern, T. and Vésteinsson, O., 2004. *Fornleifarannsóknir í Saltvík*. Reykjavik: Fornleifastofnun Íslands.

Gaborit-Chopin, D. and Bardoz M.-C., 2005. *La France romane: au temps des premiers capétiens, 987-1152, exposition Paris Musée du Louvre, 10 mars, 6 juin 2005, Paris, Musée du Louvre Ed.* Hazan, 2005, p.407.

Gabriel, I., 1988. Hof-und Sakralkultur sowie Gebrauchs- und Handelsgut im Spiegel der Kleinfunde von Starigard/Oldenburg. *Bericht der Römischen Kommission*, 69, pp.240-242.

Garcia, M.M., 2009. Mammal and Bird Bone. In: Popescu, E.S., ed. 2009. *Norwich Castle: Excavations and Historical Survey 1987-98 part II: c.1345 to modern. East Anglian Archaeology Report No.132*. Dereham: Historic Environment, Norfolk Museums and Archaeology Service.

Gardiner, M., 1997. The exploitation of Sea-Mammals in Medieval England: bones and their Social context. *Archaeological Journal*, 152, pp.173-195.

Gardiner, M., Stewart, J. and Priestley-Bell, G., 1998. Anglo-Saxon Whale exploitation: some evidence from Dengemarsh, Lydd, Kent. *Medieval Archaeology*, 42, 96-101.

Gardiner, M.F., 1990. An Anglo-Saxon and Medieval settlement at Botolphs, Bramber, West Sussex. *Archaeological Journal*, 147, pp.216-275

Gidney, L.J., 2011. Animal bone. In: Barrowman, R.C. ed. 2011. *The chapel and burial ground on St. Ninian's Isle, Shetland: Excavations past and present. London: The Society for Medieval Archaeology*. pp.147-155

Gimeno, G.-L., 1978. Hallazgo de un broche altomedieval trabajado en hueso. *Boletín del Seminario de Arte y Arqueología*, 44, pp.430-434.

Giot, P.-R. and Corgné, J., 1951. La nécropole de Saint-Urnel-en-Plomeur (Finistère). Fouilles de 1946-1950. *Gallia*, 9, pp.1-19.

Giot, P.-R. and Monnier, J.-L., 1977. Le cimetière des anciens Bretons de Saint-Urnel ou Saint-Saturnin en Plomeur (Finistère). *Gallia*, 35(1), pp.141-171.

Gisladóttir, G.A. and Vésteinsson, O., 2008. *Archaeological investigations at Sveigakot, 2006*. Reykjavik: Fornleifastofnun Íslands.

Gísladóttir, G.A., 2014. *Archaeological Fieldwork at Svalbarð, NE Iceland 2013: Baegisstaðir, Hjálmarvík, Kúðá, Svalbarð, Sjóhúsavík og Skriða*. Reykjavik: Centre d'études nordiques, U. Laval.

Goosens, T.A., 2012. *Van akker tot Hooghwerf. Onderzoek naar de bewoning van de ijzertijd, inheems-Romeinse tijd, middeleeuwen en nieuwe tijd op de haakwal van Naaldwijk (plangebied Hoogeland, gemeente Westland)*. Leiden: Archol.

Gosling, P., 1991. From Dún to Dundalk: The Topography and Archaeology of a Medieval Frontier Town A.D. c 1187-1700. *Journal of the County Louth Archaeological and Historical Society*, 22, no.3, pp.221-353.

Gowen, M., 1992. Excavation of two souterrain complexes at Marshes Upper, Dundalk, Co. Louth. *Proceedings of the Royal Irish Academy*, 92C, pp.55-121.

Graham-Campbell, J., 1994. *Cultural Atlas of het Viking World*. Oxford: Andromeda Oxford Limited.

Grahame, T., 1968. Animal bones. Hamilton, J.R.C., ed. 1968. *Excavations at Clickhimin, Shetland*. Edinburgh: Her Majesty's Stationery Office. Appendix I.

Grieg, S., 1940. Viking Antiquities in Scotland. In: Shetelig, H., ed. 1940. *Viking Antiquities in Great Britain and Ireland*. Part II. Oslo.

Grimm, J., 2006. Silence of the lambs - Archäozoologische Untersuchungen an mittelalterlichen knochenmaterial aus Emden. Beitrage zur Archäozoologie und Prähistirisch Antrhopologie, V.

Groenman-van Waateringe, W. and Wijngaarden-Bakker, L.H., ed. n.d. *Medieval archaeology and environmental research in the Netherlands*. pp.283-297.

Gustavsson, R., Hennius, A. and Ljungkvist, J., 2015. Are many Vendel and Viking Period gaming pieces made of whale bone? Forn Vännen. *Journal of Swedish Antiquarian Research*, 110(1), pp.51-54.

Halici, H., 2004. In: J. Vanden Borre and J.B. Hielkema, eds. 2004. *Een archeologische begeleiding (AB) op het kasteelterrein Caetshage te Culemborg, gemeente Culemborg (Gld.)*. Groningen: ARC publicaties.

Hall, D.W., MacDonald, A.D.S., Perry, D.R. and Terry, J., 1998. The archaeology of Elgin: excavations on Ladyhill and in the High Street, with an overview of the archaeology of the burgh. *Proceedings of the Society of Antiquity of Scotland*, 128, pp753-829.

Hallen, Y., 1994. The use of bone and antler at Foshigarry and Bac Mhic Connain, two Iron Age sites on North Uist, Western Isles. *Proceedings of the Society of Antiquaries of Scotland*, 124, pp.189-231.

Hambleton, E. and Rowley-Conwy, P., 1997. The medieval reindeer economy at Gæccevaj'njar'ga 244 B in the Varanger Fjord, North Norway. *Norwegian Archaeological Review*, 30(1), pp.55-70.

Hamilton, J.R.C., 1956. *Excavations at Jarlshof, Shetland.* Edinburgh: Her Majesty's Stationery Office.

Hamilton-Dyer, S., 2010. *Skriduklaustur Monastery, Iceland; Animal Bones 2003-2007.* Reykjavik: Skriduklaustursrannsoknir.

Hamilton-Dyer, S., McCormick, F., Murray, H.K. and Murray, J.C., 1993. In: Murray, H.K. and Murray, J.C., eds. Excavations at Rattray, Aberdeenshire. A Scottish Deserted Burgh. *Medieval Archaeology*, 37, pp.109-218.

Hamlin, A. and Lynn, C., 1988. *Pieces of the Past; Archaeological Excavations by the Department of the Environment for Northern Ireland.* Belfast: Her Majesty's Stationery Office.

Hansen, S.S, 2013. Faunal Remains from Toftanes. *Acta Archaeologica*, 84(1), pp.175-177.

Harland, J.F., 2012. Animal Husbandry: the Mammal Bone. In: Barrett, J.H., ed. 2012. Being an islander. Cambridge: McDonald Institute for Archaeological Research. Harrison, R., 2008. Möðruvellir in Hörgárdalur, N. Iceland: General Overview of the Archaeofauna Analyzed from the 2006-08 Midden Mound Excavations. HER/NORSEC Zooarchaeology Laboratory REPORT, 59. pp. 1-56.

Harrison, R., 2008. *Midden Excavation at Möðruvellir, and Prospection in Hörgárdalur; Interim Field Report Gásir Hinterlands Project 2008.* Reykjavík & New York: NABO.

Harrison, R., 2010. Small Holder Farming in Early Medieval Iceland: Skuggi in Hörgárdalur. *Archaeological Islandica*, 8, pp.51-76.

Harrison, R., 2014. *The Siglunes Archaeofauna, I. Report of the Viking Age and Medieval Faunal Remains.* NORSEC, 62.

Hatting, T., 1991. The Archaeozoology. In: Bencard, M., Jorgensen, L.B. and Madsen, H.B., eds. *Ribe Excavations 1970-76, Volume 3.* Esbjerg: Sydjysk Universitetsforlag. pp.43-58.

Hennius, A., Gustavsson, R., Ljungkvist, J. and Spindler, L., 2018. Whalebone Gaming Pieces: Aspects of Marine Mammal Exploitation in Vendel and Viking Scandinavia. *European Journal of Archaeology*, pp.1-20.

Henry, F., 1945. Remains of the Early Christian Period of Inishkea North, Co. Mayo. *The Journal of The Royal Society of Antiquaries of Ireland*, 75(3), pp.127-155.

Hicks, M.T. and Pálsdóttir, L.B., 2011. *Excavations at Skútustadir, Mývatn Northern Iceland: Preliminary Field Report After the Excavation Season June - July 2010.* Reykjavik: Fornleifastofnun Íslands.

Hicks, M.T., 2010. Skútustadir: An Interim Zooarchaeological Report following the 2009 Field Season. *NORSEC*, 4, pp.1-25

Holsworth, P., 1976. Saxon Southampton; A New Review. *Medieval Archaeology*, 20, pp.26-61.

Holthuis, L.B., Smeenk, C. and Laarman, F.J., 1998. The find of a whale barnacle, Cetopirus complanatus (Morch, 1853). *Zoologische Verhandelingen, Nationaal Natuurhistorisch Museum, Leiden*, nr 323, pp.350-363.

Hopman, M., 1993. *Een kijk op het Karolingische dierenrijk; Faunaresten van de terpen Tzummarum en Wijnaldum (Fr.).* Ph.D. Rijksuniversiteit Groningen.

Hufthammer, A.K., 2010. Analyser av osteologiske materiale fra utgravninger i 2009 og 2010 i Skjaervika og Fjellvika ved Hammerfest, Finnmark fylke. Deres journalnumre: Flellvika TS 12181 og Skjaervika TS 12160 og TS12178. In: Hendriksen, S., and Valen, C.R., 2010, eds. *Skjaervika og Fjellvika, Hammerfest Kommune. Rapport fra de arkeologiske undersokelsene 2009 og 2010.* Tromso: Tromso Museum Universitetetsmuseet, Universitetet Tromso. pp.478-479.

Hunter, J., 2007. *Vol 1: Excavations at Pool, Sanday. A multi-period settlement from Neolithic to Late Norse Times.* Oxford: Oxbow Books.

Hurley, M.F., 1995. Bone and Antler Artefacts. In: Hurley, M.F., ed. *Excavations at the North Gate, Cork.* Cork: Cork Corporation. Pp.144-148.

Ijzereef, G.F. and Laarman, F., 1986. The animal remains from Deventer (8th-19th centuries AD). *Berichten van de Rijksdienst voor het Oudheidkundig Bodemonderzoek*, 26, pp.405-443.

Ingrem, C., Mulville, J. and Cartledge, J., 2005. Animal Bone. In: Sharples, N., ed. 2005. A Norse Farmstead in the Outer Hebrides.Oxford: Oxbow Books. Oxford: Oxbow Books.

Isaksen, E., 2012. Hvalbeinsplater fra yngre jernalder: en analyse av hvalbeinplatenes kontekst og funksjon. Unpublished Master's Thesis. Universitetet i Tromsø.

Jansen, H.M., 1987. Svendborg in the Middle Ages. *Journal of Danish Archaeology*, 6, pp.198-219.

Jennbert, K., 2011. Scandinavian cave archaeology. *Cave and Karst Science*, 38(3), pp.144-150.

Jones, R.T., Reilly, K. and Pipe, A.R., 1997. The Animal Bone. In: Morley, B. and Gurney, D., ed. 1997. *Castle Rising Castle, Norfolk.* Norfolk: Archaeology and environment Division, Norfolk Museums and Archaeology Service.

Kelleher, H. 2003 *Archaeological excavation at Trench 309 and Trench 310 Washington Street.* Unpublished excavation report in the Office of the Cork City Archaeologist.

Knol, E., Prummel, W., Uytterschaut, H.T., Hoogland, M.L.P., Casparie, W.A., de Langen, G.J., Kramer, E. and Schelvis, J., 1996. The early medieval Cemetery of

Oosterbeintum (Friesland). *Palaeohistoria*, 37/38, pp. 245-416.

Krivogorskaya, Y., Perdikaris, S. and McGovern, T.H., 2005. Cleaning up the farm: a later Medieval Archaeofauna from Gjögur, a fishing farm of NW Iceland. *Archaeological Islandica*, 5, pp.381-392.

Küchelmann, C., 2010. *Vom mehr oder weniger armen Leben der Bettelmönche*. pp.83-85.

Kunst, G.K., 2014. The Animal Bones. In: Coolen, J. and Mehler, N., eds. *Excavations and Surveys at the Law Ting Holm, Tingwall, Shetland*. BAR British Series 592. pp.93-116.

Laarman, F.J., 2006. *De Botten. Intern Verslag Archeozoölogie*. Amersfoort: ROB

Lárusdottir, B., Roberts, H.M. and Þorgeirsdóttir, S., 2012. *Siglunes; Archaeological investigations in 2011*. Reykjavík: Fornleifastofnun Íslands.

Lauwerier, R.C.G.M. and Laarman, F.J., 1999. Dierlijk Botmateriaal. In: Hagers, J-K.A. and Siers, M.M., eds. 1999. *Archeologisch onderzoek in de Oosterbuurt van de gemeente Castricum*. Amersfoort: ROB.

Lauwerier, R.C.G.M., 1995. Veeteelt in Oost-Souburg. In: van Heeringen, R.A., Hendrikx, P.A. and Mars, A., eds. 1995. *Vroeg-Middeleeuwse ringwalburgen in Zeeland*. Goes & Amersfoort. pp.213-218.

Leask, H.G., 1943. Finding of Whale's Vertebra in Clochan-na-carriaige, Inishmore, Aran, Co. Galway. *The Journal of the Royal Society of Antiquaries of Ireland*, 13(1), P.24.

Lepiksaar, J. 1975. Über die Tierknochenfunde aus den mittelalterlichen Siedlungen Südschwedens. In: Clason, A.T., ed. 1975. *Archaeozoological studies*. Amsterdam. pp. 230-239,

Long, C.D., n.d. *Excavations in the Medieval City of Trondheim, Norway*.

Lynn, C.J., Collins, C., Fry, M.F, Hamond, F.W., Doggart, R., Heslip, R., 1981/1982. The Excavation of Rathmullan, a Raised Rath and Motte in County Down. *Ulster Journal of Archaeology*, Third series (44/45), pp.65-171.

MacGregor, A., 1974. The Broch of Burrain, North Ronaldsay, Orkney. *Proceedings of the Society of Antiquaries of Scotland*, 105, pp.63-118.

MacGregor, A., 1985. *Bone, Antler, Ivory & Horn*. London: Croom Helm.

MacLaren, A., 1974. A Norse House on Drimore Machair, South Uist. *Glasgow Archaeological Journal*, 3. pp.9-18.

Magendans, J.R. and Waasdorp, J.A., 1989. *Franken aan de slag Frankenslag; een vroeg-middeleeuwse nederzetting in 's-Gravenhage. 's-Gravenhage: Afdeling Verkeer en Vervoer, Openbare Werken en Monumentenzorg (VOM)*.

Magnússon, Þ., 1966. *Bátkumlið í Vatnsdal í Patreksfirði. Árbók Hins íslenzka fornleifafélags*, 63, pp.5-32.

Makowiecka, M and Makowiecki, D. 2007. Gospodarka zwietzętami we wczesnomiejskim centrum osadnic zym w Kołobtzega Budzistowie w świetle nowych badań at cheozoologicznych. In: Leciejewicz, L. and Rębkowski, M., eds. *Kołobrzeg. Wczesne miasto nad Bałtykiem*. Warszawa: Trio.

Makowiecki, D. and Makowiecka, M, 2014. The Character of Animal Exploitation and the Environment at the Polish/Prussian Frontier in the Medieval Period: A Case Study. *Archaeologia Baltica*, 20. pp.91-116.

Maltby, M., 2009. Bones: mammals, birds and fish. In: Palmer, S. and Reilly, D., eds. 2009. *Excavations of an Enigmatic Multi-Period Site on the Isle of Portland, Dorset*. Oxford: BAR Publishing.

Martín, A.E. and Lorenzo, J.N., 2007. *La Solana, estudi zooarqueologic del jaciment*. Barcelona: Universitat de Barcelona.

McCarthy, M., 1994. Appendix 2; Animal Bone Report. In: Lennon, A.M., ed. 1994. *Summary Report on Excavation of Ringfort, Raheens, No 2, near Carrigaline, Co. Cork*. Cork Historical and Archaeological Society. Pp.47-65.

McCarthy, M., 1997. Faunal Remains: Christ Church. In: Cleary, R.M., Hurley, M.F. and Shee Twohig, E., eds. 1997. *Skiddy's Castle and Christ Church, Cork; Excavations 1974-1977 by D.C. Twohig*. Cork: Cork Corporation.

McCarthy, M., 2003. 11. The faunal remains. In: Cleary, R.M. and Hurley, M.F., eds. *Cork City Excavations 1984-2000*. Cork: Cork City Council. pp.375-389.

McCarthy, M., 2008. *Appendix IV: Analysis of Animal Bone from Rosepark, Balrothery, Co. Dublin, Archaeological Excavations at Rosepark, Balrothery, Co. Dublin. Volume 1: Balrothery Excavations*. Dublin: Judith Carroll and Co. and the Heritage Council, pp.120-25.

McCormick, F. 1987. Stock rearing in Early Christian Ireland. Unpublished PhD thesis. Queen's University of Belfast.

McCormick, F. and Murphy, E., 1977. Mammal Bones. In: Walsh, C., 1997. *Archaeological Excavations at Patrick, Nicholas and Winetavern Streets*. Dingle: Brandon Book Publishers Ltd.

McCormick, F., Kerr, T., McClatchie, M. and O'Sulivan, A., 2011. *The Archaeology of Livestock and Cereal Production in Early Medieval Ireland, AD 400-1100. Early Medieval Archaeology Project (EMAP 2) Report 5:1*. Irish National Strategic Archaeological Research (INSTAR) programme 2011.

McGavin, N.A., 1983. Excavations at Kirkwall, 1978. *Proceedings of the Society of Antiquaries of Scotland*, 112, pp.392-436.

McGovern, T.H., 2003. Herding Strategies at Sveigakot, N Iceland: An Interim Report. In: Vésteinsson, O., ed. 2003. *Archaeological investigations at Sveigakot 2002*. Reykjavik: Fornleifastofnun Íslands.

McGovern, T.H., 2009. The Archaeofauna. In: Lucas, G., ed., 2009. H*ofstadir; Excavation of a Viking Age Feasting Hall in North-Eastern Iceland.* Reykjavik: Institute of Archaeology.

McGovern, T.H., Amundsen, C., Perdikaris, S, Harrison, R. and Krivogorskaya, Y., 2004. *An Interim report of a Viking-Age and Medieval Archaeofauna from Undir Junkarinsflotti, Sandoy, Faroe Islands*. NORSEC Zooarchaeology Laboratories Report, 17. NABO.

McGovern, T.H., Smiarowski, K. and Harrison, R., 2013. *Hard Times at Hofstadir? An Archaeofauna circa 1300 AD from Hofstadir in Mývatnssveit, N Iceland.* NORSEC 60.

McNeill, T., 2004. Excavations at Dunineny Castle, Co Antrim. *Medieval Archaeology*, 48, pp.167-200.

Médard, A., 2013. Botmateriaal. In: Verheist, E.M.P., 2013. *Middeleeuwse bewoningsporen in Ijzendoorn, gemeente Nederbetuwe; proefsleuven-onderzoek in het plangebied Ijzendoor-West*. Weesp: RAAP Archeologisch Adviesbureau B.V.

Méniel, P. And Arbogast, R.-M., 1989. Les restes de mammifères de la cour Napoléon du Louvre (Paris) du XIVe au XVIIIe siècle. *Revue de Paléobiologie, Museum d'Histoire Naturelle de la Ville de Geneve*, 8(2), pp.407-466.

Merples, B., 1976. The Animal Bones. In: Hassall, T.G., ed. 1976. Excavations at Oxford Castle, 1965-1973. *Oxoniensia*, 41, pp.302-304.

Moesker, T. and Cavallo, C., 2016. Dierlijk bot. In: Dijkstra, M.F.P., Verhoeven, A.A.A. and van Straten, K.C.J., 2016. eds. *Nieuw licht op Leithon. Archeologisch onderzoek naar de vroegmiddeleeuwse bewoning in plangebied Leiderdorp-Plantage*. Themata 8. Amsterdam: Universiteit van Amsterdam/Diachron UvA bv. pp.575-624.

Moffat, R., Spriggs, J. and O'Connor, S., 2008. The Use of Baleen for Arms, Armour and Heraldic Crests in Medieval Britain. *Antiquaries Journal*, 88, pp.207-215.

Mohl, U., 1961. Oversigt over dyreknogler fra Naesholm. In: La Cour., V., ed. 1961. *Naesholm*. Kobenhavn: Nationalmuseet. pp.358-424.

Morton, A.D., 1992. *Excavations at Hamwic Volume 1: excavations 1946-83, excluding Six Dials and Melbourne Street*. London: Council for British Archaeology. pp.142-153

Motteau, J., 1991. Catalogue des objets des fouilles 1973-1977, Recherches sur Tours, n° 5, Tours. *Association pour le développement des études d'archéologie urbaine à Tours*, 1991. p.138.

Mouton, D., 2005. *La Moutte d'Allemagne-en-Provence: un castrum précoce du Moyen Âge provençal*. France Bibliothèque d'Archéologie Méditerranéenne et Africaine.

Murray, E., McCormick, F. and Plunkett, G., 2004. The Food Economies of Atlantic Island Monasteries: The Documentary and Archaeo-Environmental Evidence. *Environmental Archaeology*, 9, pp. 179-189.

NADNuis, 2019. *Noordelijk Archeologisch Depot Nuis*. Available at: <http://www.nadnuis.nl/>.

Nieuwhof, A., 2007. Dieren op de wierde Englum. In: Nieuwhof, A. and Bottema-Mac Gillavry, J.N., eds. 2007. De Leege Wier van Englum: archeologisch onderzoek in Reitdiepgebied. *Jaarverslagen Vereniging voor terpenonderzoek*, 92, pp.116-159

NMS (National Museum of Scotland), 2019a. *Casket*. Available at: http://nms.scran.ac.uk/database/record.php?usi=000-100-002-043-C [Accessed 16-5-2019].

NMS (National Museum of Scotland), 2019b. *Casket of a carved whalebone*. Available at: http://nms.scran.ac.uk/database/record.php?usi=000-100-049-339-C [Accessed 16-5-2019].

Noddle, B.A., Bramwell, D. and Jones, A., 1977. In: *The Animal Bones Cardiff Archaeological Society and Contributors, 1977. Llantrithyd a Ringwork in South Glamorgan*. Cardiff: Cardiff Archaeological Society. pp.63-73.

Nordjyske, n.d. *Første fund fra middelalderens Nibe*. Available at: https://nordjyske.dk/nyheder/foerste-fund-fra-middelalderens-nibe/851c5908-4d9d-48eb-8dca-f1fc7c89ea24 [Accessed 10-05-2019].

Ó Ríordáin, B., 1973. The High Street Excavations. In: Almqvist, B. and Greene, D., 1973, eds. *Seventh Viking congress*. London: Royal Irish Academy. Pp. 135-140.

Ó Ríordáin, B., 1975. Dublin City: Christchurch Place. In: Delaney, T.G., ed. *Excavations 1974, Association of Young Irish Archaeologists, Ulster Archaeological Society and Group for the Study of Irish Historic Settlement*, pp.14-15.

O'Connor, T.P., 1989. *Bones from Anglo-Scandinavian Levels at 16-22 Coppergate*. London: Council for British Archaeology.

O'Dell, A.C., Stevenson, R.B.K., Brown, T.N., Plenderleith, H.J., and Bruce-Mitford, R.L.S., 1959. The St Ninian's Isle Silver Hoard. *Antiquity*, 33, pp.241-268.

Odner, K., 1992. *The Varanger Saami; Habitation and Economy AD 1200-1900*. Oslo: The Institute of Comparative Research in Human Culture.

Odner, K., 2001. Trade, Tribute and household responses. *Acta Borealia*, 18(1), pp.25-50.

O'Hara, 2010. An Iron Age and Early Medieval cemetery at Collierstown 1, Co. Meath. Interpreting the changing character of a burial ground. Unpublished Masters Thesis

Ó'Néill, J., 2005. Excavations of pre-Norman structures on the site of an enclosed Early Christian cemetery at Cherrywood, County Dublin. In: Duffy, S., 2005, eds. *Medieval Dublin VII. Proceedings of the Friends of Medieval Dublin Symposium 2005*. Dublin: Four Courts Press. pp.66-88.

Ørsted Brandt, L., Haase, K. and Collins, M.J., 2018. Species identification using ZooMS, with reference to the exploitation of animal resources in the medieval town of Odense. *Danish Journal of Archaeology*.

O'Sullivan, D. and Young, R., 1995. *Book of Lindisfarne, Holy Island.* London: B.T. Batsford Ltd/English Heritage.

Owen, O. and Dalland, M., 1999. *Scar: A Viking Boat Burial on Sanday, Orkney*. East Linton: Tuckwell Press.

Pálsdóttir, A.H. Gorsline, M. and McGovern, T., 2008. In: Milek, K. ed. 2008. *Vatnsfjördur 2007; Framvinduskýrslur/Interim Report*. Reykjavik: Fornleifastofnun Íslands.

Pálsdóttir, A.H., 2010. *Dýrabein frá Alþingisreit: Greining á dýrabeinum frá svæðum A, B og C (No. 2010-1)*. Íslenskar fornleifarannsóknir ehf, Reykjavík.

Pálsdóttir, L.B. and Roberts, H., 2006. *Excavations at Gásir 2005; An Interim Report/Framvinduskýrsla*. Reykjavik: Fornleifastofnun Islands.

Pálsdóttir, L.B. and Sveinbjarnarson, O.G., 2011. *Under the Glacier; 2011 Archaeological investigations on the fishing station at Gufuskálar, Snaefellnes*. Reykjavík: Fornleifastofnun Íslands.

Peets, J. and Maldre, L., 2010. *Salme paadijäänused ja luunupud. Ülle Tamla (Toim.). Ilusad asjad. Tähelepanuväärseid leide Eesti arheoloogiakogudest (47 - 88)*. Tallinn: Ajaloo Instituut

Pereira, V., 2013. *Das faunas às populações - Reflexos islâmicos do Castelo de Paderne*. Pereira Techne 1, pp.67-73.

Pestell, T., 2004. *Landscapes of Monastic Foundation: The Establishment of Religious Houses in East Anglia, ca. 650-1200*. Woodbridge: The Boydell Press. pp.91-96.

Petersen, J., 1952. *Vikingetidens Redskaper*. Oslo: I Kommisjon Hos Jacob Dybwad.

Pipe, A., 1995. Environmental Evidence. In: Mills, P., 1995. *Excavations at the Dorter Undercroft, Westminster Abbey. Transactions of the London and Middlesex Archaeological Society*. pp.69-124.

Pipe, A., 2007. The Vertebrate Remains. Miller, P. and Saxby, D., 2007. eds. *The Augustinian priory of St Mary Merton, Surrey. MOLAS Monograph 34*. Museum of London Archaeology Service. pp.246-250.

Pipe, A., Rielly, K. and Ainsley, C., 2011. Animal Bone. In: Dyson, T., Samuel, M., Steele, A. and Wright, S.M., 2011. *The Cluniac Priory and abbey of St Saviour Bermondsey, Surrey; Excavations 1984-15. MOLA Monograph 50*. London: Museum of London Archaeology.

Poole, K., 2010. Mammal and bird remains. In: Thomas, G., ed. 2010. *The later Anglo-Saxon settlement at Bishopstone: a downland mannor in the making. CBA Research Report 163*. York: Council for British Archaeology.

Prummel, W., 1983. *Excavations at Dorestad 2, Early medieval Dorestad an archaezoological study*. Amersfoort: Rijksdients for het Oudheidkunding Bodemonderzoek.

Prummel, W., 1993. *Starigard/Oldenburg. Hauptburg der Slawen in Wagrien. IV. Die Tierknochkunde unter besonderer Berücksichtigung der Beizjagd.* Neumünster: Karl Wachholtz Verlag.

Prummel, W., Gent, J.T. van, Kampanje, E.J.O., 2012. Walvisbotten uit Friese en Groninger terpen. *Paleo-aktueel*, 23, pp.41-48.

Reichstein, H. 1991. *Die wildlebenden Säugetiere von Haithabu. Berichte über die Ausgrabungen in Haithabu*, 30. Neumünster.

Reichstein, H. and Heinrick, D., 1994. *Die Saugetiere und Vögel aud der Frühgeschitlichen Wurt Elisenhof.* Frankfurt am Main, Berlin, Bern, New York, Paris, Wien, Peter Lang

Riddler, I. and Sabin, R., 2009. Whalebone Chopping Boards. In: Lucy, S., Tipper, J. and Dickens, A., eds. 2009. *The Anglo-Saxon Settlement and Cemetery at Bloodmoor Hill, Carlton Colville, Suffolk. East Anglian Archaeology, reports 131*. Cambridge: Cambridge Archaeological Unit. pp.199-201.

Riddler, I. and Trzaska-Nartowski, N., 2014. Hamwic et l'artisanat de l'os de baleine aux VIIe-IXe siecles. *Cahiers LandArc 2014* – N°6.

Riddler, I., 2006. Objects and Waste of Bone and Antler. In: Cramp, R., ed. 2006. *Wearmouth and Jarrow Monastic Sites, Volume 2*. London: English Heritage. Pp.267-281.

Riddler, I.D. and Trzaska-Nartowski, N., 2013. Lundenwic and the Middle Saxon Worked Bone Interlude. *Anglo-Saxon Studies in Archaeology and History*, 18, pp.75-98.

Riddler, I.D., 1998. Worked Whale Vertebrae. *Archaeologia Cantiana*, 118, pp.205-215.

Riddler, I.D., 2014. The Archaeology of the Anglo-Saxon Whale. In: Klein, S.S., Schipper, W. and Lewis-Simpson, S.M., eds. 2014. *The Maritime world of the Anglo-Saxons. Medieval and Renaissance texts and studies 448. Essays in Anglo-Saxon Studies 5*. Tempe: Ariz. pp.337-354

Riddler, I.D., Trzaska-Nartowski, N.I.A. and Hatton, S., in press. *An Early Medieval Craft, Object and Waste of Bone, Antler and Ivory from Ipswich Excavations*, 1974-94.

Rielly, K., 2003. The animal and fish bone. In: Malcolm, G., Bowsher, D. and Cowie, R., eds. 2003. *Excavations at the Royal Opera House 1989-99*. London: Museum of London Archaeology Service. Ch.7.20.

Rielly, K., 2015. 3.9 Animal Bone. Douglas, A., Jarrett, C., Rielly, K., Riddler, I. and Trzaska-Nartowski, N., 2015. *Medieval Craft-Working in Oxford: Excavations at 15-17 Clarendon Centre*. Wallingford: Pre-Construct Archaeology. pp.64-83.

Ritchie, G., and Welfare, H., 1983. Excavations at Ardnave, Islay. *Proceedings of the Society of Antiquaries of Scotland*, 113, pp.302-366.

Ritchie, J.N.G., 1976. Excavations of Pictish and Viking-age farmsteads at Buckquoy, Orkney. *Proceedings of the Society of Antiquaries of Scotland*, 108, pp.174-227.

Roberts, H.M., 2001. *Fornleifarannsókn a lodunum/ Archaeolgocial Excavations at Adalstraeti 14-18, 2001; A Preliminary Report/Framvinduskýrslur*. Reykjavik: Fornleifastofnun Islands.

Roberts, H.M., 2003. *Excavations at Gásir 2003 - A Preliminary Statement*. Reykjavik: Fornleifastofnun Nordurlands.

Roberts, H.M., 2003. *Gásir 2002. An Interim Report*. Reykjavik: Fornleifastofnun Islands.

Roberts, H.M., 2004a. *Excavations at Adalstraeti, 2003*. Reykjavik: Fornleifastofnun Islands.

Roberts, H.M., 2004b. *Excavations at Gásir 2003. An Interim Report*. Reykjavik: Fornleifastofnun Nordurlands.

Roberts, H.M., 2005. *Excavations at Gásir 2004; An Interim Report*. Reykjavik: Fornleifastofnun Islands.

Rodrigues, A.S.L., Charpentier, A., Bernal-Casasola, D., Gardeisen, A., Nores, C., Milan, J.A.P., McGrath, K. and Speller, C.F., 2018. Forgotten Mediterranean calving grounds of gray and North Atlantic right whales: evidence from Roman archaeological records. *Proceedings of the Royal Society B*.

Roesdahl, E. and Wilson, D.M., 1992. *From Viking to Crusader*. New York: Rizolli. p.246.

Sabin, R., Bendrey, R. and Riddler, I., 1999. Twelfth-century Porpoise Remains from Dover and Canterbury. *Archaeological Journal*, 156, pp.363-369.

Sablerolles, Y. 1990. Het dierlijk botmateriaal uit de voege Middeleeuwen afkomstig van de nederzetting op de woerd te Valkenburg (Z.H.). Een voorbeschouwing. In: Bult., E.J. and Hallewas, D.P., 1990. *Graven bij Valkenburg III*. Delft: Eburon Delft.

Schmölcke, U., 2009. Neue Bemerkungen zu alten Funden - Die Tierknochen aus der Borgsumburg ("Lembecksburg") auf Föhr. In: Abegg-Wigg, A., Meier, U.M., Segschneider, M. and Tummuscheit, A., eds. 2009. *Ringwälle und verwandte Strukturen des ersten Jahrtausends n. Chr. an Nord- und Ostsee*. Neumünster: Wacholtz Verlag.

Scott, S., 1991. The Animal Bone. In: Armstrong, P., Tomlinson, D. and Evans, D.H., eds. 1991. *Excavations at Lurk Lane Beverly 1979-82*. Dorchester: The Dorset Press. Ch.3.21.

Sidell, J., 2000. Environmental Archaeology in London 1995-1998, part 3. *London Archaeologist*, 3(7), pp.188-195.

Simpson, D.D.A., Murphy, E.M. and Gregory, R.A., 2006. *Excavations at Northon Isle of Harris*. Oxford: Archaeopress.

Sjövold, T., 1974. *The Iron Age settlement of Arctic Norway. A study in the expansion of European Iron Age culture within the Arctic Circle. Part II. Late Iron Age (Merovingian and Viking periods)*. Tromsö, Oslo, Bergen: Norwegian Universities Press.

Skomsvoll, J., 2012. Nordnorske spillsaker fra jernalder og middelalder. Unpublished Masters Thesis. Universitetet i Tromsø Våren.

Small, A., Thomas, C. and Wilson, D.M., 1973. *St. Ninian's Isle and its treasure*. Oxford: Oxford University Press. Pp123-124.

Smith, C., 2000. The animal bone. In: Perry, D.R., ed. 2000. *Castle park Dunbar*. Devon: Short Run Press. pp.194-219.

Staal, B., n.d. *Hval pa afveje*. Available at: <http://www.tv2bornholm.dk/media/74491/hval-paa-afveje_web.pdf> [Accessed 10-5-2019].

Stout, G. and Stout, M., 2008. *Excavations of Knowth Site M*. Dublin: Wordwell Ltd.

Stratford, N., 1997. *The Lewis Chessmen; and the enigma of the hoard*. London: British Museum Press.

Tate, J., Reiche, I. and Pinzari, F., 2014. The Lewis Chessmen: what can examination of the surfaces tell us? In: D.H. Caldwell and M.A. Hall, eds. 2014. *The Lewis Chessmen; new perspectives*. Edinbugh: NMS Enterprises Limited. Pp.11-28.

Teixeira, A., Venâncio, R. and Brito, C., 2014. Archaeological Remains Accounting for the Presence and Exploitation of the North Atlantic Right Whale Eubalaena glacialis on the Portuguese Coast (Peniche,

West Iberia), 16th to 17th century. *PLOS ONE*, 9(2), e85971, pp.1-12.

Telfer, A., 2003. Medieval drainage near Smithfield Market: excavations at Hosier Lane, EC1. *London Archaeologist*, 10(5), pp.115–120.

Thilderkvist, J., 2013. Ritual bones or common waste, A study of Early Medieval bone deposits in Northern Europe. Ph.D. Rijksuniversiteit Groningen. p.66.

UBAAL Unité de Bioarchéologie Animalde de Lille, n.d. *Exploitation du Rorqual commun au début de Xie s. près de Dunkerque*. [online]. Available at: <http://perso.univ-lille3.fr/~toueslati/spip.php?breve5&lang=fr> [Accessed 18 August 2014]

Ulbricht, I., 1984. *Die Verarbeitung von Knochen, Geweih und Horn im mittelalterlichen Schleswig*. Neumünster. p.58.

Universitetet i Bergen, 2019. *Paleozoologi*. Available at: <https://www.uib.no/fg/flom/63523/paleozoologi> [Accessed 09-01-2019]

Van der Plaetsen, P., 1987. Walvissen ... in het Gravensteen. Stadsarcheologie. *Bodem en Monument in Gent*, 1(1), pp.24-25.

Van Dijk, J., 1995. Zooarcheologisch onderzoek van vroeg-middeleeuws botmateriaal uit de opgraving Hoogmadeseweg te Leiderdorp. *Westerheem*, 44, pp.95-101.

Van Dijk, J., 2012. Archeozoölogisch Onderzoek. In: Williams, G.L., ed. 2012. Een Archeologische opgraving. In: *ADC ArcheoProjecten Raport*, 1785, pp.82-84

Van Dijk, J., 2018. Dierlijk bot. In: Meijlink, B.H.F.M. and Silkens, B., 2018. *Archeologisch proefsleuvenonderzoek en opgraving aan de Derringmoerweg, plangebied Hazenburg II te Arnemuiden, gemeente Middelburg*. Middelburg: Walcherse Archeologische Dienst. pp.87-91.

Van Dijk, J., Bouman, M.T.I.J., Moolhuizen, C. and Bos, J.A.A., 2011. De voedseleconomie vanaf de midden-IJzertijd tot en met de Late Middeleeuwen. In: Dijkstra, J. and Zuidhoff, F.S., eds. 2011. Kansen op de kwelder. Archeologisch onderzoek op Walcheren langs de N57. In: *ADC ArcheoProjecten Rapport* 1384. Pp.109-140.

Van Doesburg, J., and Bakker, A.M., 1999. *Archeologisch Aanvullend Onderzoek (AAO) in de binnenstad van Tiel, vindplaats Tiel-Bleekveld*. Amersfoort: Rijksdienst voor the Oudheidkunding Bodemonderzoek.

Van Neer, W. and Ervynck, A., 1993. *Archeologie en Vis*. Zellik: Instituut voor het Archeologisch Patrimonium.

Van Neer, W., Lentacker, A and Ervyck, A., 2013. Dierenresten uit veenwinningsput spoornr.27. In: De Smeldt, I. and In 't Ven, I., eds. 2013. *Het Archeologische onderzoek in Raversijde (Oostende) in de periode 1992-2005*. pp.469-470.

Vésteinsson, O., 2001. *Archaeological investigations at Sveigakot 1998-2000*. Reykjavík: Fornleifastofnun Íslands.

Vésteinsson, O., 2004. *Archaeological investigations at Sveigakot 2003*. Reykjavik: Fornleifastofnun Íslands.

Vésteinsson, O., 2008. *The church in Gásir; Interim report on excavations in 2004 and 2006*. Reykjavik: Fornleifastofnun Islands.

Wace, A.J.B., and Jehu, 1915. Cave excavations in East Fife. *Proceedings of the Society of Antiquaries of Scotland*, 49. pp.233-255.

Wallace, P.F., 2016. *Viking Dublin; The Wood Quay Excavations*. Sillins: Irish Academic Press.

Waterman, D.M. and Jope, M., 1954. Excavations at Clough Castle, Co. Down. *Ulster Journal of Archaeology*, third series, 17, pp.103-163.

Webster, L. and Backhouse, J., 1991. *The Making of England: Anglo-Saxon Art and Culture, A.D. 600-900*. London: British Museum Press.

Wilkins, B., Petts, D. and Dave, R., 2016. *Lindisfarne: The Holy Island Archaeology Project 2016, Archaeological Assessment Report*. Newgate: Digventures.

Wood, I.N., 1990. *Ripon, Francia and the Franks Casket in the Early Middle Ages. Northern History*, 26. pp 1-19.

Woollett, J. and McGovern, T.H., 2003. *Interim Report of Animal Bones from the 2002 Excavations at Gásir, Eyjafjördur, N Iceland*. CUNY Zooarchaeology Interim Report. pp.1-22.

Young, C.J., 2000. *Excavations at Carisbrooke Castle, Isle of Wight, 1921-1996. Salisbury, Trust for Wessex Archaeology Ltd. Analysis of the 11th-12th century animal bone (Pippa Smith), Bird Bones (Dale Serjeantson)*.

Zeiler, J.T., 2007. Archeozoölogisch onderzoek van de Laat - Middeleeuwse kastelen Egmond en Brederode. *ArchaeoBone Rapport*, 61, pp.1-32.

Zeiler, J.T., 2018. Vee, vogels, huisdieren en wild; Archeozoölogisch onderzoek van dierlijke resten uit de Hof ter Hille te Koksijde, België (laatste kwart 9e - eerste kwart 12e eeuw AD). *ArchaeoBone rapport*, 154. Haren: ArchaeoBone.

Zori, D. and Wake, T., n.d. *Archaeofauna from the Chieftain's Farm at Hrísbrú, 2001-2008*.

www.ingramcontent.com/pod-product-compliance
Ingram Content Group UK Ltd.
Pitfield, Milton Keynes, MK11 3LW, UK
UKHW061847041026
15591UKWH00015B/946